Neuroacanthocytosis Syndromes II

Ruth H. Walker • Shinji Saiki • Adrian Danek
Editors

Neuroacanthocytosis Syndromes II

Preface by Mark Hallett, MD
Foreword by Ginger and Glenn Irvine

Publication Sponsor:

The Advocacy for Neuroacanthocytosis Patients, London, UK

Springer

Ruth H. Walker, MB, ChB, PhD
Movement Disorders Clinic
Department of Neurology
James J. Peters Veterans Affairs Medical Center
130 W. Kingsbridge Road (127)
Bronx, NY 10468
and
Department of Neurology
Mount Sinai School of Medicine
One Gustave L. Levy Place
New York, NY 10029
USA
ruth.walker@mssm.edu

Shinji Saiki, Dr.
Department of Medical Genetics
Cambridge Institute for Medical Research
Wellcome Trust/MRC Building
Addenbrooke's Hospital
Hills Road
Cambridge CB2 2XY
UK
ss644@cam.ac.uk

Adrian Danek, Prof. Dr. med.
Neurologische Klinik
Ludwig-Maximilians-Universität
D-81366 München
Germany
danek@lmu.de

Cover illustration: The cover image consists of an ideogram, known as a kanji in Japanese, superimposed on the representation of a thorny red blood cell (acanthocyte). This kanji was designed by Teiko Dewa, a Japanese artist living in New York City, who has taught mentally- and physically-disabled children in Japan. The kanji incorporates the symbol for "star" rather than "thorn", which has negative connotations.

ISBN: 978-3-540-71692-1 e-ISBN: 978-3-540-71693-8

Library of Congress Control Number: 2005279044

Cover design: WMXDesign GmbH, Heidelberg, Germany

Printed on acid-free paper 5 4 3 2 1 0

springer.com

Preface

The differential diagnosis of a neurological patient presenting with chorea is difficult. Huntington disease (HD) is best known and can be readily established with genetic testing should clinical features not be clear. But, if not HD, there are many possibilities. These include Huntington disease-like 1 (HDL1), Huntington disease-like 2 (HDL2), chorea-acanthocytosis (ChAc), McLeod syndrome (MLS), benign hereditary chorea types 1 and 2 (BHC), familial dyskinesia and facial myokymia, pantothenate kinase-associated neurodegeneration (PKAN), neuroferritinopathy, dentatorubral-pallidoluysian atrophy (DRPLA), and spinocerebellar ataxia 17 (SCA17). All these disorders are relatively uncommon, but there are ways of making the diagnosis and the hard part is keeping them in mind. A subset of these choreas is also characterized by the presence of acanthocytes in the blood, and this fact is valuable from a diagnostic point of view, but is also an interesting clinical clue in regard to the pathophysiology. ChAc and MLS have many acanthocytes, and HDL2 and PKAN may have acanthocytes. The acanthocytes are a reminder that the manifestations of these disorders are broad and include a range of neurologic symptoms and systemic abnormalities. Since there is only little research on these entities, it is valuable to make a periodic synthesis of the state of the art, to draw people's attention to the area, and to stimulate further research. A first book on the neuroacanthocytosis syndromes came from the first international conference in 2002. There have been two further international conferences on the neuroacanthocytosis syndromes in 2005 and 2006, and the contributions from those are the chapters in the present book. Much has happened in the last five years, and we now have a better clinical characterization of the disorders and a much better understanding of the genetics and cell biology. There are also advances in therapy. This book should be helpful for anyone interested in any aspect of these disorders. Edited by the leaders in the field, the contributors are the experts, and the material should be considered authoritative.

Bethesda, MD, USA

Mark Hallett, MD

Foreword

Neuroacanthocytosis (NA) is a group of diseases that snatch away bodily control, and with it the independence and hopes of young adults around the world. As parents of a 27 year-old we shuddered to hear a compassionate neurologist tell her, "The good news is that we can name your disease; the bad news is there is nothing we can do but to try to make it easier for you." The worst news, wisely left unsaid, was that there is no stopping the degeneration of the brain and the impending life of dependence, frustration and pain.

It takes the inveterate optimist to see the positive side, that the NA diseases offer two clues to the critical causes of neurodegeneration:

- the abnormalities in the cell membranes that are manifest in the red blood cells of patients and
- the impact of mutations in a number of specific genes that may each be a step in a neurodegenerative pathway.

The anguish of patients and the compulsion to understand these clues stimulated Adrian Danek, Hans Jung, Tony Monaco, Akira Sano, François Tison, Ruth Walker and many others, including Thomas Witt, Jordan Grafman cedilla in Francois, Mitchell Brin, the late W.L. Marsh, Colvin Redman, Maria Teresa Dotti and Alexander Storch to find spare resources and time over more than 10 years to build a foundation of know-ledge and a network of collaboration to study these diseases. We have them all to thank for the momentum of the movement.

The compassion and curiosity of these champions combined with their loving conviction that personal action may make things better are contagious. They are shared by physicians and researchers as well as the families and friends of the patients whose generous donations have allowed the Advocacy for Neuroacanthocytosis Patients to support several small research projects as well as help to finance three international meetings and resulting publications. Recent events give us reason to hope that funding will be found for a multi-disciplinary, international task force.

This anguish and hope are combined in the logo created by the Japanese artist Teiko Dewa, who superimposed the Japanese kanji symbol for "star", signifying "light", "hope" and "radiance", over the symbol of the thorny cells that characterize neuroacanthocytosis.

The hospitality of Professor Eva Andermann and her institution, the Montreal Neurological Institute and Hospital, as well as Professor Masanori Nakagawa of the Kyoto Prefectural Medical University made possible the symposia (see Figs. 1 and 2) that are the foundation for the contents of this book. All of the authors of the chapters in the book as well as Joseph B. Martin, Dean of the Harvard Medical School, who opened the Montreal Symposium, have contributed substance as well as the imagination that leads to vision of where an answer may be found.

We were both immeasurably helped and encouraged by the financial support for the symposia from Mr. and Mrs. Carl H. Pforzheimer III, Mr. and Mrs. In-Suk Oh, Mr. and Mrs. Ian Mackay, High Q Foundation, Mary Kinross Charitable Trust, The Movement Disorder Society, Montreal Neurological Institute, Nippon Boehringer Ingelheim Limited and Thyssen-Stiftung. Grooms-Shaftsbury, the English charity that has worked with disabled young adults since 1862, gives us vital financial administrative support.

Money and motivation only produce results in the presence of dedication. The dedication of the leadership of the symposia that included Eva Andermann, Adrian Danek, Hans H. Jung, Shinji Saiki and Ruth Walker was vital. Enormous help in organizing the two symposia came from Teiko Dewa, Naoko Funabashi, Etsuko Morris, Meredith Stangenberg, Hazel Stoddart and Robin Warwick who all volunteered to produce the program materials as well as the many details that made the events successful.

Ruth Walker, together with Shinji Saiki, put together the Kyoto program. Ruth, who also chaired the Symposium and co-edited this book, makes a mighty contribution. She, together with Adrian Danek, whose concern and untiring effort is the inspiration and the driving force of this special collaboration, deserve our special thanks.

It is all of these people together with the patients they serve who are the Advocacy for Neuroacanthocytosis Patients.

Secretariat
The Advocacy of Neuroacanthocytosis Patients

Ginger and Glenn Irvine

Fig. 1 Attendees at Second International Neuroacanthocytosis Symposium, Montreal, Canada, April 17th–19th 2005. *Left to right Back row*: Akira Sano, MD, PhD; Massimo Avoli, MD, PhD; François Tison, MD, PhD; Masayuki Nakamura, MD, PhD; Abbas Sadikot, MD, PhD; Colvin M. Redman, PhD; Giel J.C.G..M. Bosman, PhD; Gordon W. Stewart, MD; Russell L. Margolis, MD; Steven M. Holland, MD; Karen Gale, PhD; Kailash P. Bhatia, MD, FRCP; Joseph B. Martin, MD, PhD; Adrian Danek, MD; Antonio Velayos-Baeza, PhD; Carol Dobson-Stone, MBiochem, DPhil; Renzo Guerrini, MD; Frederick Andermann, MD, FRCP(C); Robert A. Hegele, MD, FRCP(C), FACP; James Phelan, PhD; Richard A. Hardie, MD; Glenn Irvine; Juan Botas, PhD. *Front row*: Pierre Burbaud, MD; Jacqueline McIntosh, MA, MRCSLT; Hans H. Jung, MD; Mio Ichiba, MD; Robert S. Fuller, PhD; Ruth H. Walker, MB, ChB, PhD; Eva Andermann, MD, PhD, FCCGM; Clotilde Lévecque, PhD; Soohee Lee, PhD; Françoise Marie-Francoise Chesselet, MD, PhD; Michael Hayden, MB, ChB, PhD, FRCP(C), FRSC; David Rosenblatt, MD; Klaus L. Leenders, MD

Fig. 2 Attendees at Third International Neuroacanthocytosis Symposium; Kyoto, Japan, October 28th 2006. *Left* to *right; back row*: Yohei Tamura, MD; Nagato Kuriyama, MD; Tadataka Kawakami, MD; Glenn Irvine; Mio Ichiba, MD; Toshiki Mizuno, MD; Benedikt Bader, MD; Mark Walterfang, MD; Lucia de Franceschi, PhD; Gordon W. Stewart, MD; Sonia Gandhi, MD; Giel Bosman, PhD; Clotilde Léveque, PhD; Hans H. Jung MD; Antonio Velayos-Baeza, PhD; Akira Sano, MD, PhD; Adrian Danek, MD; Mark Guttman, MD; Jan Aasly, MD; Masayuki Nakamura, MD, PhD; Mikihiro Kihara MD, PhD; Yutaka Kurano, MD; Felix Geser, MD, PhD; Yasufumi Kageyama, MD; Ginger Irvine; Jun Ochiai, MD; Robin Warwick BS. *Front row*: Naoko Funabashi; Etsuko Morris; Ruth H. Walker, MB, ChB, PhD; Genjiro Hirose, MD, PhD; Shinji Saiki, MD; Mieko Matsuda, MD; Pang Ying Shih, MD; Fusako Yokochi, MD, PhD; Soohee Lee, PhD; Chiho Ishida, MD; Junko Takahashi, MD; Yoshihiko Tani, MD, PhD; Mac Ho, DPhil; Nobuo Terada, MD

Contents

List of Abbreviations

5-HIAA	5-hydroxyindole acetic acid
5-HT	5-hydroxytyramine
aa	amino acids
AOO	age of onset
ABC	ATP-binding cassette
ABL	abetalipoproteinemia
AC-PC	anterior commissure-posterior commissural [intercommissural] line
AD	autosomal dominant
AD	Alzheimer's disease
ADC	apparent diffusion coefficient
ADLs	activities of daily living
AE	anion exchanger
ALT	alanine transaminase
AOA	ataxia with ocular motor apraxia
AP	anterior-posterior
APAAP	alkaline phosphatase-anti-alkaline phosphatase
AR	autosomal recessive
ARP3	actin-related protein 3
AST	aspartate aminotransferase
ATP	adenosine triphosphate
beta-CIT	[^{123}I]β-carbomethoxy-3-β-(4-iodophenyl)-tropane
BP	blood pressure
$BP_{BB}\,II_{late}$	late phase II of beat-to-beat BP during Valsalva maneuver
BPF	brain parenchymal fraction
BHC	benign hereditary chorea
BSA	bovine serum albumen
BT	botulinum toxin
C	chorea
C	cysteine
cc	corpus callosum
C. elegans	*Caenorhabditis elegans*

Ca	caudate
CASS	composite autonomic scoring scale
CBF	cerebral blood flow
cd	caudate nucleus
cDNA	complementary DNA
CGD	chronic granulomatous disease
CHAC	chorea-acanthocytosis gene, now renamed *VPS13A,* coding for chorein
ChAc	chorea-acanthocytosis
CI	chorein immunoreactivity
CK	creatine kinase
$CMRO_2$	cerebral metabolic rate of oxygen
CNS	central nervous system
CPK	creatine phosphokinase
CSF	cerebrospinal fluid
CT	computed tomography
CUGBP1	CUG-binding protein 1
D	dystonia
D2	dopamine D2 receptor
DA	dopamine
DAP	diastolic arterial pressure
DBS	deep brain stimulation
DM1	myotonic dystrophy type 1
DNA	deoxyribonucleic acid
DRPLA	dentatorubropallido-Luysian atrophy
DSM	Diagnostic and Statistical Manual
DTI	diffusion tensor imaging
DTR	deep tendon reflexes
ECE	endothelin converting enzyme
ECG	electrocardiogram
EDS	excessive daytime somnolence
EDTA	ethylene-diamine-tetracetic acid
EHDN	European Huntington's Disease Network
EEG	electroencephalography
EM	electron microscopy
EMG	electromyography
EOG	electro-oculogram
ER	endoplasmic reticulum
ERM	ezrin radixin moesin
ES	embryonic stem
F	female
FDG	fluorodeoxyglucose
FH inherit	family history and inheritance pattern
GABA	gamma amino butyric acid
GAD	glutamic acid decarboxylase

GAPDH	glyceraldehyde 3-phosphate dehydrogenase
GBS	Guillain–Barré syndrome
GFAP	glial fibrillary acidic protein
GT	greek symbol gamma
GL	glycolipids
GM	grand mal seizure
GM	grey matter
GP	globus pallidus
GPe	globus pallidus, external segment
GPi	globus pallidus, internal segment
GTP	guanidine triphosphate
HARP	hypoprebetalipoproteinemia, acanthocytosis, retinitis pigmentosa, and pallidal degeneration
H&E	hematoxylin and eosin
HD	Huntington's disease
HDL	Huntington's disease-like
HFS	high frequency stimulation
HIV	human immunodeficiency virus
hm	homozygously
H.N.	hospital normal
Hp	haptoglobin
HPLC	high performance liquid chromatography
HPRT	hypoxanthine guanine phosphoribosyl transferase
HR	heart rate
HR_{DB}	heart rate with deep breathing
ht	heterozygously
htt	huntingtin
HVA	homovanillic acid
IBGC	idiopathic basal ganglia calcification
ICCA	benign infantile convulsions and paroxysmal choreoathetosis syndrome
IF	immunofluorescence
Ig	immunoglobulin
IP	immunoprecipitation
IQ	intelligence quotient
ISHH	in situ hybridization histochemistry
IT15	interesting transcript 15 (HD gene)
IVM	involuntary movements
JCRV	Japana Centro Revuo Medicina
JPH3	junctophilin 3 gene
kDa	kilodalton
KEL	gene coding for Kell protein
Kx	X-linked antigen of the Kell system

LDH	lactate dehydrogenase
LH	left hemisphere
M	male
MAbs	monoclonal antibodies
MAP	mean arterial pressure
MBNL1	muscleblind-like protein 1
MCV	mean cellular volume
MDT	multidisciplinary therapy
MHC-F	myosin heavy chain fast
MIM	Mendelian Inheritance in Man (http://www.ncbi.nlm.nih.gov/entrez/query.fcgi?db = OMIM)
MLS	McLeod syndrome
MMSE	mini mental status examination
MNI	Montreal Neurological Institute
MRI	magnetic resonance imaging
mRNA	messenger RNA
MRS	magnetic resonance spectroscopy
MSA	multiple system atrophy
MSLT	mean sleep latency
MUPs	motor unit potentials
MW	molecular weight
N	normal
NA	neuroacanthocytosis
n/a	not available
NADH	reduced form of nicotinamide adenine dinucleotide
NADH-TR	nicotinamide adenine dinucleotide (reduced) – tetrazolium reductase
NBIA	neurodegeneration with brain iron accumulation
NCV	nerve conduction velocity
ND	not determined, not described
NE	norepinephrine
NE	not examined
NK	not known
NL	normal trait
NMD	nonsense-mediated mRNA decay
NMDA	N-methyl-D-asparate
NO	nitric oxide
np	not performed
OCD	obsessive-compulsive disorder
OH	orthostatic hypotension
OSAS	obstructive sleep apnoea syndrome
OT	occupational therapist
P	parkinsonism

PA	phosphatidic acid
PANK2	pantothenate kinase 2 gene
PAS	periodic acid Schiff
PBS	phosphate-buffered saline
PC	phosphatidylcholine
PC	proprotein convertases
PD	Parkinson's disease
PE	phosphatidylethanolamine
PED	paroxysmal exercise-induced dyskinesia
PET	positron emission tomography
PHD	paroxysmal hypnogenic dyskinesia
PI	phosphatidylinositol
PKAN	pantothenate kinase-associated neurodegeneration
PKD	paroxysmal kinesigenic dyskinesia
PLM	periodic limb movement
PNKD	paroxysmal non-kinesigenic dyskinesia
PNs	peroneal nerves
poly-Q	polyglutamine
PS	phosphatidylserine
PSG	polysomnographic recording
PTAH	phosphotungstic acid hematoxylin
PTC	premature termination codon
Pu, put	putamen
PVC	prevacuolar compartment
PVDF	polyvinylidine difluoride
PVP	posteroventral pallidotomy
Q	glutamine
QSART	quantitative sudomotor axon reflex test
RBC	red blood cell
RBD	REM behaviour disorder
REM	rapid eye movement
RE-PED-WC	Rolandic epilepsy-PED-writer's cramp
RH	right hemisphere
RLS	restless legs syndrome
RNA	riboxynucleic acid
ROI	region-of-interest
RT	room temperature
RT-PCR	reverse transcription-polymerase chain reaction
SA	splice acceptor
SAP	systolic arterial pressure
SC	Sydenham's chorea
SCA	spinocerebellar ataxia
sCK	serum creatine kinase
SD	splice donor

SD	standard deviation
SDS–PAGE	sodium dodecyl sulphate polyacrylamide gel electrophoresis
SE	sleep efficacy
SK	small conductance calcium activated potassium channels
SL	sleep latency
SLT	speech and language therapist
SM	sphingomyelin
SN	substantia nigra
SNP	single nucleotide polymorphism
SP	substance P
SPECT	single photon emission computed tomography
SPM	statistical parametric mapping
SR	sarcoplasmic reticulum
Sz	seizure
T	tremor
TGase	transglutaminase
TGN	trans-Golgi network
TM	transmembrane
TPR	tetratrico peptide repeat
tTGase	tissue transglutaminase
TUNEL	TdT-mediated dUTP nick-end labelling
TdT	terminal deoxynucleotidyltransferase
dUTP	2′-deoxyuridine 5′-triphosphate
U/l	enzymatic units per liter
UTR	untranslated region
VAMP	vesicle-associated membrane protein
VBM	voxel-based morphometry
VPS13	gene family, named after yeast protein Vpsl3p
VR	Valsalva ratio
WAIS	Wechsler Adult Intelligence Scale
WASO	wake after sleep onset
WB	Western blotting
WM	white matter
WNL	within normal limits
X	position of an amino acid that was changed to a stop codon
XK	McLeod syndrome protein
XK	McLeod syndrome gene
Y2H	yeast-two-hybrid

Contributors

Thomas Arzberger
Ludwig-Maximilians-Universität, Zentrum für Neuropathologie und Prionforschung, Feodor-Lynen-Straße 23, D-81377 München, Germany

Benedikt Bader
Ludwig-Maximilians-Universität, Zentrum für Neuropathologie und Prionforschung, Feodor-Lynen-Straße 23, D-81377 München, Germany
benedikt.bader@med.uni-muenchen.de

Kailash P. Bhatia
Institute of Neurology, Queen Square, London, WC1N 3BG, UK
k.bhatia@ion.ucl.ac.uk

Giel J.C.G.M. Bosman
Department of Biochemistry, Nijmegen Center for Molecular Life Sciences and University Medical Center-Nijmegen, The Netherlands
g.bosman@ncmls.ru.nl

Mitchell F. Brin
Department of Neurology, University of California Irvine, and Allergan LLC, Irvine, CA

Pierre Burbaud
Department of Clinical Neurophysiology, Centre Hospitalier Pellegrin, Place Amélie-Raba Léon, 33076 Bordeaux, France

Adrian Danek
Neurologische Klinik, Ludwig-Maximilians-Universität, D-81366 München, Germany
danek@lmu.de

Carol Dobson-Stone
Prince of Wales Medical Research Institute, Barker St, Randwick, NSW 2031 and University of New South Wales, Kensington, NSW 2052, Australia
c.dobson-stone@powmri.edu.au

Leja Dolenc-Grošelj
University Institute of Clinical Neurophysiology, University Medical Centre, Ljubljana, Slovenia

Lucia de Franceschi
Section of Internal Medicine, Department of Clinical and Experimental Medicine, University of Verona, Italy
lucia.defranceschi@univr.it

Sonia Gandhi
Department of Molecular Neuroscience, Institute of Neurology, Queen Square, London WC1N 3BG, UK
s.gandhi@ion.ucl.ac.uk

Felix Geser
Center for Neurodegenerative Disease Research, University of Pennsylvania School of Medicine, Philadelphia, PA, USA and Institute of Neuropathology, Department of Pathology, University Hospital Zürich, Switzerland

Imad Ghorayeb
Service des Explorations Fonctionnelles du Système Nerveux-Hôpital Pellegrin, 33076 Bordeaux, France

Richard J. Hardie
Department of Neurology, Frenchay Hospital, Bristol BS16 1LE, UK

Arthur P. Hays
Department of Neuropathology, Columbia Presbyterian Medical Center, New York City, USA

Helmut Heinsen
Julius-Maximilians-Universität, Morphologische Hirnforschung, Würzburg, Germany

Karsten Henkel
Neurologisch-verhaltensmedizinische Schmerzklinik, D-24149 Kiel, Germany

Genjiro Hirose
Department of Neurology, Neurological Center, Asanogawa General Hospital, Japan and Naka 83, Kosaka-cho, Kanazawa City, Ishikawa Prefecture, 920-8621, Japan
ghirose@asanogawa-gh.or.jp

Mio Ichiba
Departments of Psychiatry, Kagoshima University Graduate School of Medical and Dental Sciences, Kagoshima 890-8520, Japan

Hans H. Jung
Department of Neurology, University Hospital Zürich, Frauenklinikstrasse 26, 8091 Zürich, Switzerland
hans.jung@usz.ch

Jan Kassubek
Neurologische Klinik der Universität, D-89081 Ulm, Germany

Maiko Katoh
Departments of Psychiatry, Kagoshima University Graduate School of Medical and Dental Sciences, Kagoshima 890-8520, Japan

Yuko Katoh
Departments of Psychiatry, Kagoshima University Graduate School of Medical and Dental Sciences, Kagoshima 890-8520, Japan

Yasuo Kawamura
Department of Neurology, Kawamura Hospital, Gifu, Japan

Mikihiro Kihara
Department of Neurology, Suzuki Clinic, Tokyo, Japan
mikikihara@aol.com

Jan Kobal
Department of Neurology, Division of Neurology, University Medical Centre, Ljubljana, Slovenia

Hans A. Kretzschmar
Ludwig-Maximilians-Universität, Zentrum für Neuropathologie und Prionforschung, Feodor-Lynen-Straße 23, D-81377 München, Germany

Soohee Lee
New York Blood Center, 310 East 67th Street, New York, NY 10021, USA
solee@nybloodcenter.org

Klaus L. Leenders
Department of Neurology, University Medical Centre Groningen (UMCG), University of Groningen, The Netherlands
k.l.leenders@neuro.umcg.nl

Anthony J. Lees
Department of Molecular Neuroscience, Institute of Neurology, Queen Square, London WC1N 3BG, UK and Reta Lila Weston Institute for Neurological Studies, University College London, 1 Wakefield Street, London WC1N 1PJ, UK

Clotilde Lévecque
The Wellcome Trust Centre for Human Genetics, University of Oxford, Roosevelt Drive, Headington, Oxford OX3 7BN, UK
clotilde@well.ox.ac.uk

Jacqueline McIntosh
Wolfson Neurorehabilitation Centre and Atkinson Morley Wing (Neurosciences), St George's Healthcare NHS Trust, Wimbledon, London, UK
Jacq.McIntosh@stgeorges.nhs.uk

Russell L. Margolis
Laboratory of Genetic Neurobiology, Division of Neurobiology, Department of Psychiatry and Behavioral Sciences, Johns Hopkins University School of Medicine, CMSC 8-121, 600 N. Wolfe Street, Baltimore, MD 21287, USA
rmargoli@jhmi.edu

Mieko Matsuda
Departments of Psychiatry, Kagoshima University Graduate School of Medical and Dental Sciences, Kagoshima 890-8520, Japan

Anthony P. Monaco
The Wellcome Trust Centre for Human Genetics, University of Oxford, Roosevelt Drive, Headington, Oxford OX3 7BN, UK
anthony.monaco@well.ox.ac.uk

Masayuki Nakamura
Departments of Psychiatry, Kagoshima University Graduate School of Medical and Dental Sciences, Kagoshima 890-8520, Japan
nakamu36@m.kufm.kagoshima-u.ac.jp

Manuela Neumann
Ludwig-Maximilians-Universität, Zentrum für Neuropathologie und Prionforschung, Feodor-Lyness-Straße 23, D-81377 München, Germany

Shinichi Ohno
Department of Anatomy, Interdisciplinary Graduate School of Medicine and Engineering, 1110 Shimokato, Chuo-City, Yamanashi 409-3898, Japan

Thomas Pollmächer
Max-Planck-Institute für Psychiatrie, München, Germany

Dobrila D. Rudnicki
Laboratory of Genetic Neurobiology, Division of Neurobiology, Department of Psychiatry and Behavioral Sciences, Johns Hopkins University School of Medicine, CMSC 8-121, 600 N. Wolfe Street, Baltimore, MD 21287, USA

Shinji Saiki
Department of Medical Genetics, Cambridge Institute for Medical Research, Wellcome Trust/MRC Building, Addenbrooke's Hospital, Hills Road, Cambridge CB2 2XY, UK
ss644@cam.ac.uk

Akira Sano
Departments of Psychiatry, Kagoshima University Graduate School of Medical and Dental Sciences, 8-35-1 Sakuragaoka, Kagoshima 890-8520, Japan
sano@m3.kufm.kagoshima-u.ac.jp

James D. Schmelzer
Department of Neurology, Mayo Clinic, Minnesota, USA

Susanne A. Schneider
Institute of Neurology, Queen Square, London, WC1N 3BG, UK
S.Schneider@ion.ucl.ac.uk

Quan Sha
New York Blood Center, 310 East 67th Street, New York, NY 10021, USA

Hirotoshi Shibata
Japanese Red Cross Osaka Blood Center, 2-4-43, Morinomiya Joto-ku, Osaka 536-8505, Japan

Gordon W. Stewart
Department of Medicine, University College London, Rayne Building, University Street, London WC1E 6JJ, UK
g.stewart@ucl.ac.uk

William A. Symmans (deceased)
Hamilton Medical Laboratory, Hamilton, New Zealand

Junko Takahashi
Japanese Red Cross Osaka Blood Center, 2-4-43, Morinomiya Joto-ku, Osaka 536-8505, Japan

Yohei Tamura
Department of Neurology, Jikei University School of Medicine, 3-25-8, Nishi-Shinbashi, Minato-ku, Tokyo 105-8461, Japan

Mitsunobu Tanaka
Japanese Red Cross Osaka Blood Center, 2-4-43, Morinomiya Joto-ku, Osaka 536-8505, Japan

Yoshihiko Tani
Japanese Red Cross Osaka Blood Center, 2-4-43, Morinomiya Joto-ku, Osaka 536-8505, Japan
tani@osaka.bc.jrc.or.jp

Nobuo Terada
Department of Anatomy, Interdisciplinary Graduate School of Medicine and Engineering, 1110 Shimokato, Chuo-City, Yamanashi 409-3898, Japan

François Tison
Service de Neurologie-Hôpital du Haut Lévêque, 33604 Pessac, France

Marcus Tolnay
Institute of Pathology, Division of Neuropathology, University Hospital Basel, Switzerland

Shu-Ichi Ueno
Department of Psychiatry, Course of Integrated Brain Sciences, University of Tokushima School of Medicine, Japan

Antonio Velayos-Baeza
The Wellcome Trust Centre for Human Genetics, University of Oxford, Roosevelt Drive, Headington, Oxford OX3 7BN, UK
avelayos@well.ox.ac.uk

Dennis Velakoulis
Melbourne Neuropsychiatry Centre, University of Melbourne, and Neuropsychiatry Unit, Royal Melbourne Hospital, Melbourne, Australia

Ruth H. Walker
Movement Disorders Clinic, Department of Neurology, James J. Peters Veterans Affairs Medical Center, 130 W. Kingsbridge Road (127), Bronx, NY 10468, and Mount Sinai School of Medicine, New York, NY, USA
ruth.walker@mssm.edu

Mark Walterfang
Melbourne Neuropsychiatry Centre, University of Melbourne, and Neuropsychiatry Unit, Royal Melbourne Hospital, Melbourne, Australia
mark.walterfang@mh.org.au

Stephanie M.S. Wilmore
Department of Medicine, University College London, Rayne Building, University Street, London WC1E 6JJ, UK

Fusako Yokochi
Department of Neurology, Tokyo Metropolitan Neurological Hospital, 2-6-1 Musashidai, Fiuchru, Tokyo, Japan
fyokochi-tmnh@umin.ac.jp

Kurano Yutaka
Departments of Psychiatry, Kagoshima University Graduate School of Medical and Dental Sciences, Kagoshima 890-8520, Japan

Xiang Zhu
New York Blood Center, 310 East 67th Street, New York, NY 10021, USA

Part I
The Neuroacanthocytosis Syndromes

Neuroacanthocytosis Syndromes – A Current Overview

R.H. Walker(✉), S. Saiki, and A. Danek

Abstract Neuroacanthocytosis syndromes are characterized by the presence of "thorny" red blood cells and neurodegeneration of the basal ganglia, along with peripheral neuromuscular findings, seizures, and a variety of neuropsychiatric features. In recent years significant progress has been made in understanding the molecular pathophysiology of these disorders; cases are now identified as autosomal recessive chorea-acanthocytosis, X-linked McLeod syndrome, or more rarely, pantothenase kinase-associated neurodegeneration or Huntington's disease-like 2. Molecular analysis of classic reports of neuroacanthocytosis will clarify nomenclature and improve understanding of genotype-phenotype correlations. In addition, there are issues of atypical inheritance patterns which remain to be elucidated. A relatively high incidence of chorea-acanthocytosis in Japan may indicate a genetic founder effect, and has led to significant developments from Japanese researchers.

R.H. Walker
Departments of Neurology, James J. Peters Veterans Affairs Medical Center, Bronx, NY and Mount Sinai School of Medicine, New York, NY, USA
ruth.walker@mssm.edu

R.H. Walker et al. (eds.), *Neuroacanthocytosis Syndromes II.*

1 Advances in Neuroacanthocytosis

This volume follows up on "Neuroacanthocytosis syndromes" [2], which summarized the proceedings of a symposium which took place in Seeon, Bavaria, in May 2002. That meeting brought together a diverse group of researchers from around the world, including movement disorder neurologists, molecular biologists, hematologists, and neurogeneticists, and many others involved in studying this group of rare diseases. The group explored "New perspectives for the study of basal ganglia degeneration", following the discovery by scientists in England and Japan of the genetic basis of the core neuroacanthocytosis (NA) syndromes, McLeod syndrome (MLS) and chorea-acanthocytosis (ChAc) [43, 100, 142].

The collaborations initiated at this meeting led to the second Neuroacanthocytosis Symposium (Montreal Neurological Institute, Montreal, Canada, April 2005) and the third symposium, a satellite meeting of the 10th International Congress of Parkinson's disease and Movement Disorders, organised by the Movement Disorder Society, in Kyoto, Japan, October 2006 (Figs. 1 and 2 of the Foreword). An interim meeting was convened at the Third International Congress on Vascular Dementia in Prague, Czech Republic, October 2003 [27].

The present volume expands on the abstracts of the Kyoto [111] and Montreal [7] symposia. Major scientific developments since the first meeting in 2002 have included the publication of the reference method for acanthocyte detection [127], the description of *XK* mutations without full-blown MLS [54, 150], and the discovery of additional members, *XPLAC* and *XTES*, of the *XK* family [18]. The genes related to *CHAC* were found to belong to a conserved gene family involved in vacuolar protein sorting, thus leading to the renaming of *CHAC* as *VPS13A* [143]. The use of chorein antibodies for diagnosis of ChAc using Western blot assay [30] was a major clinical advance, obviating the need for molecular analysis of the large *VPS13A* gene for diagnosis in most cases.

Major developments from Japan included the observation of cellular inclusions in ChAc muscle [133] and the first animal model for ChAc [137], mice with the *VPS13A* deletion-mutation found in the Ehime province [71, 87].

The nosology of NA syndromes has evolved since the Seeon meeting. A family reported there with autosomal dominant NA [148] was identified as having Huntington's disease-like 2 [144, 146, 147, 149] (see chapter by Margolis).

The fourth "core" NA syndrome, pantothenate-kinase associated neurodegeneration (PKAN), has been better delineated with the availability of genetic testing for *PANK2* mutations. The so-called HARP syndrome (hypoprebetalipoproteinemia, acanthocytosis, retinitis pigmentosa, pallidal degeneration), of which only two cases have been reported, was proven to be allelic with PKAN [26, 46, 95]. The occurrence of acanthocytes as a feature of PKAN – in at least 8% of patients – has lately become more appreciated [39, 97].

Here we discuss the history of NA in Japan and in the English medical literature, including the appropriateness of the use of the "Levine–Critchley" eponym and the confusion in diagnosis of NA syndromes prior to the molecular era. We discuss recent developments in ChAc and issues related to the genetics of this disorder (advances in MLS are discussed in the chapter by Jung). Therapy for NA syndromes is at present in its infancy and is limited to symptom management, however patients with these debilitating disorders may nevertheless derive significant benefit from medical and non-medical interventions, which are summarised below.

2 Neuroacanthocytosis in Japan

A historical perspective of NA (probable ChAc, but never confirmed using current methods) reveals the large number case reports in Japanese, starting with Shimizu's report of 1974 [122] and continuing into the 1990s, most of which were initially presented at local meetings of the Japanese Society of Neurology [9, 13, 35, 45, 48, 50–53, 55, 57–59, 63, 65, 66, 68, 81, 91, 92, 106, 114, 119, 120, 123, 128–130, 132, 136, 151]. According to a personal communication by Prof. Hirose (and see his chapter in this volume), the topic of NA syndromes was selected for a symposium at the 1980 Annual Meeting by the then president (Prof. Kameyama, Kyoto) of the Japanese Society of Neurology [1]. Much of what was presented there [10, 41, 64, 67, 88, 107, 112, 116, 118, 153] had already been reported locally and, since funds had been devoted to the study of NA by the Ministry of Health and Welfare of Japan to Prof. Toyokura of Tokyo University, these cases were also collected as brief reports for a government record of these activities [138]. All relevant studies were reported in more detail in journals with English abstracts such as "Rinsho Shinkeigaku" (Clinical Neurology), "Shinkeinaika" (Neurological Medicine), "No To Shinkei" (Brain and Nerve), or in international journals [12, 42, 60, 62, 72, 89, 93, 94, 96, 113, 115, 117, 121, 124, 126, 131, 134, 135, 140, 141, 152].

An exact count of case numbers in Japan is difficult because of inaccessibility of material to English speakers and duplicate publications, but the number of 71 patients mentioned by Prof. Hirose in his chapter appears to be a conservative estimate. Thus Japan, in contrast to the few reports from other parts of Asia [5, 15, 17, 28, 37, 61, 69, 83, 84, 104, 139] and the approximately 150–200 cases from the rest of the world [25], appears to have the highest diagnosis rate in the world.

3 Levine–Critchley Syndrome

To date, confirmation of the identity of the patients with the original NA "Levine–Critchley syndrome" has not been possible. Attempts to trace the New England family described since 1960 by Irving M. Levine (Fig. 1) and others [33, 34, 70, 74–76, 103] have not yet been rewarding. The description of the proband is consistent with ChAc or MLS, particularly the descriptions that "his gait was lurching in character with long strides and somewhat ataxic because of quick involuntary knee buckling movements. … Speech was moderately inarticulate because it was interrupted by involuntary tongue and facial movements." [75] However, the clinical features of other family members and the pedigree diverge to some extent from the

Fig. 1 Irving M. Levine (*right*) with Mitchell F. Brin, circa 1985. Photograph courtesy of Dr Mitchell F. Brin

features of patients with molecularly identified ChAc. There was consanguinity five generations previous to the affected family members. Examination of the pedigree shows that chorea was present in one of the proband's siblings, a maternal aunt, and a cousin who had *chorea gravidarum* [75]. All of these subjects had acanthocytosis. The neurological findings in many of the other reported family members from four generations are challenging to interpret, and consist predominantly of various combinations of hyporeflexia, mild weakness and muscle atrophy. Acanthocytosis was found mostly in family members with these findings, but did not necessarily cosegregate with the neurological abnormalities. There is apparent male–male transmission excluding the diagnosis of MLS and the inheritance pattern in general appears to be autosomal dominant. However, the variability of the neurological findings, and the presence of acanthocytes in a girl with epilepsy on the father's side of the family, which was otherwise not felt to be affected, make it difficult to draw a conclusion as to the nature of the disorder being described.

The patients reported by Edmund Critchley from two families, one from East Kentucky, USA [22, 23], the other from Lancashire, UK [16, 21], however, display the typical clinical features of genetically confirmed cases [25]. Edmund Critchley (Fig. 2) nicely has summarized his findings [20], most recently in an enjoyable "Neurologist's Tale" [19].

Fig. 2 Edmund Critchley. Photograph courtesy of Dr E.M.R. Critchley

While working at the Department of Neurology of David Barrett Clark in Lexington, Kentucky, it was his task to provide neurologic care in the small towns of the hinterland: "I went about weekly to a regional clinic. We would hire a State limousine, seating at least six people and holding an EEG set, and travel down a state highway at about 90 mph (looking out for helicopters) to a National Park, where we stay the night before an early start treating epilepsy, genetic diseases and other neurological disorders referred to us." [19] In these isolated regions he discovered his "prize patient", for whom he uses the alias Terry. "The poor were white, the Scots-Irish, the inbred white people living along the 'creeks' or narrow river valleys, the Appalachian poor. … Forced to work on the plantations of Virginia to pay their passage to America, many chose to escape and live rough as frontiersmen. They found a creek with a bit of land, safe from other feuding wild men … married and produced offspring. The progeny moved further along the creek to poorer and poorer land, intermarrying with their own kith and kin" [19].

"Terry had presented to Abe Wikler, Professor of Psychiatry, Neurology and Pharmacy, who had attempted to try to control his movement disorder to no avail. Dave Clark was determined that he should be thoroughly investigated and I was the man to do so .... Terry was a 29-year-old white male. When first seen at the age of 26, he exhibited involuntary movements and had a grossly swollen, raw, bitten tongue. He had had 15–20 similar episodes of tongue, lip and cheek biting, which often occurred at night. The episodes started 6 years earlier on a background of increased generalized weakness, nervousness, 'fits and jerks', and had increased in frequency and severity. The involuntary movements included finger-snapping, grimacing, dystonic and choreiform movements, hyperextension of the trunk, twisting movements of his shoulders, sucking noises, plosive sounds and drooling. ... There had been times when he could not speak plainly: 'the inside of his mouth would draw', he would 'snap at his lips, and his stomach would stick'. When he ate, his tongue would involuntarily push food out on to his plate. For 4 years he had preferred to retire to a separate room to eat.

"He showed no psychotic or hallucinatory behaviour, but on his later admissions, appeared somewhat disinhibited sexually. Over 2 years, he had two episodes of 'passing out' preceded by abnormal noises and shaking or tremor of the abdomen and outstretched extremities, 'drawing up of the legs'. These episodes, which lasted for 30 min, were followed by confusion and a 'wild look' which persisted for approximately 1 h. On admission in 1967, his involuntary movements were so intense that he could not walk without assistance. He was alert, well-orientated, had no gross memory defects, and was disturbed by his own repulsive appearance. He slurred and stuttered when talking, was often indistinct and had occasional inappropriate laughter. Despite the involuntary movements, there was no ataxia and co-ordination tests were intact. He had generalized hypotonia, flexor plantar responses, and loss of deep tendon reflexes. There was a suspicion of thinning with coarse twitching of his calves. His IQ was 72 (WAIS), verbal 81, and performance 61. ... He was the 10th child.

"The eldest died of seizures, bit her tongue and had involuntary limb movements. She became forgetful, emaciated and bedfast with violent shaking of her limbs. Two others also died about the age of 26. The fourth child had an illness of 2 years duration, with passing out spells and rejection of food from the mouth. The fifth gave birth to an unaffected child and soon afterwards became bedfast and emaciated. The ninth remained healthy till aged 31, before developing choreic movements and grand mal fits. She has always refused admission to hospital but did permit an examination on herself and her normal 17-month child. I witnessed what appeared to be an attack of hystero-epilepsy with partial loss of consciousness, opisthotonus (bending the back in hyperextension) and drawing and grasping movements of all four limbs. Speech and swallowing were normal but she showed facial grimacing sans distal choreic movements. The deep tendon reflexes were absent and the diagnosis of acanthocytosis confirmed" [19].

Critchley provides us with the vivid picture of a severe and progressive neurological condition starting in the third decade in possibly seven male and female siblings (out of ten) from a genetically isolated background favoring intermarriage and manifestation of autosomal recessive traits. He clearly describes severe orofacial dyskinesia with a tongue that would involuntarily push food out on to the plate, generalized chorea severely interfering with gait, sudden trunk flexion and hyperextension, some neuropsychiatric and cognitive impairment and possible epileptic seizures. Critchley also noted the resemblance to the tics of the Gilles de la Tourette syndrome [20].

It is commonly said that "chance favors the prepared mind" and this was certainly true for Edmund Critchley when he happened to see a single patient with an almost identical movement disorder, her parents unrelated, coming from different counties [20]. "Within a year of my return to England, a 30-year old patient was referred to me … her husband had noticed … that her walk was ungainly and had avoided drinking with her in company because of her tendency to slobber … increasingly irritated by grunting noises, of which at first she was unaware. … She started to lose weight … complaining of intermittent difficulty in swallowing and a tightness in the throat. … Most disturbing were the wide variety of oro-facial tics … continually present during waking hours, though fluctuating in severity from day to day … Her speech was dysarthric, partly broken by involuntary movements. She would also grunt, suck, make repetitive sounds, bite her tongue and lower lip, or struggle to control the pooling of saliva. The involuntary limb movements were sometimes dystonic and choreiform with hyperextension and flexion of her trunk and throwing out of an arm or leg … she took her own life … A coroner's post-mortem was cursorily performed …" [19].

All of Critchley's cases appear to have had a phenotype identical to that seen in patients with a molecular diagnosis of ChAc [25]. However, in the absence of genetic testing, we may never know whether Levine–Critchley syndrome truly was ChAc or some other related neurodegenerative disease, making use of the eponym rather imprecise.

4 Subsequent Neuroacanthocytosis Reports

The question of diagnosis is illustrated by the paper of Hardie and collaborators from Queen Square [38]. This often-cited NA case series has become problematic since its group of 19 patients has since been shown to be genetically heterogeneous (see chapter by Gandhi et al.). Cases 1–4 (family H) were found to be compound heterozygotes for mutations in the *VPS13A* gene (family CHAC 01 [105]). Cases 5–10 (family L) carry a McLeod mutation [44]. This family, which was originally reported as having a benign condition [79], is unique because of the severe disease expression in a female mutation carrier. Her neuropathological findings are described in chapters by Danek et al. and Geser et al. Cases 11–12 (family B) were only found to have one *VPS13A* mutation (family CHAC 09 [100]). A few of Hardie's cases have not yet been diagnosed using molecular methods.

Another report in which the diagnosis remains to be clarified is that of two Russian-Jewish brothers who presented in their 50's with proximal muscle weakness and sensory neuropathy in addition to chorea and orofacial dyskinesia [4]. Based on an observation from Israel, where homozygous deletions of *VPS13A* (6059delC) had been found in two unrelated patients, it was postulated that these cases provide support for a founder effect of ChAc in Ashkenazi Jews [78]. However, gender, age at onset and clinical manifestations are more consistent with a diagnosis of MLS than of ChAc, as is the pedigree structure of the two brothers [4]. Motor restlessness, seizures and neuropathy in the mother would not be expected in the autosomal recessive condition of ChAc and could be better explained if she were a manifesting MLS mutation carrier.

5 Chorea-Acanthocytosis – Recent Developments

5.1 Molecular Studies

Autosomal recessive chorea-acanthocytosis (ChAc) is due to mutations of the large gene *VPS13A* [100, 25, 29, 142] which encodes for the protein chorein that appears to be fairly widely expressed in brain [11, 14, 71] (see chapter by Bader et al.). The function of its protein product, chorein, is not known, apart from that it is likely to be a sorting protein, and may be involved in intracellular protein trafficking (see chapter by Velayos-Baeza et al.). A large number of different mutations, including frameshifts (small deletions or insertions), nonsense, splice-site, deletions, and missense mutations, and deletions of whole exons have been found throughout the entire gene, making screening a challenge [29, 100, 142].

5.2 Clinical Features

The typical onset of ChAc is in young adulthood, but may occur at younger ages. Initial presentation with a variety of neuropsychological syndromes is being increasingly recognized, including depression, psychosis, obsessive-compulsive symptoms, trichotillomania [78], tourettism [90, 108], anxiety and agitation (see chapter by Sano). Self-mutilation due to lip- and tongue-biting may be due to apparent obsessive behaviors or to involuntary movements [145]. Behavioral changes suggesting a frontal lobe-type syndrome, including disinhibition and self-neglect, may present prior to development of the movement disorder. Dementia is frequently seen, with deficits primarily in memory and executive skills. The use of psychiatric medications may obscure the diagnosis for several years, as the involuntary movements are attributed to tardive dyskinesia.

The involuntary movements of ChAc are usually chorea and dystonia, but occasionally Parkinsonism. Orofacial and lingual dyskinesias are typically prominent and troublesome, and may result in marked difficulty with eating [99].

Seizures are seen in approximately 40% of patients [99], but were more common in a cluster of French-Canadian families (70–80%; discussed in more detail below). Most attacks are described as generalized, probably of temporal origin, although mesial temporal sclerosis is not documented [3]. The presence of cortical abnormalities, either structural or functional [11, 32, 98], is under discussion (see chapter by Leenders and Jung).

Neuromuscular signs comprise peripheral neuropathy and myopathy [99]. Serum creatine kinase (CK) elevation may be detected before the appearance of neurologic signs or symptoms [78]. Muscle atrophy and weakness, as well as CK elevation, may be regarded as secondary to chronic denervation. However, a primary myopathic process is suggested by nemaline rods [133] or accumulations of tissue transglutaminase [85] on muscle biopsy, and a predisposition to rhabdomyolysis [102]. In symptomatic carriers of apparently a single mutation of chorein, accumulations of chorein may be seen along the muscle cell membrane [110] (see chapter by Saiki and Tamura).

Involvement of the autonomic nervous system is occasionally noted in ChAc (see chapter by Kihara et al.).

Non-neurologic findings such as hepatosplenomegaly are sometimes, but not invariably, observed [99]. Hepatosplenomegaly does not seem to be a significant cause of morbidity, as long as its etiology is recognized, and it does not prompt invasive investigations.

Neuroimaging in ChAc is typically reported as strongly resembling that in HD [73], although quantitative studies [40] indicate that the head of the caudate nucleus is particularly vulnerable (see chapter by Henkel et al.).

Polysomnographic studies in several ChAc patients reveal sleep fragmentation, poor sleep efficiency, short total sleep time and a high rate of wakefulness after sleep onset [125] (see chapter by Ghorayeb et al.). These findings likely explain the daytime somnolence in these patients. The observations resemble those in HD and thus may be due to a common pathophysiological mechanism.

6 Inheritance Patterns

Although the inheritance of ChAc is generally accepted as being autosomal recessive, and that of MLS as being X-linked, there are reports of apparent NA families in which the inheritance pattern does not conform to either of these. As discussed above, inheritance has been suggested as being autosomal dominant in Levine's cases [75] and the same has been speculated about in Critchley's Kentucky family [23]. The daughter of the eldest sibling (who was similarly affected to her brother "Terry") also had severe neurological abnormalities, however, her syndrome was quite dissimilar (with a Friedreich's ataxia-like phenotype and developmental delay) and Critchley himself doubted a relation of the two conditions [19].

A French-Canadian case cluster, centered around Saguenay-Lac St. Jean, Quebec, a farming region with a small founding population, was initially thought to show dominant inheritance [6]. Most cases originally presented with epilepsy as mentioned above [3]. Molecular analysis permitted clarification of the rather complex pattern of inheritance in four French Canadian pedigrees with eleven affected patients [31]. Three out of the four pedigrees were homozygous for a deletion of the four terminal exons of *VPS13A*. Two females in the fourth family were homozygous for a splicing mutation (4242 + 1 G > T). The affected males in this highly consanguineous pedigree were compound heterozygotes for the two mutations just mentioned. Identification of a common haplotype associated with the terminal deletion in four apparently unrelated families implied a founder effect for ChAc in French Canadians. Thus, the recessive transmission of *VPS13A* mutations was obscured by the high degree of consanguinity in the Saguenay-Lac St. Jean population, with the misleading impression of dominant inheritance [6].

Autosomal dominant inheritance of NA was suggested for an Italian family, although consanguinity of the parents suggests the possibility of pseudo-dominance [80]. Acanthocytes were also observed in members otherwise unaffected by NA symptoms from a further family [86]. Molecular analysis of these cases has not been reported.

Some of the descriptions of Japanese families with autosomal dominant transmission of NA syndromes could be due to isolation of the population and intermarriage. A founder effect has been shown for ChAc in Japan, in the form of the deletion mutation of *VPS13A* from the Ehime province [142]. One family has been reported with possibly autosomal dominant inheritance of ChAc, prior to the advent of genetic testing, although MLS or HDL2 (the latter unlikely in this population) was not excluded [62].

In two Japanese families apparent autosomal dominant effects of *VPS13A* mutations have been described. In one family of an affected woman with a homozygous point mutation (3889C > T causing R1297X), several relatives who carried a single copy of the diseased gene were said to display symptoms and signs characteristic for ChAc [47]. However, it is not completely clear that the neurological findings in the family could not have resulted from alternative causes.

In another family a single mutation was found of one allele of the gene, a heterozygous G > A transversion at the last nucleotide of *VPS13A* exon 57 in a brother and sister [49, 108, 109]. Their unaffected mother, not consanguineous with the father, showed a homozygous wild type sequence at that site. The father and grandfather were not available for examination, but it was presumed that the mutation had been introduced from the paternal side. Although genetic analysis was not performed, ChAc was also felt to be present in three subjects with acanthocytosis and chorea from successive generations in side branches of the pedigree, fitting a pattern of autosomal dominant inheritance. It is possible that, comparable to initial findings in the French-Canadians, additional mutations were present in this pedigree. A second mutation of the index siblings may have been missed due to technical reasons, as was not felt to be unusual in our analyses of ChAc, where the second mutation had remained undetected in altogether 12 patients from 66 pedigrees [25]. Assay of chorein levels and further genetic analysis will help clarify this issue.

7 Treatment of Neuroacanthocytosis Syndromes

7.1 *Pharmacological Therapy*

To date treatment for the NA syndromes is purely symptomatic. As in other choreiform disorders, reduction of involuntary movements may be achieved in principle by reducing dopaminergic neurotransmission, using atypical antipsychotic agents or tetrabenazine. However, as in HD, this may not necessarily result in functional improvement. Anticonvulsants such as levetiracetam [24, 101] and topiramate [36] can be beneficial in secondary choreas, and may be considered in NA. Levetiracetam has been reported as providing specific benefit to truncal tics in ChAc [77], yet anecdotal experience in a single case has not confirmed this effect. Carbamazepine and lamotrigine may worsen involuntary movements [8].

Neuropsychiatric issues are often a major cause of morbidity and mortality, as suicide is not infrequent (see chapter by Sano). Depression should be aggressively treated. Selective serotonin-reuptake inhibitors and tricyclic antidepressants may be useful for depression, in addition to mood-stabilizing medications such as anticonvulsants. The second-generation neuroleptics, especially clozapine and quetiapine, may improve both mood disorders and chorea. Classical neuroleptics should be avoided in order to avoid potential induction of tardive movement disorders.

7.2 *Neurosurgery*

With the development of experience in functional neurosurgery for other movement disorders, both hypo-and hyperkinetic, a small number of patients with NA syndromes have undergone either deep brain stimulation or ablative procedures (see chapter by Yokochi and Burbaud). The results are somewhat promising, but need to be interpreted within the context of a multi-symptomatic, progressive neurodegenerative disorder.

7.3 *Other Therapeutic Issues*

A multidisciplinary team approach to issues of mobility, communication, nutrition, swallowing, and psychosocial aspects is vital to address the many issues which may arise with these complex disorders. Adjunctive non-medical therapies are invaluable and must be individualised to each patient's needs and goals [82] (see chapter by McIntosh).

In MLS, clinicians should be aware of potential blood transfusion reactions. Hemolysis may result from the production of anti-Kx and anti-Kell antibodies, and autologous blood should be banked (see chapter by Tani).

Cardiac myopathies and dysrhythmias may occur in MLS, and very occasionally in ChAc [56], thus, patients should be monitored regularly and treated as indicated. Autonomic dysfunction (see chapter by Kihara et al.) may also play a role in sudden death. Sleep disorders have as yet received little therapeutic attention (see chapter by Ghorayeb et al.).

8 Conclusions

Developments in molecular biology have facilitated the diagnosis of NA syndromes and, most importantly, the provision of accurate genetic counseling. Molecular diagnosis has permitted confirmation of disease identity for patients who may present with atypical syndromes, and has revealed the diversity of phenotypic variations. As with current thinking for other, more common, movement disorders such as Parkinson's and Huntington's disease, we are increasingly recognizing non-motor aspects of NA syndromes, enabling us to address these additional clinical features in our patients.

Recent studies of neuroimaging and neuropathology are providing us with insights in basal ganglia circuitry, and may shed light upon the unique psychiatric features of NA syndromes.

Animal models and cell culture studies of protein function are vital to understand the effects of protein mutation at the level of the cell and the system. We hope that the small steps presented in this volume will ultimately lead to therapies for our patients with these neglected syndromes.

References

1. Neurological disease with acanthocytosis (1980). Clin Neurol (Tokyo) 20:1056–1084
2. Al-Asmi A, Jansen A, Badhwar A et al. (2005) Familial temporal lobe epilepsy as a presenting feature of choreoacanthocytosis. Epilepsia 46:1256–1263
3. Aminoff MJ (1972) Acanthocytosis and neurological disease. Brain 95:749–760
4. Anand IS, Khwaja GA, Gupta M (1994) Choreoacanthocytosis. J Assoc Physicians India 42:495–496
5. Andermann E, Badhwar A, Al Asmi A et al. (2002) Chorea-acanthocytosis in a large French-Canadian kindred with dominant or pseudo-dominant inheritance and variable expressivity. Am J Hum Genet 71:530
6. Andermann E, Danek A, Irvine G et al. (2005) Second international neuroacanthocytosis symposium: expanding the spectrum of choreatic syndromes: Montreal neurological hospital and institute, April 17–19 2005. Mov Disord 20:1673–1684
7. Andermann F, Jansen A, Al-Asmi A et al. (2006) Epilepsy in neuroacanthocytosis. Mov Disord 20:1680
8. Araki J, Tatsumi Y, Sannomiya Y et al. (1983) Chorea-acanthocytosis and deformability of erythrocytes [Japanese]. Jpn J Clin Hematol 24:1055–1059
9. Arima S, Mori R, Kato N (1980) A case of the familial chorea with acanthocytosis [Japanese]. Clin Neurol (Tokyo) 20:1069–1070

10. Arzberger T, Heinsen H, Buresch N et al. (2006) The neuropathology of chorea-acanthocytosis: from stereology to an immunohistochemical detection of chorein. Mov Disord 20:1679
11. Asano K, Osawa Y, Yanagisawa N et al. (1985) Erythrocyte membrane abnormalities in patients with amyotrophic chorea with acanthocytosis. Part 2. Abnormal degradation of membrane proteins. J Neurol Sci 68:161–173
12. Asano K, Oshima M (1990) Protein extractabilities from the erythrocyte membranes in patients with chorea-acanthocytosis. Jpn J Exp Med 60:355–358
13. Bader B, Arzberger T, Dobson-Stone C et al. (2007) Chorein expression in post-mortem brain tissue of chorea-acanthocytosis patients and non-affected controls. J Neural Transm 114:XXVIII
14. Behari M, Saha P, Prasad K et al. (1991) Choreoacanthocytosis with marked dysphagia and laryngeal dystonia. J Assoc Physicians India 39:967–968
15. Betts JJ, Nicholson JT, Critchley EM (1970) Acanthocytosis with normolipoproteinaemia: biophysical aspects. Postgrad Med J 46:702–707
16. Bharucha EP, Bharucha NE (1989) Choreo-acanthocytosis. J Neurol Sci 89:135–139
17. Calenda G, Peng J, Redman CM et al. (2006) Identification of two new members, XPLAC and XTES, of the XK family. Gene 370:6–16
18. Critchley E (2001) A neurologist's tale. The Memoir Club, Spennymoor
19. Critchley EMR (1971) Acanthocytosis associated with tics and involuntary movements. Z Neurol 200:336–340
20. Critchley EMR, Betts JJ, Nicholson JT et al. (1970) Acanthocytosis, normolipoproteinemia and multiple tics. Postgrad Med J 46:698–701
21. Critchley EMR, Clark DB, Wikler A (1967) An adult form of acanthocytosis. Trans Am Neurol Assoc 92:132–137
22. Critchley EMR, Clark DB, Wikler A (1968) Acanthocytosis and neurological disorder without betalipoproteinemia. Arch Neurol 18:134–140
23. D'Amelio M, Callari G, Gammino M et al. (2005) Levetiracetam in the treatment of vascular chorea: a case report. Eur J Clin Pharmacol 60:835–836
24. Danek A (ed) (2004) Neuroacanthocytosis syndromes. Springer, Dordrecht, The Netherlands
25. Danek A, Dobson-Stone C, Velayos-Baeza A et al. (2006) The phenotype of chorea-acanthocytosis: a review of 106 patients with *VPS13A* mutations. Mov Disord 20:1678
26. Danek A, Hegele A (2004) Compound heterozygous *PANK2* mutations confirm HARP and Hallervorden–Spatz syndromes are allelic (correspondence). Neurology On-line material: http://www.neurology.org/cgi/eletters/61/10/1423
27. Danek A, Jung HH, Melone MAB et al. (2005) Neuroacanthocytosis: new developments in a neglected group of dementing disorders. J Neurol Sci 229–230:171–186
28. Dixit SN, Sharma S, Behari M et al. (1993) Neuroacanthocytosis with pure chorea. J Assoc Physicians India 41:613
29. Dobson-Stone C, Danek A, Rampoldi L et al. (2002) Mutational spectrum of the *CHAC* gene in patients with chorea-acanthocytosis. Eur J Hum Genet 10:773–781
30. Dobson-Stone C, Velayos-Baeza A, Filippone LA et al. (2004) Chorein detection for the diagnosis of chorea-acanthocytosis. Ann Neurol 56:299–302
31. Dobson-Stone C, Velayos-Baeza A, Jansen A et al. (2005) Identification of a *VPS13A* founder mutation in French Canadian families with chorea-acanthocytosis. Neurogenetics 6:151–158
32. Dydak U, Mueller S, Sandor PS et al. (2006) Cerebral metabolic alterations in McLeod syndrome. Eur Neurol 56:17–23
33. Engel WK, Levy RI, Watkins SP et al. (1967) Discussion: an adult form of acanthocytosis. Trans Am Neurol Assoc 92:135–137
34. Estes JW, Morley TJ, Levine IM et al. (1967) A new hereditary acanthocytosis syndrome. Am J Med 42:868–881
35. Fujimoto Y, Isozaki E, Yokochi F et al. (1997) A case of chorea-acanthocytosis successfully treated with posteroventral pallidotomy [Japanese]. Clin Neurol (Tokyo) 37:891–894

36. Gatto EM, Uribe RC, Raina G et al. (2004) Vascular hemichorea/hemiballism and topiramate. Mov Disord 19:836–838
37. Gunatilake SB, Janaka De Silva H, Seneviratne S et al. (1997) Neuroacanthocytosis. Ceylon Med J 42:183–184
38. Hardie RJ, Pullon HWH, Harding AE et al. (1991) Neuroacanthocytosis: a clinical, haematological and pathological study of 19 cases. Brain 114:13–49
39. Hayflick SJ, Westaway SK, Levinson B et al. (2003) Genetic, clinical, and radiographic delineation of Hallervorden–Spatz Syndrome. N Engl J Med 348:33–40
40. Henkel K, Danek A, Grafman J et al. (2006) Head of the caudate nucleus is most vulnerable in chorea-acanthocytosis: a voxel-based morphometry study. Mov Disord 21:1728–1731
41. Hirayama K, Yamada T (1980) Familial neuro-acanthocytosis in adult. Special consideration on symptomatology of involuntary movements [Japanese]. Clin Neurol (Tokyo) 20:1064–1066
42. Hirayama M, Hamano T, Shiratori M et al. (1997) Chorea-acanthocytosis with polyclonal antibodies to ganglioside GM1. J Neurol Sci 151:23–24
43. Ho M, Chelly J, Carter N et al. (1994) Isolation of the gene for McLeod syndrome that encodes a novel membrane transport protein. Cell 77:869–880
44. Ho MF, Chalmers RM, Davis MB et al. (1996) A novel point mutation in the McLeod syndrome gene in neuroacanthocytosis. Ann Neurol 39:672–675
45. Hosokawa T, Omoto K, Kanaseki T et al. (1992) Studies on the erythrocyte membrane skeleton in a patient with chorea-acanthocytosis. Theoretical speculation on the mechanism of neurological involvement. Brain and Nerve (Tokyo) 44:739–744
46. Houlden H, Lincoln S, Farrer M et al. (2003) Compound heterozygous *PANK2* mutations confirm HARP and Hallervorden–Spatz syndromes are allelic. Neurology 61:1423–1426
47. Ichiba M, Muroya S, Tomemori Y et al. (2005) Inheritance pattern of ChAc in a pedigree in Kagoshima, Japan. Mov Disord 20:1678–1679
48. Imoto K (1983) A case report. Clin Neurol (Tokyo) 23:86
49. Ishida C, Makifuchi T, Saiki S et al. (2007) An autopsied case of chorea-acanthocytosis with a single detected mutation of *VPS13A*. Mov Disord 22:VI
50. Itoga E, Kito S, Fujita H (1978) A case of amyotrophic chorea with acanthocytosis. J Hiroshima Med Assoc 31:772–775
51. Itoga E, Kito S, Fujita H et al. (1979) A pedigree of amyotrophic chorea with acanthocytosis. Saishin Igaku (Osaka) 308–313
52. Itoga E, Kito S, Funakawa W et al. (1981) Amyotrophic chorea with acanthocytosis. Neurol Med (Tokyo) 15:8–15
53. Iwata M, Fuse S, Sakuta M et al. (1984) Neuropathological study of chorea-acanthocytosis. Jpn J Med 23:118–122
54. Jung HH, Hergersberg M, Vogt M et al. (2003) McLeod phenotype associated with a *XK* missense mutation without hematologic, neuromuscular, or cerebral involvement. Transfusion 43:928–938
55. Kageyama Y, Kodama Y, Tadano M et al. (2000) A case of chorea-acanthocytosis with dilated cardiomyopathy and myopathy [Japanese]. Clin Neurol (Tokyo) 40:816–820
56. Kageyama Y, Matsumoto K, Ichikawa K et al. (2007) A new phenotype of chorea-acanthocytosis with dilated cardiomyopathy and myopathy. Mov Disord 22:1669–1670
57. Kamakura K, Shimizu T, Shimada I et al. (1979) Involuntary movements associated with acanthocytosis. Report of two cases. Saishin Igaku (Osaka) 34:304–307
58. Kamakura K, Shimizu T, Toyokura Y et al. (1981) Chorea-acanthocytosis (Levine–Critchley syndrome). Report of a family with biochemical study. Neurol Med (Tokyo) 15:1–7
59. Kamiya K, Sugino F, et al. (1983) A case report of chorea-acanthocytosis. Clin Neurol (Tokyo) 23:632
60. Kawakami T, Takiyama Y, Sakoe K et al. (1999) A case of McLeod syndrome with unusually severe myopathy. J Neurol Sci 166:36–39
61. Khwaja GA (1999) Neuroacanthocytosis. J Assoc Physicians India 47:656

62. Kito S, Itoga E, Hiroshige Y et al. (1980) A pedigree of amyotrophic chorea with acanthocytosis. Arch Neurol 37:514–517
63. Kobayashi J (1983) A case of myoclonus epilepsy, mitochondrial myopathy and acanthocytosis. Clin Neurol (Tokyo) 23:86
64. Kohriyama T, Yoshida Y, Nakamura N et al. (1980) Pathogenesis of chorea with acanthocytosis. Clin Neurol (Tokyo) 20:1076–1078
65. Kohriyama T, Yoshida Y, Nakamura N et al. (1981) Chorea-acanthocytosis. Report of two cases with study on membrane-abnormality. Neurol Med (Tokyo) 15:30–35
66. Kohriyama T, Yoshida Y, Nakamura N et al. (1980) Progressive chorea with acanthocytosis. A report of two cases and biochemical studies. Nihon Naikagakkai-shi 69:26–32
67. Kondo K, Ueno E, Makishita H et al. (1980) Clinical studies of 4 cases of chorea with acanthocytosis. Clin Neurol (Tokyo) 20:1073–1075
68. Kondo K, Yanagisawa N (1981) Familial chorea with neuropathy and acanthocytosis: report of six cases with special reference to its clinical variety. Neurol Med (Tokyo) 15:118–127
69. Kumari KTS, Ravikumar A, Nair RA et al. (1997) Changes in the glycosaminoglycans, glycoproteins and lipid components of erythrocyte membrane in neuroacanthocytosis. Neurology (India) 45:23–27
70. Kuo PT, Bassett DR (1962) Blood and tissue lipids in a family with hypo-beta-lipoproteinemia. Circulation 26:660
71. Kurano Y, Nakamura M, Ichiba M et al. (2007) In vivo distribution and localization of chorein. Biochem Biophys Res Commun 353:431–435
72. Kuroiwa Y, Ohnishi A, Sato Y et al. (1984) Chorea acanthocytosis: clinical pathological and biochemical aspects. Int J Neurol 18:64–74
73. Kutcher JS, Kahn MJ, Andersson HC et al. (1999) Neuroacanthocytosis masquerading as Huntington's disease: CT/MRI findings. J Neuroimaging 9:187–189
74. Levine IM (1964) An hereditary neurological disease with acanthocytosis. Neurology 14:272
75. Levine IM, Estes JW, Looney JB (1968) Hereditary neurological disease with acanthocytosis, a new syndrome. Arch Neurol 19:403–409
76. Levine IM, Yettra M, Stefanini M (1960) A hereditary neurological disorder with acanthocytosis. Neurology 10:425
77. Lin FC, Wei LJ, Shih PY (2006) Effect of levetiracetam on truncal tic in neuroacanthocytosis. Acta Neurol Taiwan 15:38–42
78. Lossos A, Dobson-Stone C, Monaco AP et al. (2005) Early clinical heterogeneity in choreoacanthocytosis. Arch Neurol 62:611–614
79. Lowe D, Ibbotson RM, Loizou A (1977) Benign normolipoproteinaemic acanthocytosis. Br Med J 2:99–100
80. Marson AM, Bucciantini E, Gentile E et al. (2003) Neuroacanthocytosis: clinical, radiological, and neurophysiological findings in an Italian family. Neurol Sci 24:188–189
81. Matsumoto H (1986) Meal-time assistance of a patient with chorea-acanthocytosis [Japanese]. Kurinikaru Sutadi 7:888–893
82. McIntosh J (2006) Speech/Occupational therapy for neuroacanthocytosis. Mov Disord 20:1682
83. Meenakshi-Sundaram S, Arun Kumar MJ, Sridhar R et al. (2004) Neuroacanthocytosis misdiagnosed as Huntington's disease: a case report. J Neurol Sci 219:163–166
84. Mehndiratta MM, Malik S, Kumar S et al. (2000) McLeod syndrome (a variant of neuroacanthocytosis). J Assoc Physicians India 48:356–357
85. Melone MA, Di Fede G, Peluso G et al. (2002) Abnormal accumulation of tTGase products in muscle and erythrocytes of chorea-acanthocytosis patients. J Neuropathol Exp Neurol 61:841–848
86. Metzer WS (1990) Neuroacanthocytosis with autosomal dominant inheritance, normal serum CK and preserved myotatic reflexes. Mov Disord 5(Suppl 1):92
87. Mizuno E, Nakamura M, Agemura A et al. (2007) Brain-specific transcript variants of 5′ and 3′ ends of mouse *VPS13A* and *VPS13C*. Biochem Biophys Res Commun 353:902–907

88. Mori K, Ide Y, Haku R (1980) A case with choreic involuntary movement and acanthocytosis, who responded to steroid treatment. Clin Neurol (Tokyo) 20:1071–1072
89. Mukoyama M, Kazui H, Sunohara N et al. (1986) Mitochondrial myopathy, encephalopathy, lactic acidosis, and stroke-like episodes with acanthocytosis: a clinicopathological study of a unique case. J Neurol 233:228–232
90. Müller-Vahl KR, Berding G, Emrich HM et al. (2007) Chorea-acanthocytosis in monozygotic twins: clinical findings and neuropathological changes as detected by diffusion tensor imaging, FDG-PET and (123)I-beta-CIT-SPECT. J Neurol 254:1081–1088
91. Nagashima T, Iwashita H, Kuroiwa Y et al. (1979) Chorea-acanthocytosis. A report of a family [Japanese]. Clin Neurol (Tokyo) 19:609–615
92. Nakanishi K, et al. (1983) A case of chorea-athetosis and poikilocytosis. Proceedings of 24th Japanese Neurology Meeting 62
93. Ohnishi A, Sato Y, Nagara H et al. (1981) Neurogenic muscular atrophy and low density of large myelinated fibres of sural nerve in chorea-acanthocytosis. J Neurol Neurosurg Psychiatry 44:645–648
94. Okamoto K, Ito J, Furusawa T et al. (1997) CT and MR findings of neuroacanthocytosis. J Comput Assist Tomogr 21:221–222
95. Orrell RW, Amrolia PJ, Heald A et al. (1995) Acanthocytosis, retinitis pigmentosa, and pallidal degeneration: a report of three patients, including the second reported case with hypoprebetalipoproteinemia (HARP syndrome). Neurology 45:187–192
96. Oshima M, Osawa Y, Asano K et al. (1985) Erythrocyte membrane abnormalities in patients with amyotrophic chorea with acanthocytosis. Part 1. Spin labeling studies and lipid analyses. J Neurol Sci 68:147–160
97. Pellecchia MT, Valente EM, Cif L et al. (2005) The diverse phenotype and genotype of pantothenate kinase-associated neurodegeneration. Neurology 64:1810–1812
98. Rafalowska J, Drac H, Jamrozik S (1996) Neuroacanthocytosis. Review of literature and case report. Folia Neuropathol 34:178–183
99. Rampoldi L, Danek A, Monaco AP (2002) Clinical features and molecular bases of neuroacanthocytosis. J Mol Med 80:475–491
100. Rampoldi L, Dobson-Stone C, Rubio JP et al. (2001) A conserved sorting-associated protein is mutant in chorea-acanthocytosis. Nat Genet 28:119–120
101. Recio MV, Hauser RA, Louis ED et al. (2005) Chorea in a patient with cerebral palsy: treatment with levetiracetam. Mov Disord 20:762–764
102. Robinson D, Smith M, Reddy R (2004) Neuroacanthocytosis. Am J Psychiatry 161:1716
103. Rovito DA, Pirone FJ (1963) Acanthrocytosis associated with schizophrenia. Am J Psychiatry 120:182–185
104. Roy MK, Saha SP, Gangopadhya PK et al. (1999) A case of chorea with neuroacanthocytosis. J Assoc Physicians India 47:242–243
105. Rubio JP, Danek A, Stone C et al. (1997) Chorea-acanthocytosis: genetic linkage to chromosome 9q21. Am J Hum Genet 61:899–908
106. Ryu S, Awasaki H, Yamaguchi A (1982) A case report on chorea acanthocytosis. Clin Neurol (Tokyo) 22:734–734
107. Sahashi I, Takahashi A, Mitsuma A et al. (1980) Familial onset of abnormal erythrocytes, Shy–Drager syndrome, and involuntary movements. Clinical and hereditary studies. Clin Neurol (Tokyo) 20:1082–1084
108. Saiki S, Hirose G, Sakai K et al. (2004) Chorea-acanthocytosis associated with Tourettism. Mov Disord 19:833–836
109. Saiki S, Sakai K, Kitagawa Y et al. (2003) Mutation in the *CHAC* gene in a family of autosomal dominant chorea-acanthocytosis. Neurology 61:1614–1616
110. Saiki S, Sakai K, Murata KY et al. (2007) Primary skeletal muscle involvement in chorea-acanthocytosis. Mov Disord 22:848–852
111. Saiki S, Sano A, Danek A et al. (2007) Third international neuroacanthocytosis symposium: the Asian perspective; Kyoto Prefectural University of Medicine, October 28 2006. Mov Disord 22:I–VI

112. Saito T, Hasegawa I, Kowa H (1980) Symposium: neurological disorder with acanthocytosis. Clin Neurol (Tokyo) 20:1079–1081
113. Sakai T, Antoku Y, Iwashita H et al. (1991) Chorea-acanthocytosis: abnormal composition of covalently bound fatty acids of erythrocyte membrane proteins. Ann Neurol 29:664–669
114. Sakai T, Goto I, Kuroiwa Y et al. (1981) Chorea-acanthocytosis and Huntington's chorea [Japanese]. Neurol Med (Tokyo) 15:24–29
115. Sakai T, Iwashita H, Kakugawa M (1985) Neuroacanthocytosis syndrome and choreo-acanthocytosis (Levine–Critchley syndrome). Neurology 35:1679
116. Sakai T, Iwashita H, Sato Y et al. (1980) Chorea-acanthocytosis: familial and sporadic case [Japanese]. Clin Neurol (Tokyo) 20:1062–1063
117. Sakai T, Mawatari S, Iwashita H et al. (1981) Choreoacanthocytosis. Clues to clinical diagnosis. Arch Neurol 38:335–338
118. Sakuta M, Takemura T, Kamakura K et al. (1980) An autopsy case of chorea-acanthocytosis [Japanese]. Clin Neurol (Tokyo) 20:1059–1061
119. Sato Y, Ohnishi A, Tateishi J et al. (1984) An autopsy case of chorea-acanthocytosis. Special reference to the histopathological and biochemical findings of basal ganglia. Brain and Nerve (Tokyo) 36:105–111
120. Senda Y, Koike Y, Sugimura K et al. (1987) Chorea-acanthocytosis with catecholamine abnormality and orthostatic hypotension: a case report [Japanese]. Clin Neurol (Tokyo) 27:898–903
121. Shibasaki H, Sakai T, Nishimura H et al. (1982) Involuntary movements in chorea-acanthocytosis: a comparison with Huntington's chorea. Ann Neurol 12:311–314
122. Shimizu T, Inoue S, Sugita H et al. (1974) Self-mutilation, choreoathetosis, muscular hypotonia, absence of deep tendon reflexes and normouricemia, report of adult case [Japanese]. Neurol Med (Tokyo) 1:135–136
123. Shiokawa T, Suga S, Beppu H et al. (1982) Familial cases in chorea-acanthocytosis [Japanese]. Clin Neurol (Tokyo) 22:735
124. Shizuka M, Watanabe M, Aoki M et al. (1997) Analysis of the McLeod syndrome gene in three patients with neuroacanthocytosis. J Neurol Sci 150:133–135
125. Silvestri R, Raffaele M, De Domenico P et al. (1995) Sleep features in Tourette's syndrome, neuroacanthocytosis and Huntington's chorea. Neurophysiol Clin 25:66–77
126. Sobue G, Mukai E, Fujii K et al. (1986) Peripheral nerve involvement in familial chorea-acanthocytosis. J Neurol Sci 76:347–356
127. Storch A, Kornhass M, Schwarz J (2005) Testing for acanthocytosis: a prospective reader-blinded study in movement disorder patients. J Neurol 252:84–90
128. Sugimura K, Katsui Y, Nakazhima K et al. (1987) Peripheral nerve involvement in chorea-acanthocytosis [Japanese]. Clin Neurol (Tokyo) 27:211–217
129. Tabira T (2000) Chorea-acanthocytosis [Japanese]. Ryoikibetsu Shokogun Shirizu 188–189
130. Takahashi Y, Kojima T, Atsumi Y et al. (1983) A case of chorea-acanthocytosis with various psychotic symptoms. Seishin Shinkeigaku Zasshi 85:457–472
131. Takashima H, Sakai T, Iwashita H et al. (1994) A family of McLeod syndrome, masquerading as chorea-acanthocytosis. J Neurol Sci 124:56–60
132. Takiyama Y, Saito T, Kaneko J et al. (1986) An unusual case associated with atrophy of the caudate nucleus, acanthocytosis and severe muscular wasting–neuroacanthocytosis syndrome [Japanese]. Clin Neurol (Tokyo) 26:1130.1137
133. Tamura Y, Matsui K, Yaguchi H et al. (2005) Nemaline rods in chorea-acanthocytosis. Muscle Nerve 31:516–519
134. Tanaka M, Hirai S, Kondo S et al. (1998) Cerebral hypoperfusion and hypometabolism with altered striatal signal intensity in chorea-acanthocytosis: a combined PET and MRI study. Mov Disord 13:100–107
135. Terada N, Fujii Y, Ueda H et al. (1999) Ultrastructural changes of erythrocyte membrane skeletons in chorea-acanthocytosis and McLeod syndrome revealed by the quick-freezing and deep-etching method. Acta Haematol 101:25–31

136. Terao S, Sobue G, Takahashi M et al. (1995) Disturbances of hypothalamic-pituitary hormone secretion in familial chorea-acanthocytosis [Japanese]. Brain and Nerve (Tokyo) 47:57–61
137. Tomemori Y, Ichiba M, Kusumoto A et al. (2005) A gene-targeted mouse model for chorea-acanthocytosis. J Neurochem 92:759–766
138. Toyokura Y, Kamakura K, Shimada Y (1982) Familial chorea-acanthocytosis (the Levine–Critchley syndrome): a review of the reported cases in Japan. In: The ministry of health and welfare of Japan (ed.) Annual report of the research committee of CNS degenerative diseases (Tokyo) pp 335–351
139. Tsai C-H, Chen R-S, Chang H-C et al. (1997) Acanthocytosis and spinocerebellar degeneration: a new association?. Mov Disord 12:456–459
140. Uchida K, Nakajima K, Shima H et al. (1992) The first example of the McLeod phenotype in a Japanese baby with chronic granulomatous disease. Transfusion 32:691
141. Ueno E, Oguchi K, Yanagisawa N (1982) Morphological abnormalities of erythrocyte membrane in the hereditary neurological disease with chorea, areflexia and acanthocytosis. A study with freeze fracture electron microscopy. J Neurol Sci 56:89–97
142. Ueno S-I, Maruki Y, Nakamura M et al. (2001) The gene encoding a newly discovered protein, chorein, is mutated in chorea-acanthocytosis. Nat Genet 28:121–122
143. Velayos-Baeza A, Vettori A, Copley RR et al. (2004) Analysis of the human *VPS13* gene family. Genomics 84:536–549
144. Walker RH, Jankovic J, O'Hearn E et al. (2003) Phenotypic features of Huntington's disease-like 2. Mov Disord 18:1527–1530
145. Walker RH, Liu Q, Ichiba M et al. (2006) Self-mutilation in chorea-acanthocytosis: manifestation of movement disorder or psychopathology? Mov Disord 21:2268–2269
146. Walker RH, Melnick A, Walsh M et al. (1997) Dominantly inherited chorea-parkinsonism-dementia-acanthocytosis with red cell protein membrane abnormalities. Ann Neurol 42:407–408
147. Walker RH, Morgello S, Davidoff-Feldman B et al. (2002) Autosomal dominant chorea-acanthocytosis with polyglutamine-containing neuronal inclusions. Neurology 58:1031–1037
148. Walker RH, Morgello S, Davidoff-Feldman B et al. (2004) Autosomal-dominant chorea-acanthocytosis: report of a family and neuropathology. In: Danek A (ed) Neuroacanthocytosis syndromes. Springer, Dordrecht, pp 55–65
149. Walker RH, Rasmussen A, Rudnicki D et al. (2003) Huntington's disease-like 2 can present as chorea-acanthocytosis. Neurology 61:1002–1004
150. Walker RH, Danek A, Uttner I et al. (2007) McLeod phenotype without the McLeod syndrome. Transfusion 47:299–305
151. Yamada T, Hirayama K, Akai J (1986) One autopsy case of Levine–Critchley syndrome: clinico-pathological study [Japanese]. Clin Neurol (Tokyo) 26:156–161
152. Yamamoto T, Hirose G, Shimazaki K et al. (1982) Movement disorders of familial neuroacanthocytosis syndrome. Arch Neurol 39:298–301
153. Yamamoto T, Hirose G, Takado M et al. (1980) A familial neurological disorder with acanthocytosis. Special reference to movement disorders [Japanese]. Clin Neurol (Tokyo) 20:1067–1068

Differential Diagnosis of Chorea

S.A. Schneider, R.H. Walker, and K.P. Bhatia(✉)

Abstract Causes of chorea are multifold and include inherited (genetic) and acquired forms. Among the genetic disorders, Huntington's disease remains most important. However, other disorders with an indistinguishable clinical presentation are being increasingly recognized, and are referred to as the Huntington's disease like (HDL) syndromes. These include HDL1 due to mutation of the *PRNP* gene (chr. 20p12); HDL2 due to mutation of the *JPH3* gene (chr. 16q24); HDL3 mapped to chr. 4 (gene unknown) and HDL4/SCA17 due to mutation of the *TBP* gene (chr. 6q27). Other disorders discussed here include dentatorubropallidoluysian atrophy; the spinocerebellar ataxias; neuroferritinopathy; pantothenate-kinase associated neurodegeneration; chorea-acanthocytosis and McLeod syndrome. We review the growing list of recognized genetic (and acquired) disorders and discuss the clinical approach and useful factors to arrive at the etiological diagnosis.

K.P. Bhatia
Institute for Neurology, Queen Square, London, WC1N 3BG, UK
k.bhatia@ion.ucl.ac.uk

R.H. Walker et al. (eds.), *Neuroacanthocytosis Syndromes II.*

1 Definition of Chorea

Chorea is characterized by involuntary movements which flit and flow unpredictably from one body part to another. Three other forms of hyperkinetic disorders, namely tics, dystonia and myoclonus have to be differentiated from chorea. Tics, unlike chorea, are stereotyped and suppressible with a build up of an inner urge or tension. Dystonia is characterized by sustained muscle contractions causing abnormal postures, and myoclonic jerks are of a brief shock-like nature. Chorea can occur with other movement disorders in certain neurological syndromes.

Chorea is derived from Greek for "dance" and hearkens back to a medieval dancing procession called St. Vitus' chorea [73]. Medical practitioners including Meige and Charcot were fascinated by this dance-like "procession of hysteric nature". This procession still continues annually in Echternach, Luxembourg, when pilgrims gather at the grave of St. Willibrord on Whit Tuesdays.

Causes of chorea are multifold and include genetic and acquired forms, such as immune-mediated, infectious, toxic, metabolic and vascular causes. In view of the wide range of differential diagnoses, clinical clues are useful to facilitate diagnosis. In particular, information about disease onset including the onset age, type of onset (sudden or insidious), progression, and family history as well as examination findings help to arrive at the diagnosis.

In this chapter, we will first give descriptions of the main genetic (Table 1) and acquired non-genetic (Table 2) causes, and then give a clinical approach to a patient with chorea regarding the differential diagnosis.

2 Genetic Causes of Chorea

2.1 Autosomal Dominant Inheritance

2.1.1 Huntington's Disease

The most important cause of genetic chorea is Huntington's disease (HD), a neurodegenerative autosomal dominant disorder due to mutation of the *huntingtin* gene (*htt; IT15*) on chromosome 4 [1]. HD is due to expansion of a physiological polyglutamine stretch which ranges from 27 to 35 in healthy individuals. Penetrance is incomplete with between 36 and 39 repeats, and individuals may or may not develop disease. Ranges of 40 or more repeats eventually cause HD, and the longer the stretch the more severe the clinical phenotype. Disease onset is inversely related to the number of repeats [46, 119].

HD prevalence has been found to be high in regions of Venezuela and Scotland, and relatively low in Japan, Finland and Norway [57, 123]. In Europe and North America the prevalence is about 4–8 per 100,000.

Onset of classic HD is around age 40 with a combination of personality changes, generalized chorea, and cognitive decline. However, in children or adolescents, HD tends to present as an akinetic rigid variant (Westphal form), rather than with chorea. This young-onset variant is more commonly inherited from the father, due to meiotic instability with an increased risk of expansion. Other features in both adult and young onset forms include eye movement abnormalities (impersistence of gaze, difficulty initiating saccades), dysarthria, dysphagia, pyramidal signs, and ataxia resulting in imbalanced gait and reduced postural stability. Dystonia, myoclonus, tics and tremor can also occur as part of the clinical spectrum. Progression is inexorable with death after 15–20 years. Brain imaging reveals progressive atrophy of the caudate nuclei, present even before onset of symptoms. The diagnosis is based on genetic testing. Positive testing has major implications for the entire family and genetic counselling should be offered to explain the risk of inheriting the disease (50%) and penetrance (100% when the polyglutamine stretch exceeds 40 repeats) [75, 119].

The underlying mechanisms of HD are not fully understood. Aggregation of mutant protein fragments, interference with transcription factors and gene expression, abnormal levels of nerve growth factors and calcium, mitochondrial dysfunction, and reduction of synaptic transmission resulting in interference of signaling pathways all seem to play a role [35, 46, 93].

Therapy remains symptomatic. Tetrabenazine reduces chorea through presynaptic dopamine depletion and mild D2-receptor blockage. Typical and atypical neuroleptics also reduce chorea. However, chorea is usually more bothersome for relatives and carers than for patients. Furthermore, if medication is excessive, motor abilities may markedly deteriorate as chorea becomes replaced by disabling hypokinesia, hence, the primary aim should be to lessen chorea rather than fully suppress it.

Depression can be treated with classical antidepressants. Deep brain stimulation of the globus pallidus internus has been performed experimentally, and has been reported to provide temporary benefit in single cases, but it cannot stop neurodegeneration [59]. Multiple drug trials are currently being carried out in the hope of finding neuroprotective agents.

2.1.2 Dentatorubral-Pallidoluysian Atrophy

Dentatorubral-pallidoluysian atrophy (DRPLA) is in many respects similar to HD. It is an autosomal dominant neurodegenerative disorder [101] due to expansion of trinucleotide (CAG) repeats [67]. The mutated *atrophin-1* gene was mapped to chromosome 12p13.31 [131]. The repeat size ranges from 8 to 25 repeats in healthy subjects and from 49 to 88 repeats in affected patients [71, 100]. Instability in transmission has been reported, with an average change in repeat length of 4 repeats for paternal transmission and a decrease of 1 during maternal transmission [71]. Similarly to HD, repeat size correlates with age of onset and disease severity. Three clinical phenotypes have been described; presentation with prominent chorea similar to HD; prominent ataxia; and prominent myoclonus. Usual age of onset is in the third decade with death in the forties. However, early onset disease with severe progressive myoclonic epilepsy and cognitive decline is reported. Late onset disease may present with mild cerebellar ataxia [5].

DRPLA is relatively prevalent in Japan. However, Caucasian cases have been reported [11, 78, 137]. Recently, four Portuguese families with DRPLA [92] were

Table 1 Summary of genetic causes of chorea

Condition	Inheritance	Chromosomal location	Gene	Age of onset	Clinical characteristics
Huntington's disease	AD	4p15	*IT15/Huntingtin*	20–40	Chorea, personality changes, dementia
DRPLA	AD	12p13	*Atrophin-1*	20–30s	Ataxia, chorea, myoclonus
HDL1	AD	20p12	*PRNP*	20–40	HD phenocopy with prominent psychiatric features
HDL2	AD	16q24	*JPH3*	35	HD phenocopy, in black Africans
HDL3	AR	4p15	n.k.	Childhood	Extrapyramidal, pyramidal, ataxia, dementia (single family)
HDL4/SCA17	AD	6q27	*TBP*	25–40	Ataxia or HD phenocopy
SCA 1	AD	6p23	*ATXN1*	30–40	Ataxia, extrapyramidal features including chorea
SCA 2	AD	12q24	*ATXN2*	5–45	Ataxia, extrapyramidal features, neuropathy, dementia
SCA 3	AD	14q32.1	*MJD-1*	25–45	Ataxia, chorea, dystonia, parkinsonism
Neuroferritinopathy	AD	19q13	*FTL*	40	Chorea, dystonia, oromandibular involvement, parkinsonism, dysarthria
Benign hereditary chorea	AD	14q13	*TITF-1*	Childhood	Non-progressive chorea, thyroid, pulmonary abnormalities
Chorea-acanthocytosis	AR	9q21	*VPS13A*	30	Chorea, dystonia, oromandibular involvement, self-mutilation, neuropathy
PKAN	AR	20p13	*PANK2*	Childhood	Dystonia, oromandibular involvement, parkinsonism, chorea, dementia
Karak syndrome	AR	22q12-q13	*PLA2G6*	Childhood	Ataxia, chorea
Wilson's disease	AR	13q14	*ATP7B*	20–30	Extrapyramidal, psychiatric, Kayser-Fleischer ring; hepatic involvement
Aceruloplasminemia	AR	3q23	*CP*	20–50	Ataxia, spasticity, dystonia, retinal degeneration, diabetes mellitus
Friedreich's ataxia	AR	9p13	*FXN*	Childhood	Ataxia, spasticity, chorea, dystonia, myoclonus
Ataxia-telangiectasia	AR	11q22.3	*ATM*	Childhood	Ataxia, oculomotor apraxia, chorea, dysarthria
Ataxia with oculomotor apraxia I	AR	9p13.3	*APTX*	Childhood	Ataxia, oculomotor apraxia, chorea, neuropathy
Ataxia with oculomotor apraxia 2	AR	9q34	*SETX*	Childhood	Ataxia, oculomotor apraxia, chorea, neuropathy

McLeod	X-linked	Xp21	*XK*	40–50	Chorea, dystonia, parkinsonism, tremor, psychiatric features, neuropathy
Lesch–Nyhan	X-linked	Xq26	*HPRT*	Childhood	Dystonia, spasticity, mental retardation, self -mutilation
Lubag	X-linked	Xq13.3	*DYT3*	10–40	Dystonia, parkinsonism, chorea, tremor
PKD	–	Hetereogenous	n.k.	7–15	Paroxysmal dyskinesias triggered by movement
PNKD	AD	2q33	*MR-1*	2–79	Paroxysmal dyskinesias triggered by alcohol or coffee
PED	–	Heterogeneous	n.k.	2–30	Paroxysmal dyskinesias triggered by exercise
Paroxysmal choreoathetosis with spasticity	AD	1p21	n.k.	Childhood	Dystonia, choreoatherosis, dysarthria, spasticity triggered by stress etc
Episodic ataxia 1	AD	12p3	*KCNA1*	Childhood	Episodic ataxia, myokymia, dysarthria

AD autosomal dominant, *AR* autosomal recessive, *n.k.* not known, PKD Paroxysmal Kinesigenic Dyskinesias; PNKD Paroxysmal Non-Kinesigenic Dyskinesias; PED Paroxysmal Exercise-induced Dyskinesias;

Table 2 Non-genetic causes of chorea

Immune-mediated	Sydenham's chorea
	Systemic lupus erythematosis
	Antiphospholipid syndrome
	Paraneoplastic syndromes
Drug-induced	Dopamine-blocking agents
	L-dopa
	Dopamine agonists
	Psychostimulants including cocaine
	Antiepileptic drugs including carbamazepine, valproic acid
	Tricyclic antidepressants
	Baclofen
	Calcium channel blockers
	Lithium
	Steroids including oral contraceptives, estrogen replacement therapy
	Theophylline
	Digoxin
	Cyclosporine
Infections	HIV
	Creutzfeld–Jakob disease
	Tuberculosis
	Measles
	Mycoplasma pneumoniae
	Parvovirus
Metabolic/Hormonal	Hyperthyroidism
	Hypo- and hyperglycemia
	Pregnancy
Vascular	Ischemia
	Haemorrhagic stroke
	Vasculitis including moyamoya disease
	Post-pump chorea

found to have two intragenic SNPs in introns 1 and 3, in addition to the CAG repeat, of the DRPLA gene. Notably, all four Portuguese families shared the same haplotype, which was also identical to Japanese DRPLA chromosomes. The fact that this haplotype is the most frequent in Japanese normal alleles, but rare in Portuguese controls, may explain the relatively frequency of DRPLA in Japan compared to Europe.

2.1.3 Huntington's Disease-like (HDL) Syndromes

Recently, it has been recognized that clinically diagnosed HD is genetically heterogeneous, as demonstrated by a report of 618 patients [126]. Only 93% of those with a clinical phenotype of HD were found to have the HD-causing *IT15* gene expansion. For the remaining cases, with a clinical picture of HD in the absence of an *IT15* mutation, the term "Huntington's disease-like" (HDL) syndromes was coined. Mutations in unrelated genes have recently been identified [10, 66, 89, 142]. Among HDL phenocopies, HDL 1 and HDL 4 seem to be most common. HDL 3 is autosomal recessive, and has only been reported in one family to date. HDL 3 is discussed in Sect. 2.2.5 under the recessive disorders.

HDL 1

HDL 1 is an autosomal dominant adult-onset progressive neurodegenerative disorder with prominent psychiatric features, now known to be a prion disease [76]. Similar to HD, the clinical picture is that of abnormal movements, difficulty in coordination, dementia, personality changes and psychiatric symptoms. Seizures have been also described [142]. In a French family the clinical picture resembled Gerstmann-Straussler-Schenker disease [76]. Mean age at onset is 20–45 years. Atrophy of the basal ganglia, the frontal and temporal lobes [142], and the cerebellum with kuru and multicentric plaques labeled with anti-prion antibodies [76] was demonstrated by neuropathological exam. Despite the clinical suggestion of spongiform encephalopathy, spongiosis was not prominent. Linkage to chromosome 20p12 was reported [142]. A 192-nucleotide insertion in the region of the prion protein gene (*PRNP*) encoding an octapeptide repeat in the prion protein was detected [97, 127].

HDL 2

HDL 2 is a rare cause of HD phenocopies, representing only about 2% of patients without the *IT15* mutation [126, 127]. The frequency of HDL 2 is relatively high in black South Africans [9, 90]. North American and Mexican families with African origins have been described [88]. However, so far it has not been reported in Japanese or Caucasians.

Onset is in the fourth decade with a clinical picture resembling classic HD but similarities to the juvenile-onset variant have also been described (pedigree W), but with the absence of seizures and often normal eye movements [89].

Pathological examination showed a picture indistinguishable from classic HD [90]. The causal mutation is a CTG/CAG expansion on chromosome 16q24.3 in the *junctophilin-3* gene [61]. Similar to HD, there is a correlation between age of onset and repeat length. The function of junctophilin is not fully understood but a role in junctional membrane structures and in the regulation of calcium has been suggested [70]. In the normal population, the repeat length ranges from 6 to 27 CTG/CAG triplets [70]. Pathological repeat expansions range from 43 to 57 triplets, with length instability in maternal transmission [90]. To date, the impact of alleles with 36–39 triplets is uncertain [90]. Approximately 10% of HDL2 patient have acanthocytosis on peripheral blood smear [136]. HDL2 is discussed in more detail in the chapter by Margolis.

HDL 4/SCA17

HDL 4 or SCA17 is an autosomal dominant triplet repeat disorder. The mutated gene on chromosome 6q27, *TBP*, encodes for the TATA box-binding protein, an important general transcription initiation factor. Normal CAG repeat stretches range from 25 to 42 in Caucasians, with larger repeats considered pathological. Onset age is between 19 and 48 years with rare childhood onset [87]. Although cerebellar ataxia is the most common feature, the phenotype is markedly heterogeneous and extrapyramidal, pyramidal, epilepsy, dementia, or psychiatric disturbances may be prominent. A clinical picture indistinguishable from classical HD has been reported in both heterozygous and homozygous mutation carriers [10, 72]. Although within most families HD-like presentation is observed only sporadically or in solitary individuals, HDL phenotypic homogeneity in all members of a SCA17 family has also been described [117]. The broad spectrum of clinical manifestations correlates with the neuropathological findings of the cerebellar pathology, involvement of the cerebral cortex, basal ganglia, and hippocampus [113].

2.1.5 Spinocerebellar Ataxias

Chorea may be seen in other autosomal dominant cerebellar ataxias including SCA 1 [102], SCA 2 [50], and SCA 3 (Machado-Joseph disease).

2.1.6 Neuroferritinopathy

This autosomal dominant condition due to mutation of the *FTL* gene on chromosome on 19q13, coding for ferritin light chain, presents with extrapyramidal features including chorea, dystonia with prominent oromandibular dyskinesias and parkinsonism (without tremor). Abnormal aggregates of ferritin and iron in the brain can be detected by imaging. Other features include dysarthria, spasticity, cerebellar signs, frontal lobe syndrome and dementia which are variably present [24, 27]. Average age of onset is 40 years. The condition is more common in the Cumbrian region of England due to a founder effect. A French family, possibly sharing a common ancestor, with exactly the same mutation, has also been described [24].

Patients have low serum ferritin levels. MRI may reveal cystic changes in the basal ganglia and bilateral pallidal necrosis [24, 85]. Abnormalities of the mitochondrial respiratory chain were demonstrated in skeletal muscle biopsy [24].

2.1.7 Benign Hereditary Chorea (BHC)

Following the first description of benign hereditary chorea in 1967 [53] of two brothers from Mississippi with early-onset non-progressive chorea, a number of cases have been reported [17, 24, 118, 139]. For several years there was debate about whether BHC is a syndrome [118] or a true entity. However, recently BHC was linked to chromosome 14q encoding for the *TITF-1* gene (*NKX2-1* gene), a homeodomain-containing transcription factor essential for the organogenesis of the lung, thyroid and the basal ganglia [14, 33, 74].

Onset of disease is in early infancy with focal or generalized chorea with both intra- and interfamilial variability. Pulmonary and thyroid dysfunction may be present. Atypical additional features include intention tremor [109], dysarthria and gait disturbances [25] and mental impairment [80]. However, as many reports date back to the era before genetic testing, the diagnosis is questionable.

Pathological study of a genetically-proven case revealed no significant abnormalities using standard methods [70]. However, immunohistochemical staining showed loss of most TITF-1-immunoreactive striatal interneurons as compared to four matched control brains [69, 70].

Treatment is not needed in most cases, but favorable responses to levodopa [6], haloperidol, chlorpromazine, and prednisone [139] has been reported.

2.1.8 Idiopathic Basal Ganglia Calcification

Idiopathic basal ganglia calcification (IBGC; Fahr's disease) is a heterogeneous group of disorders in which there is deposition of calcium in the basal ganglia and other cerebral regions, particularly the deep cerebellar nuclei. The clinical picture may include dystonia, parkinsonism, chorea, ataxia, cognitive impairment and behavioural changes. IBGC is genetically heterogeneous, and may occur sporadically. In one family, linkage was demonstrated to 14q (IBC1) [49] although in other families with autosomal dominant inheritance, linkage to this locus was excluded [15, 106, 141]. In some families, the pattern of inheritance and additional clinical features suggest mitochondrial inheritance [112, 145].

2.2 *Autosomal Recessive Forms of Chorea*

2.2.1 Pantothenate Kinase-Associated Neurodegeneration (PKAN)

Hallervorden and Spatz [54] first described five sisters with progressive dysarthria and dementia in 1922. Extrapyramidal symptoms dominate, particularly generalized dystonia with oromandibular involvement and parkinsonism, in addition to spasticity, behavioral changes followed by dementia, and pigmentary retinal degeneration, usually with onset in childhood [58]. However, chorea as the main feature has also been reported in a late adult-onset pathologically proven case [52].

Pathological examination of the original cases [54] revealed brown discoloration of the globus pallidus and substantia nigra, which is now known to be due to iron deposition most abundantly in the globus pallidus interna. The synonym "neurodegeneration with brain iron accumulation type 1" (NBIA-1) reflects these findings. Axonal spheroids, and gliosis of the pallidum and substantia nigra are also seen. MRI detects the pallidal abnormalities and the iron deposits. T2-weighted images show a central hyperintensity (probably representing fluid accumulation or edema) of the globus pallidus with a rim of signal hypointensity (iron deposition). This 'eye of the tiger' sign [121] is usually seen early in disease course [58] in the vast majority of cases. Acanthocytosis is found in about 8% of patients with PKAN.

PKAN is caused by mutation of the pantothenate kinase (*PANK2*) gene on chromosome 20p13-p12.3.

Recently, patients with a clinical picture similar to PKAN and iron deposition on MRI (in absence of the classic eye of the tiger sign) but without *PANK2* mutations, were found to harbour mutations the *PLA2G6* gene, encoding for a calcium-independent phospholipase A_2, on chromosome 22q13 (Karak syndrome), a form of neuroaxonal dystrophy [58, 99].

2.2.2 Chorea-Acanthocytosis

Chorea-acanthocytosis is an autosomal recessive neurodegenerative disorder due to mutation of the *VPS13A* (*CHAC*) gene on chromosome 9q21 [114, 111]. Clinical features include chorea, dystonia with prominent orofacial involvement and self mutilation, tics, parkinsonism, eye movement abnormalities suggestive of brain stem involvement, seizures, subcortical dementia and psychiatric features with impairment of frontal lobe function [31, 51]. Myopathy and neuropathy are often present.

Blood tests reveal presence of acanthocytosis in the blood smear and elevated creatine kinase.

MRI demonstrates progressive caudate atrophy. Extensive neuronal loss and gliosis affecting the striatum, pallidum, and substantia nigra were found on post-mortem examination [56]. Various aspects of chorea-acanthocytosis are discussed in detail elsewhere in this volume.

2.2.3 Wilson's Disease

Even though Wilson's disease most commonly presents with parkinsonian features and dystonia and tremor, chorea may be present and should therefore be considered [84].

2.2.4 Aceruloplasminemia

Deficiency of ceruloplasmin is due to inheritance of mutations in the gene for ceruloplasmin and results in iron deposition in the cerebellum and basal ganglia [96, 143] and thus belongs to the NBIA spectrum. Typical presentation is with diabetes mellitus and retinal degeneration in the 20's. In middle age neurological signs appear, usually ataxia and spasticity. Movement disorders including dystonia, especially orofacial, parkinsonism and chorea may develop. Dementia may manifest in later years. Symptomatic heteroplasmic carriers have also been reported.

Ceruloplasmin functions as a ferroxidase, thus iron oxidation from Fe^{2+} to Fe^{3+} is impaired, and neurons are more vulnerable to oxidative stress. Neuropathologically, astrocytes and neurons laden with iron are found in the cerebellum, basal ganglia, and cortex [96, 143].

2.2.5 HDL3

A Saudi-Arabian family presenting with early onset mental deterioration, dysarthria, extrapyramidal (dystonia) and pyramidal signs was reported as autosomal recessive variant of HD by Al-Tahan et al. [2] and Kambouris et al. [66]. Onset age was 3–4 years with extrapyramidal symptoms, ataxia, gait impairment, spasticity and intellectual decline. Brain imaging revealed progressive atrophy of the caudates bilaterally and the frontal cortex. HDL-3 was mapped to chromosome 4p15.3 [66], however, weakness of the evidence has been emphasized by Lesperance and Burmeister [81].

2.2.6 Friedreich Ataxia

Although progressive gait instability or general clumsiness, neuropathy and cardiomyopathy scoliosis are the classic presentation [38], chorea [55] and myoclonus [39] have also rarely been described. Friedreich ataxia, caused by expansion in the frataxin gene on chromosome 9, affects about 40,000 individuals in Europe and is very rare among black Africans.

2.2.7 Ataxia Telangiectasia

The wide range of clinical phenotypes in ataxia telangiectasia includes early-onset truncal ataxia, ocular motor apraxia, peripheral neuropathy, dysarthria and extrapyramidal features including facial hypomimia and dystonia. Chorea was present in the majority (68 of 70) of patients [140].

2.2.8 Ataxia with Ocular Motor Apraxia

The clinical spectrum of ataxia with ocular motor apraxia type 1 (AOA1), due to mutation in the *aprataxin* gene on chromosome 9p13.3 comprises oculomotor apraxia, axonal sensorimotor neuropathy, hypoalbuminaemia and hypercholesterolaemia. Choreic movements are frequent at onset (80%), but usually disappear in the course of the disease.

Similarly, AOA2 may present with a combination of eye signs (gaze nystagmus, strabismus, impaired smooth pursuit), ataxia, extrapyramidal features, including chorea, dystonia and tremor, dysphagia and neuropathy with onset in childhood or adolescence [77]. Findings of elevated cholesterol, creatine kinase, and alpha-fetoprotein support the diagnosis. In some populations AOA2 is the most common inherited cause of ataxia after Friedreich's ataxia. Similar to AOA1 the mutant protein may be involved in RNA repair [98].

2.2.9 Infantile Bilateral Striatal Necrosis

This condition causes chorea or dystonia in infants, and may be inherited in an autosomal recessive manner [8], but also appears to be due to mitochondrial mutations [32]. The diagnostic findings are of bilateral lesions in the striatum.

2.2.10 Other Inherited Metabolic Disorders

Movement disorders may be present as part of a constellation of neurological abnormalities in a number of inherited metabolic disorders. Autosomal recessive inheritance of mutations of critical synthetic enzymes is usually the etiology. Dystonia appears to be more common than chorea, possibly due to the co-existence of rigidity or spasticity, due to pyramidal tract involvement. Diagnosis is by assaying blood and urine for amino acids, by enzyme assays in lymphocytes, or genetic testing, as indicated.

Glutaric acidura typically presents with generalized dystonia and encephalopathy, and occasionally chorea [44, 105, 135]. On MRI dilation of the sylvian fissures can be seen and lesions of the putamen. Chorea, typically mild, can be seen in propionic academia, due to propionic-CoA carboxylase deficiency [122, 129]. Other aminoacidopathies in which chorea may occasionally be seen include 3-methylglutaconic academia [47], and succinic semialdehyde dehydrogenase deficiency.

2.3 X-linked Disorders

2.3.1 McLeod Syndrome

McLeod neuroacanthocytosis syndrome is phenotypically very similar to autosomal recessive chorea-acanthocytosis, but is characterized by absence of the Kx antigen and reduced expression of Kell antigens on red blood cells. It is due to mutation of the XK gene [29, 30] and is discussed in detail in the chapters by Jung, Geser, and Lee.

2.3.2 Lesch–Nyhan Disease

Lesch–Nyhan syndrome is caused by mutation in the hypoxanthine guanine phosphoribosyltransferase (*HPRT*) gene on chromosome Xq26-q27.2 resulting in virtually complete deficiency of HPRT, with residual activity of less than 1.5%. Main clinical features include mental retardation, spastic cerebral palsy, extrapyramidal features with dystonia and chorea, uric acid urinary stones, and self-mutilation behavior with biting of fingers and lips [64].

2.3.3 Lubag

This X-linked disorder is characterized by dystonia and parkinsonism, and is found solely amongst Filipinos from province of Capiz on the island of Panay [79]. Presentation is typically in the teens with focal dystonia or parkinsonism, with most patients developing mixed features [40]. However a range of movement disorders have been reported, including chorea, tremor and myoclonus [40]. Occasional affected carrier females have been reported with chorea [41, 138].

The gene, *DYT3*, has been identified as coding for a multiple transcript system whose function is not yet known [104].

This diagnosis should be considered in any Filipino, even females, presenting with any movement disorder and a detailed family history taken so that appropriate counseling can be given [40].

3 Paroxysmal Forms of Chorea

Intermittent or episodic sudden attacks of dystonia, chorea and ballismus or their combination, with preserved consciousness and without abnormalities between attacks, are referred to as paroxysmal movement disorders [116]. Four different forms can be distinguished depending on the main triggering factors, as classified by Demirkiran and Jankovic [34]. These include paroxysmal kinesigenic dyskinesias (PKD) typically triggered by sudden movements; paroxysmal non-kinesigenic dyskinesias (PNKD) classically triggered by alcohol, coffee, hunger, fatigue and emotions; paroxysmal exercise-induced dyskinesias (PED) which are induced by prolonged or sustained exercise; and paroxysmal hypnogenic dyskinesias (PHD) during non-REM sleep. Duration of attacks varies from seconds (PKD and PHD) to hours (PNKD and PED). Age of onset is between 6 months and 79 years. Recent genetic studies revealed mutation in the *MR-1* gene on chromosome 2q33–35 encoding for the myofibrillogenesis regulator gene in the PNKD syndrome [110], and mutations on chromosomes 20q13 and 15q24 encoding for subunits of acetylcholine receptors in the autosomal dominant variant of PHD [125]. PKD and PED have been linked to chromosome 16 [130] but hetereogeneity and overlap with other syndromes like the ICCA (benign infantile convulsions and paroxysmal choreoathetosis) syndrome and the Rolandic epilepsy-PED-Writer's cramp (RE-PED-WC) syndrome have been suggested [124, 134].

Patients with episodic ataxia 1, due to point mutation of a potassium channel gene KCNA1 on 12p13 [16, 28] may sometimes manifest choreoathetosis during ataxic episodes [45]. More typical features are dysarthria and persistent myokymia [82, 83]. Genetic heterogeneity is suggested by the report of a family also with stress-induced episodic ataxia and chorea, and facial myokymia in which this mutation was excluded [42].

Paroxysmal choreoathetosis with spasticity was localized to 1p.21 in one family [7]. Episodes of dystonia, choreoatherosis, dysarthria, spasticity and imbalance were precipitated by exercise, stress, alcohol consumption and sleep deprivation. In between attacks, spasticity was persistent.

4 Non-genetic Causes of Chorea

4.1 Immune-Mediated

Sydenham's chorea is the prototype for an autoimmune-mediated chorea following infectious disease. Onset is typically in childhood around age 5–15 and 4–8 weeks after pharyngitis caused by group A β-haemolytic streptococcus. Girls are more commonly affected. Typical presentation is with generalized chorea, although hemichorea is seen in one fifth of patients. Obsessive-compulsive behaviour, hyperactivity and attention deficit can be seen [86]. Other features associated with rheumatic fever like cardiac involvement, particularly mitral valve dysfunction, and arthritis may be present [21]. However, chorea may also remain the only feature [22].

Sydenham's chorea is usually self-limiting after 8–9 months. However, studies showed that chorea was still present and had re-occurred after 2 years in about half of patients [23]. Secondary prophylaxis with penicillin is therefore recommended. In the acute setting, sodium valproate and carbamazepine are effective [48]. Dopamine-blocking agents are not recommended as first choice as they carry the risk of inducing parkinsonism and dystonia.

Other immune-mediated conditions causing chorea include systemic lupus erythematosis (SLE), antiphospholipid antibody syndrome [91, 115], vasculitis, coeliac disease [107] and paraneoplastic syndromes, particularly in small-cell lung carcinoma associated with anti-Hu, anti-Yo and anti-CRMP5 antibodies [36, 68].

4.2 Drugs and Toxins

A number of drugs and toxins are known to potentially cause chorea (see table) [62, 110, 120], therefore a detailed drug history is part of the history. Chorea may be a chronic adverse effect, i.e. tardive dyskinesia due to typical neuroleptics or other dopamine-blocking agents, or may be reversible and directly related to a drug effect.

4.3 Infectious Chorea

HIV and its complications can cause movement disorders including chorea [20, 103, 133]. Hemiballism-hemichorea and tremor are the most common hyperkinesias in HIV positive patients. This may be due to a direct effect of the virus, via opportunistic infections like toxoplasmosis causing space-occupying lesions, or from the use of drugs.

Other infections which may cause chorea include Creutzfeld–Jakob disease [12, 94] and tuberculosis [3, 65].

In children, striatal necrosis may occur as a complication of measles encephalitis [18], *Mycoplasma pneumoniae* infection [146], or following undefined febrile illness [144]. Chorea has also been reported in the setting of encephalopathy due to parvovirus infection [43].

4.4 Metabolic and Hormonal Causes

Acquired chorea due to metabolic dysfunction may be caused by hyperthyroidism [63, 128], hypo- and hyperglycaemia [13, 60, 132]. Chorea during pregnancy, referred to as chorea gravidarum, most commonly occurs during the first or second trimester [19]. This may also been seen with hormone replacement therapy or with oral contraceptives.

4.5 Vascular

Vascular lesions resulting in damage of the caudate-putamen and subthalamus may result in chorea. Overall, however, the risk of developing chorea seems relatively low as concluded by a large study of 1500 stroke patients, 56 (0.04%) of which developed movement disorders up to 1 year after the stroke. In this series chorea was most common and tended to affect the eldest [4]. Other authors have attributed about 40% of sporadic chorea cases to cerebral vasculopathy [108].

Rarer causes include moyamoya disease, which is characterized by spontaneous occlusion of one or usually both internal carotid arteries [147]. In children, chorea has been observed following heart surgery, also referred to as postpump chorea [26, 37, 95].

4.6 Neoplasm and Paraneoplastic

Similar to vasculopathies, neoplastic brain disease can cause strategic lesions, involving the basal ganglia or adjacent areas, resulting in focal or hemi-chorea. Paraneoplastic syndromes may also result in chorea, as mentioned above.

5 Approach to a Patient with Chorea

As discussed above, there are numerous causes of chorea. A clinical approach keeping in mind the type of onset, the age, presence of a family history, antecedent illness or drug exposure, the distribution of chorea (focal or generalized) and presence of associated features are the key points. Additional features may include other extrapyramidal and pyramidal signs or cerebellar signs. The presence of additional neuropsychiatric findings (cognitive decline, mood disturbance, behavioural changes) and features of the general physical examination (skin changes, heart, liver) are also very important in arriving at an etiological diagnosis.

For example, in a child presenting with chorea, one would not consider HD as early-onset HD presents as akinetic-rigid syndrome. In this situation, metabolic or acquired disorders (e.g. Wilson's disease or Sydenham's chorea) are more likely. On the other hand, in a middle-aged person with insidious onset after age 30 and generalized distribution, HD is more likely and would need to be excluded. The HDL phenocopies would be considered next. Finally, acute onset in an elderly person suggests a vascular cause. Associated clinical features like eye movements can be helpful. Gaze abnormalities are most common in HD, whilst a vertical supranuclear palsy hints towards Nieman-Pick type C or a storage disorder. Similarly, apraxia is seen in ataxia telangiectasia, polyneuropathy in chorea-acanthocytosis or McLeod syndrome, and retinitis pigmentosa in PKAN. Clues like these help to narrow down the list of differential diagnoses, so that appropriate tests can be ordered and the correct diagnosis can be made.

References

1. The Huntington's Disease Collaborative Research Group (1993) A novel gene containing a trinucleotide repeat that is expanded and unstable on Huntington's disease chromosomes. Cell 72(6):971–983
2. Al-Tahan AY, Divakaran MP, Kambouris M, Bohlega S, Salih M, Ogunniyi A et al. (1999) A novel autosomal recessive 'Huntington's disease-like' neurodegenerative disorder in a Saudi family. Saudi Med J 20:85–89
3. Alarcon F, Duenas G, Cevallos N, Lees AJ (2000) Movement disorders in 30 patients with tuberculous meningitis. Mov Disord 15(3):561–569
4. Alarcon F, Zijlmans JC, Duenas G, Cevallos N (2004) Post-stroke movement disorders: report of 56 patients. J Neurol Neurosurg Psychiatry 75(11):1568–1574
5. Aoki M, Abe K, Kameya T, Watanabe M, Itoyama Y (1994) Maternal anticipation of DRPLA. Hum Mol Genet 3(7):1197–1198
6. Asmus F, Horber V, Pohlenz J, Schwabe D, Zimprich A, Munz M et al. (2005) A novel TITF-1 mutation causes benign hereditary chorea with response to levodopa Neurology 64(11):1952–1954
7. Auburger G, Ratzlaff T, Lunkes A, Nelles HW, Leube B, Binkofski F et al. (1996) A gene for autosomal dominant paroxysmal choreoathetosis/spasticity (CSE) maps to the vicinity of a potassium channel gene cluster on chromosome 1p, probably within 2 cM between D1S443 and D1S197. Genomics 31(1):90–94
8. Basel-Vanagaite L, Straussberg R, Ovadia H, Kaplan A, Magal N, Shorer Z et al. (2004) Infantile bilateral striatal necrosis maps to chromosome 19q. Neurology 62(1):87–90
9. Bauer I, Gencik M, Laccone F, Peters H, Weber BH, Feder EH et al. (2002) Trinucleotide repeat expansions in the junctophilin-3 gene are not found in Caucasian patients with a Huntington's disease-like phenotype. Ann Neurol 51(5):662
10. Bauer P, Laccone F, Rolfs A, Wullner U, Bosch S, Peters H et al. (2004) Trinucleotide repeat expansion in SCA17/TBP in white patients with Huntington's disease-like phenotype. J Med Genet 41(3):230–232
11. Becher MW, Rubinsztein DC, Leggo J, Wagster MV, Stine OC, Ranen NG et al. (1997) Dentatorubral and pallidoluysian atrophy (DRPLA). Clinical and neuropathological findings in genetically confirmed North American and European pedigrees. Mov Disord 12(4):519–530
12. Bowen J, Mitchell T, Pearce R, Quinn N (2000) Chorea in new variant Creutzfeldt–Jacob disease. Mov Disord 15(6):1284–1285

13. Branca D, Gervasio O, Le Piane E, Russo C, Aguglia U (2005) Chorea induced by non-ketotic hyperglycaemia: a case report. Neurol Sci 26(4):275–277
14. Breedveld GJ, van Dongen JW, Danesino C, Guala A, Percy AK, Dure LS et al. (2002) Mutations in TITF-1 are associated with benign hereditary chorea. Hum Mol Genet 11(8):971–979
15. Brodaty H, Mitchell P, Luscombe G, Kwok JJ, Badenhop RF, McKenzie R et al. (2002) Familial idiopathic basal ganglia calcification (Fahr's disease) without neurological, cognitive and psychiatric symptoms is not linked to the IBGC1 locus on chromosome 14q. Hum Genet 110(1):8–14
16. Browne DL, Gancher ST, Nutt JG, Brunt ER, Smith EA, Kramer P et al. (1994) Episodic ataxia/myokymia syndrome is associated with point mutations in the human potassium channel gene, KCNA1. Nat Genet 8(2):136–140
17. Bruyn GW, Myrianthopoulos NC (1986) Chronic juvenile hereditary chorea (benign hereditary chorea of early onset). In: Vinken PJ, Bruyn GW, Klawans HL (eds) Extrapyramidal disorders. Elsevier Science, Amsterdam, pp 335–348
18. Cambonie G, Houdon L, Rivier F, Bongrand AF, Echenne B (2000) Infantile bilateral striatal necrosis following measles. Brain Dev 22(4):221–223
19. Cardoso F (2002) Chorea gravidarum. Arch Neurol 59(5):868–870
20. Cardoso F (2002) HIV-related movement disorders: epidemiology, pathogenesis and management. CNS Drugs 16(10):663–668
21. Cardoso F, Eduardo C, Silva AP, Mota CC (1997) Chorea in fifty consecutive patients with rheumatic fever. Mov Disord 12(5):701–703
22. Cardoso F, Seppi K, Mair KJ, Wenning GK, Poewe W (2006) Seminar on choreas. Lancet Neurol 5(7):589–602
23. Cardoso F, Vargas AP, Oliveira LD, Guerra AA, Amaral SV (1999) Persistent Sydenham's chorea. Mov Disord 14(5):805–807
24. Chinnery PF, Curtis AR, Fey C, Coulthard A, Crompton D, Curtis A et al. (2003) Neuroferritinopathy in a French family with late onset dominant dystonia. J Med Genet 40(5):e69
25. Chun RW, Daly RF, Mansheim BJ Jr, Wolcott GJ (1973) Benign familial chorea with onset in childhood. JAMA 225(13):1603–1607
26. Curless RG, Katz DA, Perryman RA, Ferrer PL, Gelblum J, Weiner WJ (1994) Choreoathetosis after surgery for congenital heart disease. J Pediatr 124(5 Pt 1):737–739
27. Curtis AR, Fey C, Morris CM, Bindoff LA, Ince PG, Chinnery PF et al. (2001) Mutation in the gene encoding ferritin light polypeptide causes dominant adult-onset basal ganglia disease. Nat Genet 28(4):350–354
28. D'Adamo MC, Imbrici P, Sponcichetti F, Pessia M (1999) Mutations in the KCNA1 gene associated with episodic ataxia type-1 syndrome impair heteromeric voltage-gated K(+) channel function. FASEB J 13(11):1335–1345
29. Danek A, Rubio JP, Rampoldi L, Ho M, Dobson-Stone C, Tison F et al. (2001) McLeod neuroacanthocytosis: genotype and phenotype. Ann Neurol 50(6):755–764
30. Danek A, Tison F, Rubio J, Oechsner M, Kalckreuth W, Monaco AP (2001) The chorea of McLeod syndrome. Mov Disord 16(5):882–889
31. Danek A, Walker RH (2005) Neuroacanthocytosis. Curr Opin Neurol 18(4):386–392
32. De Meirleir L, Seneca S, Lissens W, Schoentjes E, Desprechins B (1995) Bilateral striatal necrosis with a novel point mutation in the mitochondrial ATPase 6 gene. Pediatr Neurol 13(3):242–246
33. de Vries BB, Arts WF, Breedveld GJ, Hoogeboom JJ, Niermeijer MF, Heutink P (2000) Benign hereditary chorea of early onset maps to chromosome 14q. Am J Hum Genet 66(1):136–142
34. Demirkiran M, Jankovic J (1995) Paroxysmal dyskinesias: clinical features and classification. Ann Neurol 38(4):571–579
35. DiFiglia M, Sapp E, Chase KO, Davies SW, Bates GP, Vonsattel JP et al. (1997) Aggregation of huntingtin in neuronal intranuclear inclusions and dystrophic neurites in brain. Science 277(5334):1990–1993

36. Dorban S, Gille M, Kessler R, Pieret F, Declercq I, Sindic CJ (2004) [Chorea-athetosis in the anti-Hu syndrome]. Rev Neurol (Paris) 160(1):126–129
37. du Plessis AJ, Bellinger DC, Gauvreau K, Plumb C, Newburger JW, Jonas RA et al. (2002) Neurologic outcome of choreoathetoid encephalopathy after cardiac surgery. Pediatr Neurol 27(1):9–17
38. Durr A (2002) Friedreich's ataxia: treatment within reach. Lancet Neurol 1(6):370–374
39. Durr A, Cossee M, Agid Y, Campuzano V, Mignard C, Penet C et al. (1996) Clinical and genetic abnormalities in patients with Friedreich's ataxia. N Engl J Med 335(16):1169–1175
40. Evidente VG, Advincula J, Esteban R, Pasco P, Alfon JA, Natividad FF et al. (2002) Phenomenology of "Lubag" or X-linked dystonia-parkinsonism. Mov Disord 17(6):1271–1277
41. Evidente VG, Nolte D, Niemann S, Advincula J, Mayo MC, Natividad FF et al. (2004) Phenotypic and molecular analyses of X-linked dystonia-parkinsonism ("lubag") in women. Arch Neurol 61(12):1956–1959
42. Fernandez M, Raskind W, Wolff J, Matsushita M, Yuen E, Graf W et al. (2001) Familial dyskinesia and facial myokymia (FDFM): a novel movement disorder. Ann Neurol 49(4):486–492
43. Fong CY, de Sousa C (2006) Childhood chorea-encephalopathy associated with human parvovirus B19 infection. Dev Med Child Neurol 48(6):526–528
44. Friedman JR, Thiele EA, Wang D, Levine KB, Cloherty EK, Pfeifer HH et al. (2006) Atypical GLUT1 deficiency with prominent movement disorder responsive to ketogenic diet. Mov Disord 21(2):241–245
45. Gancher ST, Nutt JG (1986) Autosomal dominant episodic ataxia: a heterogeneous syndrome. Mov Disord 1(4):239–253
46. Gardian G, Vecsei L (2004) Huntington's disease: pathomechanism and therapeutic perspectives. J Neural Transm 111(10–11):1485–1494
47. Gascon GG, Ozand PT, Brismar J (1994) Movement disorders in childhood organic acidurias. Clinical, neuroimaging, and biochemical correlations. Brain Dev 16(Suppl):94–103
48. Genel F, Arslanoglu S, Uran N, Saylan B (2002) Sydenham's chorea: clinical findings and comparison of the efficacies of sodium valproate and carbamazepine regimens. Brain Dev 24(2):73–76
49. Geschwind DH, Loginov M, Stern JM (1999) Identification of a locus on chromosome 14q for idiopathic basal ganglia calcification (Fahr disease). Am J Hum Genet 65(3):764–772
50. Geschwind DH, Perlman S, Figueroa CP, Treiman LJ, Pulst SM (1997) The prevalence and wide clinical spectrum of the spinocerebellar ataxia type 2 trinucleotide repeat in patients with autosomal dominant cerebellar ataxia. Am J Hum Genet 60(4):842–850
51. Gradstein L, Danek A, Grafman J, Fitzgibbon EJ (2005) Eye movements in chorea-acanthocytosis. Invest Ophthalmol Vis Sci 46(6):1979–1987
52. Grimes DA, Lang AE, Bergeron C (2000)Late adult onset chorea with typical pathology of Hallervorden–Spatz syndrome. J Neurol Neurosurg Psychiatry 69(3):392–395
53. Haerer AF, Currier RD, Jackson JF (1967) Hereditary nonprogressive chorea of early onset. N Engl J Med 276(22):1220–1224
54. Hallervorden J, Spatz H (1992) Eigenartige Erkrankung im extrapyramidalen System mit besonderer Beteiligung des Globus pallidus und der Substantia nigra: Ein Beitrag zu den Beziehungen zwischen diesen beiden Zentren. Z Ges Neurol Psychiat(79):254–302
55. Hanna MG, Davis MB, Sweeney MG, Noursadeghi M, Ellis CJ, Elliot P et al. (1998) Generalized chorea in two patients harboring the Friedreich's ataxia gene trinucleotide repeat expansion. Mov Disord 13(2):339–340
56. Hardie RJ, Pullon HW, Harding AE, Owen JS, Pires M, Daniels GL et al. (1991) Neuroacanthocytosis. A clinical, haematological and pathological study of 19 cases. Brain 114(Pt 1A):13–49
57. Harper PS (1992) The epidemiology of Huntington's disease. Hum Genet 89(4):365–376
58. Hayflick SJ (2006) Neurodegeneration with brain iron accumulation: from genes to pathogenesis. Semin Pediatr Neurol 13(3):182–185
59. Hebb MO, Garcia R, Gaudet P, Mendez IM (2006) Bilateral stimulation of the globus pallidus internus to treat choreathetosis in Huntington's disease: technical case report. Neurosurgery 58(2):E383

60. Hefter H, Mayer P, Benecke R (1993) Persistent chorea after recurrent hypoglycemia. A case report. Eur Neurol 33(3):244–247
61. Holmes SE, O'Hearn E, Rosenblatt A, Callahan C, Hwang HS, Ingersoll-Ashworth RG et al. (2001) A repeat expansion in the gene encoding junctophilin-3 is associated with Huntington's disease-like 2. Nat Genet 29(4):377–378
62. Jain KK (1996) Drug -induced movement disorders. In: Jain KK (ed) Drug-induced neurological disorders. Hogrefe & Huber, Seatle, WA, 137–164
63. Javaid A, Hilton DD (1988) Persistent chorea as a manifestation of thyrotoxicosis. Postgrad Med J 64(756):789–790
64. Jinnah HA, Visser JE, Harris JC, Verdu A, Larovere L, Ceballos-Picot I et al. (2006) Delineation of the motor disorder of Lesch–Nyhan disease. Brain 129(Pt 5):1201–1217
65. Kalita J, Ranjan P, Misra UK, Das BK (2003) Hemichorea: a rare presentation of tuberculoma. J Neurol Sci 208(1–2):109–111
66. Kambouris M, Bohlega S, Al Tahan A, Meyer BF (2000) Localization of the gene for a novel autosomal recessive neurodegenerative Huntington-like disorder to 4p15.3. Am J Hum Genet 66(2):445–452
67. Kanazawa I (1998) Dentatorubral-pallidoluysian atrophy or Naito-Oyanagi disease. Neurogenetics 2(1):1–17
68. Kinirons P, Fulton A, Keoghan M, Brennan P, Farrell MA, Moroney JT (2003) Paraneoplastic limbic encephalitis (PLE) and chorea associated with CRMP-5 neuronal antibody Neurology 61(11):1623–1624
69. Kleiner-Fisman G, Calingasan NY, Putt M, Chen J, Beal MF, Lang AE (2005) Alterations of striatal neurons in benign hereditary chorea. Mov Disord 20(10):1353–1357
70. Kleiner-Fisman G, Rogaeva E, Halliday W, Houle S, Kawarai T, Sato C et al. (2003) Benign hereditary chorea: clinical, genetic, and pathological findings. Ann Neurol 54(2):244–247
71. Koide R, Ikeuchi T, Onodera O, Tanaka H, Igarashi S, Endo K et al. (1994) Unstable expansion of CAG repeat in hereditary dentatorubral-pallidoluysian atrophy (DRPLA). Nat Genet 6(1):9–13
72. Koide R, Kobayashi S, Shimohata T, Ikeuchi T, Maruyama M, Saito M et al. (1999) A neurological disease caused by an expanded CAG trinucleotide repeat in the TATA-binding protein gene: a new polyglutamine disease? Hum Mol Genet 8(11):2047–2053
73. Krack P (1999) Relicts of dancing mania: the dancing procession of Echternach. Neurology 53(9):2169–2172
74. Krude H, Schutz B, Biebermann H, von Moers A, Schnabel D, Neitzel H et al. (2002) Choreoathetosis, hypothyroidism, and pulmonary alterations due to human NKX2-1 haploinsufficiency. J Clin Invest 109(4):475–480
75. Langbehn DR, Brinkman RR, Falush D, Paulsen JS, Hayden MR (2004) A new model for prediction of the age of onset and penetrance for Huntington's disease based on CAG length. Clin Genet 65(4):267–277
76. Laplanche JL, Hachimi KH, Durieux I, Thuillet P, Defebvre L, Delasnerie-Laupretre N et al. (1999) Prominent psychiatric features and early onset in an inherited prion disease with a new insertional mutation in the prion protein gene. Brain 122(Pt 12):2375–2386
77. Le BI, Bouslam N, Rivaud-Pechoux S, Guimaraes J, Benomar A, Chamayou C et al. (2004) Frequency and phenotypic spectrum of ataxia with oculomotor apraxia 2: a clinical and genetic study in 18 patients. Brain 127(Pt 4):759–767
78. Le BI, Camuzat A, Castelnovo G, Azulay JP, Genton P, Gastaut JL et al. (2003) Prevalence of dentatorubral-pallidoluysian atrophy in a large series of white patients with cerebellar ataxia. Arch Neurol 60(8):1097–1099
79. Lee LV, Pascasio FM, Fuentes FD, Viterbo GH (1976) Torsion dystonia in Panay, Philippines. Adv Neurol 14:137–151
80. Leli DA, Furlow TW Jr, Falgout JC (1984) Benign familial chorea: an association with intellectual impairment. J Neurol Neurosurg Psychiatry 47(5):471–474
81. Lesperance MM, Burmeister M (2000) Interpretation of linkage data for a Huntington-like disorder mapping to 4p15.3. Am J Hum Genet 67(1):262–263

82. Litt M, Kramer P, Browne D, Gancher S, Brunt ER, Root D et al. (1994) A gene for episodic ataxia/myokymia maps to chromosome 12p13. Am J Hum Genet 55(4):702–709
83. Lubbers WJ, Brunt ER, Scheffer H, Litt M, Stulp R, Browne DL et al. (1995) Hereditary myokymia and paroxysmal ataxia linked to chromosome 12 is responsive to acetazolamide. J Neurol Neurosurg Psychiatry 59(4):400–405
84. Machado A, Fen CH, Mitiko DM, Cancado E, Soares AR, Scaff M et al. (2006) Neurological manifestations in Wilson's disease: report of 119 cases. Mov Disord 21(12):2192–2196
85. Maciel P, Cruz VT, Constante M, Iniesta I, Costa MC, Gallati S et al. (2005) Neuroferritinopathy: missense mutation in FTL causing early-onset bilateral pallidal involvement. Neurology 65(4):603–605
86. Maia DP, Teixeira AL Jr, Quintao Cunningham MC, Cardoso F (2005) Obsessive compulsive behavior, hyperactivity, and attention deficit disorder in Sydenham chorea. Neurology 64(10):1799–1801
87. Maltecca F, Filla A, Castaldo I, Coppola G, Fragassi NA, Carella M et al. (2003) Intergenerational instability and marked anticipation in SCA-17. Neurology 61(10):1441–1443
88. Margolis RL, Holmes SE, Rosenblatt A, Gourley L, O'Hearn E, Ross CAet al. (2004) Huntington's Disease-like 2 (HDL2) in North America and Japan. Ann Neurol 56(5):670–674
89. Margolis RL, O'Hearn E, Rosenblatt A, Willour V, Holmes SE, Franz ML et al. (2001) A disorder similar to Huntington's disease is associated with a novel CAG repeat expansion. Ann Neurol 50(6):373–380
90. Margolis RL, Rudnicki DD, Holmes SE (2005) Huntington's disease like-2: review and update. Acta Neurol Taiwan 14(1):1–8
91. Martino D, Chew NK, Mir P, Edwards MJ, Quinn NP, Bhatia KP (2006) Atypical movement disorders in antiphospholipid syndrome. Mov Disord 21(7):944–949
92. Martins S, Matama T, Guimaraes L, Vale J, Guimaraes J, Ramos L et al. (2003) Portuguese families with dentatorubropallidoluysian atrophy (DRPLA) share a common haplotype of Asian origin. Eur J Hum Genet 11(10):808–811
93. McGowan DP, Roon-Mom W, Holloway H, Bates GP, Mangiarini L, Cooper GJ et al. (2000) Amyloid-like inclusions in Huntington's disease. Neuroscience 100(4):677–680
94. McKee D, Talbot P (2003) Chorea as a presenting feature of variant Creutzfeldt–Jakob disease. Mov Disord 18(7):837–838
95. Medlock MD, Cruse RS, Winek SJ, Geiss DM, Horndasch RL, Schultz DL et al. (1993) A 10-year experience with postpump chorea. Ann Neurol 34(6):820–826
96. Miyajima H (2003) Aceruloplasminemia, an iron metabolic disorder. Neuropathology 23(4):345–350
97. Moore RC, Xiang F, Monaghan J, Han D, Zhang Z, Edstrom L et al. (2001) Huntington's disease phenocopy is a familial prion disease. Am J Hum Genet 69(6):1385–1388
98. Moreira MC, Klur S, Watanabe M, Nemeth AH, Le B, I, Moniz JC et al. (2004) Senataxin, the ortholog of a yeast RNA helicase, is mutant in ataxia-ocular apraxia 2. Nat Genet 36(3):225–227
99. Morgan NV, Westaway SK, Morton JE, Gregory A, Gissen P, Sonek S et al. (2006) PLA2G6, encoding a phospholipase A2, is mutated in neurodegenerative disorders with high brain iron. Nat Genet 38(7):752–754
100. Nagafuchi S, Yanagisawa H, Ohsaki E, Shirayama T, Tadokoro K, Inoue T et al. (1994) Structure and expression of the gene responsible for the triplet repeat disorder, dentatorubral and pallidoluysian atrophy (DRPLA). Nat Genet 8(2):177–182
101. Naito H, Oyanagi S (1982) Familial myoclonus epilepsy and choreoathetosis: hereditary dentatorubral-pallidoluysian atrophy. Neurology 32(8):798–807
102. Namekawa M, Takiyama Y, Ando Y, Sakoe K, Muramatsu SI, Fujimoto KI et al. (2001) Choreiform movements in spinocerebellar ataxia type 1. J Neurol Sci 187(1–2):103–106
103. Nath A, Hobson DE, Russell A (1993) Movement disorders with cerebral toxoplasmosis and AIDS. Mov Disord 8(1):107–112
104. Nolte D, Niemann S, Muller U (2003) Specific sequence changes in multiple transcript system DYT3 are associated with X-linked dystonia parkinsonism. Proc Natl Acad Sci U S A 100(18):10347–10352

105. Ojwang PJ, Pegoraro RJ, Deppe WM, Sankar R, McKerrow N, Varughese L et al. (2001) Biochemical and molecular diagnosis of glutaric aciduria type 1 in a black South African male child: case report. East Afr Med J 78(12):682–685
106. Oliveira JR, Spiteri E, Sobrido MJ, Hopfer S, Klepper J, Voit T et al. (2004) Genetic heterogeneity in familial idiopathic basal ganglia calcification (Fahr disease). Neurology 63(11):2165–2167
107. Pereira AC, Edwards MJ, Buttery PC, Hawkes CH, Quinn NP, Giovannoni G et al. (2004) Choreic syndrome and coeliac disease: a hitherto unrecognised association. Mov Disord 19(4):478–482
108. Piccolo I, Defanti CA, Soliveri P, Volonte MA, Cislaghi G, Girotti F (2003) Cause and course in a series of patients with sporadic chorea. J Neurol 250(4):429–435
109. Pincus JH, Chutorian A (1967) Familial benign chorea with intention tremor: a clinical entity. J Pediatr 70(5):724–729
110. Rainier S, Thomas D, Tokarz D, Ming L, Bui M, Plein E et al. (2004) Myofibrillogenesis regulator 1 gene mutations cause paroxysmal dystonic choreoathetosis. Arch Neurol 61(7):1025–1029
111. Rampoldi L, Dobson-Stone C, Rubio JP, Danek A, Chalmers RM, Wood NW et al. (2001) A conserved sorting-associated protein is mutant in chorea-acanthocytosis. Nat Genet 28(2):119–120
112. Reske-Nielsen E, Jensen PK, Hein-Sorensen O, Abelskov K (1988) Calcification of the central nervous system in a new hereditary neurological syndrome. Acta Neuropathol (Berl) 75(6):590–596
113. Rolfs A, Koeppen AH, Bauer I, Bauer P, Buhlmann S, Topka H et al. (2003) Clinical features and neuropathology of autosomal dominant spinocerebellar ataxia (SCA17). Ann Neurol 54(3):367–375
114. Rubio JP, Danek A, Stone C, Chalmers R, Wood N, Verellen C et al. (1997) Chorea-acanthocytosis: genetic linkage to chromosome 9q21. Am J Hum Genet 61(4):899–908
115. Sanna G, Bertolaccini ML, Cuadrado MJ, Laing H, Khamashta MA, Mathieu A et al. (2003) Neuropsychiatric manifestations in systemic lupus erythematosus: prevalence and association with antiphospholipid antibodies. J Rheumatol 30(5):985–992
116. Schneider SA, Bhatia K (2006) Paroxysmal dyskinesias – an overview. In: Jankovic J, Tolosa E (eds) Parkinson's disease and movement disorders. Lippincott Williams & Wilkins, Philadelphia, pp 459–467
117. Schneider SA, van de Warrenburg BP, Hughes TD, Davis M, Sweeney M, Wood N et al. (2006) Phenotypic homogeneity of the Huntington's disease-like presentation in a SCA17 family. Neurology 67(9):1701–1703
118. Schrag A, Quinn NP, Bhatia KP, Marsden CD (2000) Benign hereditary chorea – entity or syndrome? Mov Disord 15(2):280–288
119. Semaka A, Creighton S, Warby S, Hayden MR (2006) Predictive testing for Huntington's disease: interpretation and significance of intermediate alleles. Clin Genet 70(4):283–294
120. Sethi KD (2006) Drug-induced movement disorders. New York, Basel, Marcel Dekker.
121. Sethi KD, Adams RJ, Loring DW, el Gammal T (1988) Hallervorden–Spatz syndrome: clinical and magnetic resonance imaging correlations. Ann Neurol 24(5):692–694
122. Sethi KD, Ray R, Roesel RA, Carter AL, Gallagher BB, Loring DW et al. (1989) Adult-onset chorea and dementia with propionic academia. Neurology 39(10):1343–1345
123. Simpson SA, Johnston AW (1989) The prevalence and patterns of care of Huntington's chorea in Grampian Br J Psychiatry 155:799–804
124. Spacey SD, Valente EM, Wali GM, Warner TT, Jarman PR, Schapira AH et al. (2002) Genetic and clinical heterogeneity in paroxysmal kinesigenic dyskinesia: evidence for a third EKD gene. Mov Disord 17(4):717–725
125. Steinlein OK, Mulley JC, Propping P, Wallace RH, Phillips HA, Sutherland GR et al. (1995) A missense mutation in the neuronal nicotinic acetylcholine receptor alpha 4 subunit is associated with autosomal dominant nocturnal frontal lobe epilepsy. Nat Genet 11(2):201–203

126. Stevanin G, Camuzat A, Holmes SE, Julien C, Sahloul R, Dode C et al. (2002) CAG/CTG repeat expansions at the Huntington's disease-like 2 locus are rare in Huntington's disease patients. Neurology 58(6):965–967
127. Stevanin G, Fujigasaki H, Lebre AS, Camuzat A, Jeannequin C, Dode C et al. (2003) Huntington's disease-like phenotype due to trinucleotide repeat expansions in the TBP and JPH3 genes Brain 126(Pt 7):1599–1603
128. Sudo K, Tashiro K (1998) Hyperthyroidism-associated chorea. Lancet 352(9123):239
129. Surtees RA, Matthews EE, Leonard JV (1992) Neurologic outcome of propionic academia. Pediatr Neurol 8(5):333–337
130. Swoboda KJ, Soong B, McKenna C, Brunt ER, Litt M, Bale JF Jr et al. (2000) Paroxysmal kinesigenic dyskinesia and infantile convulsions: clinical and linkage studies. Neurology 55(2):224–230
131. Takano T, Yamanouchi Y, Nagafuchi S, Yamada M (1996) Assignment of the dentatorubral and pallidoluysian atrophy (DRPLA) gene to 12p 13.31 by fluorescence in situ hybridization. Genomics 32(1):171–172
132. Taurin G, Golfier V, Pinel JF, Deburghgraeve V, Poirier JY, Edan G et al. (2002) Choreic syndrome due to Hashimoto's encephalopathy. Mov Disord 17(5):1091–1092
133. Tse W, Cersosimo MG, Gracies JM, Morgello S, Olanow CW, Koller W (2004) Movement disorders and AIDS: a review Parkinsonism. Relat Disord 10(6):323–334
134. Valente EM, Spacey SD, Wali GM, Bhatia KP, Dixon PH, Wood NW et al. (2000) A second paroxysmal kinesigenic choreoathetosis locus (EKD2) mapping on 16q13-q22.1 indicates a family of genes which give rise to paroxysmal disorders on human chromosome 16. Brain 123(Pt 10):2040–2045
135. Voll R, Hoffmann GF, Lipinski CG, Trefz FK, Weisser J (1993) [Glutaric acidemia/glutaric aciduria I as differential chorea minor diagnosis]. Klin Padiatr 205(2):124–126
136. Walker RH, Rasmussen A, Rudnicki D, Holmes SE, Alonso E, Matsuura T et al. (2003) Huntington's disease-like 2 can present as chorea-acanthocytosis. Neurology 61(7):1002–1004
137. Warner TT, Williams L, Harding AE (1994) DRPLA in Europe. Nat Genet 6(3):225
138. Waters CH, Takahashi H, Wilhelmsen KC, Shubin R, Snow BJ, Nygaard TG et al. (1993) Phenotypic expression of X-linked dystonia-parkinsonism (lubag) in two women. Neurology 43(8):1555–1558
139. Wheeler PG, Weaver DD, Dobyns WB (1993) Benign hereditary chorea. Pediatr Neurol 9(5):337–340
140. Woods CG, Taylor AM (1992) Ataxia telangiectasia in the British Isles: the clinical and laboratory features of 70 affected individuals. Q J Med 82(298):169–179
141. Wszolek ZK, Baba Y, Mackenzie IR, Uitti RJ, Strongosky AJ, Broderick DF et al. (2006) Autosomal dominant dystonia-plus with cerebral calcifications. Neurology 67(4):620–625
142. Xiang F, Almqvist EW, Huq M, Lundin A, Hayden MR, Edstrom L et al. (1998) A Huntington's disease-like neurodegenerative disorder maps to chromosome 20p. Am J Hum Genet 63(5):1431–1438
143. Xu X, Pin S, Gathinji M, Fuchs R, Harris ZL (2004) Aceruloplasminemia: an inherited neurodegenerative disease with impairment of iron homeostasis. Xu Ann N Y Acad Sci 1012:299–305
144. Yamamoto K, Chiba HO, Ishitobi M, Nakagawa H, Ogawa T, Ishii K (1997) Acute encephalopathy with bilateral striatal necrosis: favourable response to corticosteroid therapy. Eur J Paediatr Neurol 1(1):41–45
145. Younes-Mhenni S, Thobois S, Streichenberger N, Giraud P, Mousson-de-Camaret B, Montelescaut ME et al. (2002) [Mitochondrial encephalomyopathy, lactic acidosis and stroke-like episodes (Melas) associated with a Fahr disease and cerebellar calcifications]. Rev Med Interne 23(12):1027–1029
146. Zambrino CA, Zorzi G, Lanzi G, Uggetti C, Egitto MG (2000) Bilateral striatal necrosis associated with *Mycoplasma pneumoniae* infection in an adolescent: clinical and neuroradiologic follow up. Mov Disord 15(5):1023–1026
147. Zheng W, Wanibuchi M, Onda T, Liu H, Koyanagi I, Fujimori K et al. (2006) A case of moyamoya disease presenting with chorea. Childs Nerv Syst 22(3):274–278

An Update on the Hardie Neuroacanthocytosis Series

S. Gandhi(✉), R.J. Hardie, and A.J. Lees

Abstract In 1991 Hardie described the clinical and pathological features of 19 cases of neuroacanthocytosis, resulting in the largest series reported with this rare disorder. During the past 15 years, there have been many advances in our understanding of the neuroacanthocytosis syndrome, including the identification of several different molecular causes. We have revisited the original Queen Square series in an attempt to correlate the clinical picture and natural history of each case with the new genetic findings.

S. Gandhi
Department of Molecular Neuroscience, Institute of Neurology, Queen Square, London, WC1N 3BG, UK
s.gandhi@ion.ucl.ac.uk

R.H. Walker et al. (eds.), *Neuroacanthocytosis Syndromes II*.

1 Introduction

Our knowledge of neuroacanthocytosis (NA) represents a good example of the stages which occur in the development of understanding any rare genetic neurological disease. A distinctive clinical syndrome is first described in a number of case reports and is then followed by discoveries which shed light on the underlying molecular basis of the disease. In this chapter we use the latest developments in the field of NA and apply them to a previously well-described, and often cited clinical series of patients reported from the National Hospital for Neurology and Neurosurgery, Queen Square, London.

The original association between the haematological abnormality of acanthocytosis and a neurological syndrome was recognised in 1950 [2]. The earliest descriptions linked a neurological syndrome, acanthocytosis and an accompanying disturbance in lipid metabolism (such as abetalipoproteinaemia). From 1967, case reports of individuals and families with hereditary acanthocytosis and a neurological syndrome without lipid abnormalities also appeared [4, 5]. Finally, reports in the 1980s confirmed another association, between a rare blood group characterised by weak expression of the Kell system and absent expression of Kx antigen (described in 1961 [1]), acanthocytosis and neurological abnormalities. This X-linked genetic disorder was termed 'McLeod syndrome'. 'Neuroacanthocytosis' was now divided into three broad groups consisting of (1) NA with lipid abnormalities (2) hereditary NA without lipid abnormalities (3) X-linked NA without lipid abnormalities and with the McLeod blood group.

In 1991 Hardie and colleagues reported a clinical, haematological, and pathological series of 19 patients with NA [6], consisting of 12 cases from three different families and seven non-familial cases. All cases were included in the study if they exhibited acanthocytosis, neurological symptoms or signs, and normal plasma lipoproteins. The series included two cases of the McLeod phenotype in a single family.

It was concluded that NA was a progressive disease with a mean age of onset of 32 years (range 8–62). The movement disorder was characterised by chorea, although dystonia, tics, and parkinsonism were also reported. Orofaciolingual involuntary movements were also common. Two cases had no movement disorder. Over half the patients had cognitive impairment or psychiatric features. Depressed or absent tendon reflexes were found in over two thirds, and in three cases a sural nerve biopsy confirmed a chronic axonal neuropathy. Creatine kinase (CK) levels were raised in over half of the cases. The authors concluded that there was marked phenotypic variation, and that the genetic basis for NA was probably heterogeneous.

Since the original paper was published, many developments in understanding the molecular basis of this disease have occurred. These developments have enabled us to apply genetic diagnoses to a number of the cases within it. This has the advantage of redefining a subgroup phenotype by its molecular aetiology, and also of permitting more informative genotype–phenotype correlations.

2 Refining the Diagnosis

2.1 *Autosomal Recessive Chorea-Acanthocytosis*

Rubio et al. performed linkage analysis of 11 families with presumed chorea-acanthocytosis (ChAc) [17]. A clinical diagnosis of ChAc in each family was based on a neurological syndrome characterized by an abnormal movement disorder, myopathy, neuropathy, seizures, neuropsychiatric abnormalities, and the presence of acanthocytosis, and the absence of McLeod syndrome or abetalipoproteinaemia. Two of the families in the linkage study were described in Hardie's series of 1991 (CHAC family 1 and CHAC family 9). In 1997, Rubio et al. reported linkage of the genetic abnormality in these 11 families to a 6-cM region on chromosome 9q21 [17]. The results confirmed that in all 11 cases ChAc was a homogeneous autosomal recessive disorder.

In 2001, Rampoldi and colleagues succeeded in identifying the *CHAC* gene [14] (now *VPS13A*). Sixteen distinct mutations were identified in the 11 ChAc families which co-segregated with other affected family members. Of the 11 families, three had homozygous mutations of *VPS13A* in affected members, six had compound heterozygous mutations in *VPS13A* and two had a single heterozygous mutation in *VPS13A*. From the original Hardie series, a compound heterozygous mutation 269T→A (exon 4) and 6404-6405insT (exon 48) was identified in ChAc family 1. A single heterozygous mutation 8162A→G (exon 27) was identified in ChAc family 9. This has raised the question as to whether a second mutation in *VPS13A* is present and was not detected, or whether a single mutation in *VPS13A* is sufficient to cause disease in certain cases.

2.2 *McLeod Syndrome*

McLeod syndrome is a condition defined on the basis of abnormal expression of the Kell blood group antigens and the absence of a red blood cell surface antigen called Kx (see chapter by Jung). It is an X-linked multi-system disorder with central nervous system, neuromuscular, psychiatric, cardiac and haematological abnormalities. The clinical syndrome overlaps with ChAc, although in the latter there is an entirely normal expression of the Kell antigens. Ho et al. reported the identification of the gene responsible for McLeod syndrome in 1994, the *XK* gene [8]. Following this, the same authors performed mutation screening on a family from the Hardie series that appeared to have two affected members with the McLeod blood group. The mutation identified was a novel 1-bp deletion in exon 2 at codon 90, which creates a frameshift, alters the amino acid sequence, and is predicted to result in premature termination of translation generating a protein with 128 amino acids instead of 444 [9]. The mutation was not present in any of the unaffected members of the family. Three females in the family were heterozygous for the mutation. Female carriers of the McLeod gene showed blood group mosaicism in the Kell system and some erythrocyte morphological abnormalities, but do not manifest the central nervous system disease. However in this family there was a severely affected heterozygous female. Interestingly, analysis for skewed X inactivation in her tissues confirmed that there was skewed inactivation of the X-chromosome carrying the normal XK locus in all tissues including the brain. The unbalanced inactivation of the X chromosome would account for why the heterozygous female manifested the disease.

2.3 Pantothenate-Kinase Associated Degeneration (PKAN) and HARP Syndrome

PKAN is a rare autosomal recessive neurodegenerative disease with onset during childhood [13]. It is characterized by early onset dystonia, spasticity and dementia. Pathologically it has features of bilateral globus pallidus and substantia nigra degeneration with deposition of iron in the affected regions. Approximately 10% of cases of PKAN are associated with acanthocytes [7]. In 2001, Zhou et al. reported that mutations in the *PANK2* gene were the cause of PKAN (formerly known as Hallervorden–Spatz syndrome) [18].

HARP, a syndrome of hypoprebetalipoproteinaemia, acanthocytosis, retinitis pigmentosa, and pallidal degeneration, shares many common clinical and pathological features with PKAN. Two cases of HARP syndrome have been reported to have PANK2 mutations [3, 10]. However HARP also shares certain characteristic features (acanthocytosis and orolingual dystonia and dysarthria) with the NA phenotype and this may lead to difficulties in making an accurate diagnosis.

In the original Hardie NA series was a case of an 18-year-old male with onset of dystonia at the age of 10, progressing to generalized and orofacial and lingual dystonia (case 17). He had evidence of 4% acanthocytes and was diagnosed with NA. Reinvestigation 4 years later revealed a bilateral pigmentary retinopathy and an MRI demonstrated features of the 'eye-of-the-tiger' sign characteristic of PKAN disease. Due to these additional atypical features this case was reclassified in 1995 as a variant of HARP syndrome, although without hypoprebetalipoproteinaemia [12]. The underlying pathogenetic mutation in this case remains unknown.

3 Genotype-Clinical Phenotype Correlation

The cases from the Hardie series with known genetic mutations were reanalyzed according to their molecular diagnosis. The details of the individuals' clinical and pathological features are outlined in Table 1.

Table 1 Clinical, haematological and pathological features in cases with known mutations

Case	AOO	Sex	Movement disorder	Orofacial dyskinesia	Dysarthria	Biting	Psychiatric features	Cognitive features	Seizures	Neuropathy	Acanthocytes %	Pathology regions of neuronal loss
ChAc family 1 [269T→A (exon 4) + 6404-6405insT (exon 48)]												
1	37	M	P	–	+	–	+	+	+	+	0	Ca, Pu, GP,SN
2	39	F	CP	+	+	–	+	+	–	+	0	
3	40	F	CPDT	–	+	–	–	+	–	+	10	
4	44	F	CPT	–	+	–	+	+	–	+	10	
ChAc family 9 [8162A→G (exon 27)]												
11	12	F	PDT	+	+	+	+	+	–	+	30	
12	8	M	D	–	+	–	+	+	–	–	10	
McLeod family [1-bp deletion codon 90 (exon 2)]												
5	51	F	CT	+	+	–	+	+	+	+	30	Ca, Pu, GP
6	–	F	C	–	–	–	–	–	–	–	10	
7	24	M	C	–	–	–	+	–	+	+	NK	
8	22	M	–	–	–	–	–	–	+	+	NK	
9	27	M	–	–	–	–	–	–	–	+	20	
10	–	F	–	–	–	–	–	–	–	–	5	

Case refers to case number in Hardie et al. [6]
AOO age of onset, *M* male, *F* female
C chorea, *P* parkinsonism, *D* dystonia, *T* tremor
+ = present, – absent, *NK* not known
Neuropathy: + = hypo/areflexia or electrophysiological evidence of neuropathy
Pathology: *Ca* caudate, *Pu* putamen, *GP* globus pallidus, *SN* substantia nigra

3.1 *NA Secondary to* VPS13A (CHAC) *Mutation*

Six cases from two families (family 1 and family 9) were identified from the Hardie series with a segregating mutation in *VPS13A*.

The mean age of onset was 25 years (range 8–44). In fact the age of onset was consistent within each family [family 1: 40 years (37–44); family 9: 10 years (8–12)] and therefore the specific type of mutation may have an effect on the age of onset of the condition. The duration of the disease was 23 years (22–24) calculated from age of onset to age of death in two cases from family 1. This may also therefore display interfamily variation.

A movement disorder was the presenting feature in two thirds of cases, and cognitive/psychiatric features in the remainder. Progressive parkinsonism (bradykinesia and rigidity) was present in all but one case, and chorea, tics and dystonia occurred in half the cases. Orolingual dyskinesias were present in one third of cases. Biting of the lip, tongue or cheek was observed in only one out of the six cases with a known *VPS13A* mutation (although was also present in two of the sporadic cases reported by Hardie et al.). Two thirds of the patients had limitation of upgaze or definite supranuclear gaze palsy. Dysarthria, caused by involuntary choreiform movements and oromandibular and lingual dystonia, was present and progressive in all cases, resulting in severe dysarthria in one third. In addition to the movement disorder, all six cases demonstrated psychiatric features including emotional lability, antisocial behaviour, depression and obsessionality. All six cases demonstrated cognitive impairment. Formal neuropsychometry was performed on three, all of whom showed signs of predominant impaired executive function, suggesting frontal lobe dysfunction [11]. Five out of the six cases were associated with either hyporeflexia or electrophysiological evidence of a peripheral neuropathy. Seizures were present in one case and CK was raised in one case. Neuroimaging demonstrated caudate atrophy in three out of five cases studied. The level of acanthocytes was 10–30% in two thirds of cases while acanthocytes were not detected in the remainder.

3.2 *NA Secondary to McLeod Gene Mutation*

Four affected members of the family carried the frameshift mutation in the *XK* gene: one manifesting female heterozygote, and three male carriers. Two further heterozygote females with the mutation were also examined. The mean age of onset was 24.3 years (22–27) for the affected males, and 51 years for the affected female. The duration of the disease calculated from age of onset to age of death was 8 years (7–9) in the affected males. Seizures were the presenting features in three out of four cases, and chorea/tics in one case. Chorea and tics were present during the course of the disease in three out of the four cases. Two cases showed mild psychiatric features such as personality change and distractibility. Cognitive decline, predominantly frontal lobe dysfunction, was noted in one. All four cases had clinical or electophysiological evidence of neuropathy and elevated CK levels.

The level of acanthocytes was recorded at 20–30% in affected members of the family and 5–10% in asymptomatic female carriers.

The two further female carriers were asymptomatic but on examination had areflexia and mild choreiform movements associated with acanthocytosis and raised CK levels.

4 Genotype-Pathological Phenotype Correlation

The neuropathological features of postmortem brain in three of the cases from the Hardie series were reported in two papers in 1994 [15, 16]. Of these three cases, one is reported from ChAc family 1 associated with a compound heterozygous mutation in *VPS13A*. Another case described the affected female case from the family carrying the mutation in the McLeod gene. The third case described in these papers was a sporadic case of NA from the cohort, whose genetic status is unpublished.

4.1 VPS13A (CHAC) *Mutation [269T→A (Exon 4) + 6404-6405insT (Exon 48)] (Hardie's Case 2)*

Macroscopically there was atrophy of the caudate nucleus, putamen, and pallidum [16]. Histological examination showed severe neuronal loss and gliosis in the caudate nucleus, with less severe involvement of the putamen and external and internal globus pallidus. There was sparing of the thalamus, subthalamic nucleus, brainstem, cerebellum, and spinal cord. Neuronal loss, in the absence of Lewy body formation, was observed in the substantia nigra. Quantification of the loss of pigmented neurons and tyrosine hydroxylase-immunoreactive neurons demonstrated significant reduction in neurons throughout the substantia nigra, although most notably affecting the ventrolateral part. The neuronal density in this case was within the range of the neuronal loss seen in postmortem brains of patients with idiopathic Parkinson's disease [15]. This patient had evidence of parkinsonism during life, and had a disease duration of 22 years.

4.2 McLeod (XK) *Mutation [1-bp Deletion Codon 90 (Exon 2)] (Hardie's Case 5 – Female)*

Macroscopic examination showed dilated lateral ventricles and severe atrophy of the caudate and putamen nuclei [16]. Microscopic examination confirmed neuronal loss and gliosis most severely affecting the posterior part of the caudate nucleus, but also affecting the putamen to a less extent. Gliosis was evident in the globus pallidus. The brainstem, thalamus, subthalamic nuclei, cerebellum and cerebral cortex were spared. Quantification of neurons in the substantia nigra confirmed that the neuronal count was at the lower limit of the control range, and that there was no clear neuronal loss [15]. This patient did not suffer from parkinsonism during life, and had a disease duration of 6 years. (This case is also discussed in chapter by Danek et al. and Geser et al.).

5 Conclusion

We conclude that the clinical phenotype of NA secondary to *VPS13A* mutations consists of a progressive disease of mean onset 25 years, and characterized predominantly by a movement disorder of parkinsonism, chorea, dystonia, orolingual dyskinesias, cognitive features related to frontal lobe dysfunction, psychiatric features, and neuropathy. Pathologically these features are associated with severe neuronal loss in the caudate and putamen, and neuronal loss in the substantia nigra. The clinical phenotype of NA secondary to mutations in the *XK* gene consists of a progressive disease of mean onset 24.2 years and characterized by a predominance of seizures, a movement disorder of chorea and tics, and a neuropathy. Cognitive and psychiatric abnormalities are more variable. Pathologically this mutation was associated with identical features to the case of ChAc, except for sparing of the substantia nigra.

It is difficult to draw definitive divisions between the two phenotypes with such small numbers. However it is apparent from the subset of cases described in this chapter that there is significant heterogeneity of the underlying molecular diagnosis of the NA series. At the current time, the genetic basis of 12 out of the 19 cases of the Hardie series is known and published. Further work will be directed at identifying the molecular basis of the rest of cases. This work, together with the descriptions of new cases, will enable us to understand the specific phenotype associated with a particular mutation, the variable expressivity of the mutation and phenotype within any one family, and the allelic heterogeneity of the mutations giving rise to inter-family phenotypic variation. Clinical and pathological follow-up of the cases from this cohort will also be valuable in order to draw further correlations between the aetiology of the disease and the natural history of its progression.

References

1. Allen FH Jr, Krabbe SM, Corcoran PA (1961) A new phenotype (McLeod) in the Kell blood-group system. Vox Sang 6:555–560
2. Bassen FA, Kornzweig AL (1950) Malformation of the erythrocytes in a case of atypical retinitis pigmentosa. Blood 5:381–387
3. Ching KH, Westaway SK, Gitschier J Higgins JJ, Hayflick SJ (2002) HARP syndrome is allelic with pantothenate kinase-associated neurodegeneration. J Neurol 58:1673–1674
4. Critchley EM, Clark DB, Wikler A (1968) Acanthocytosis and neurological disorder without betalipoproteinemia. Arch Neurol 18:134–140

5. Estes JW, Morley TJ, Levine IM, Emerson CP (1967) A new hereditary acanthocytosis syndrome. Am J Med 42:868–881
6. Hardie RJ, Pullon HW, Harding AE, Owen JS, Pires M, Daniels, GL, Imai Y, Misra VP, King RH, Jacobs JM (1991) Neuroacanthocytosis. A clinical, haematological and pathological study of 19 cases. Brain 114(Pt 1A):13–49
7. Hayflick SJ, Westaway SK, Levinson B, Zhou B, Johnson MA, Ching KH, Gitschier J (2003) Genetic, clinical, and radiographic delineation of Hallervorden–Spatz syndrome. N Engl J Med 348:33–40
8. Ho M, Chelly J, Carter N, Danek A, Crocker P, Monaco AP (1994) Isolation of the gene for McLeod syndrome that encodes a novel membrane transport protein. Cell 77:869–880
9. Ho MF, Chalmers RM, Davis MB, Harding AE, Monaco AP (1996) A novel point mutation in the McLeod syndrome gene in neuroacanthocytosis. Ann Neurol 39:672–675
10. Houlden H, Lincoln S, Farrer M, Cleland PG, Hardy J, Orrell RW (2003) Compound heterozygous PANK2 mutations confirm HARP and Hallervorden–Spatz syndromes are allelic. Neurology 61:1423–1426
11. Kartsounis LD, Hardie RJ (1996) The pattern of cognitive impairments in neuroacanthocytosis. A frontosubcortical dementia. Arch Neurol 53:77–80
12. Orrell RW, Amrolia PJ, Heald A, Cleland PG, Owen JS, Morgan-Hughes JA, HardingAE, Marsden CD (1995) Acanthocytosis, retinitis pigmentosa, and pallidal degeneration: a report of three patients, including the second reported case with hypoprebetalipoproteinemia (HARP syndrome). Neurology 45:487–492
13. Pellecchia MT, Valente EM, Cif L, Salvi, S, Albanese, A, Scarano, V, Bonuccelli, U, Bentivoglio, AR, D'Amico, A, Marelli, C, Di, Giorgio A, Coubes, P, Barone, P, and Dallapiccola, B (2005) The diverse phenotype and genotype of pantothenate kinase-associated neurodegeneration. Neurology 64:1810–1812
14. Rampoldi L, Dobson-Stone C, Rubio JP, Danek, A, Chalmers, RM, Wood, NW, Verellen, C, Ferrer, X, Malandrini, A, Fabrizi, GM, Brown, R, Vance, J, Pericak-Vance, M, Rudolf, G, Carre, S, Alonso, E, Manfredi, M, Nemeth, AH, Monaco, AP (2001) A conserved sorting-associated protein is mutant in chorea-acanthocytosis. Nat Genet 28:119–120
15. Rinne JO, Daniel SE, Scaravilli F, Harding, AE, Marsden, CD (1994) Nigral degeneration in neuroacanthocytosis. Neurology 44:1629–1632
16. Rinne JO, Daniel SE, Scaravilli F, Pires, M, Harding, AE, Marsden, CD (1994) The neuropathological features of neuroacanthocytosis. Mov Disord 9:297–304
17. Rubio JP, Danek A, Stone C, Chalmers, R, Wood, N, Verellen, C, Ferrer, X, Malandrini, A, Fabrizi, GM, Manfredi, M, Vance, J, Pericak-Vance, M, Brown, R, Rudolf, G, Picard, F, Alonso, E, Brin, M, Nemeth, AH, Farrall, M, Monaco, AP (1997) Chorea-acanthocytosis: genetic linkage to chromosome 9q21. Am J Hum Genet 61:899–908
18. Zhou B, Westaway SK, Levinson B, Johnson MA, Gitschier J, Hayflick SJ (2001) A novel pantothenate kinase gene (PANK2) is defective in Hallervorden–Spatz syndrome. Nat Genet 28:345–349

Update on McLeod Syndrome

H.H. Jung

Abstract McLeod syndrome is an X-linked neuroacanthocytosis syndrome caused by mutations of the *XK* gene. Central nervous system manifestations resemble Huntington's disease, and include a choreatic movement disorder, dysexecutive cognitive deficits, psychiatric abnormalities, and generalized seizures. Neuromuscular manifestations include myopathy, sensory-motor axonal neuropathy, and cardiomyopathy. In the recent years, McLeod syndrome has increasingly recognized in various countries. Several studies demonstrate a high phenotypic variability. Most mutations in the *XK* gene predict an absent or truncated XK protein, and no clear genotype–genotype correlation has been found. Missense mutations are rare and may be associated with a milder phenotype. Imaging studies demonstrate striatal atrophy and subtle cerebral metabolic abnormalities. Neuropathological studies reveal striatal neuronal loss and gliosis without specific features. There is an ongoing search for a suitable animal model to disclose the pathogenetic mechanisms of the disorder and search for possible therapeutic targets.

H.H. Jung
Department of Neurology, University Hospital Zürich, Frauenklinikstrasse 26, 8091 Zürich, Switzerland
hans.jung@usz.ch

R.H. Walker et al. (eds.), *Neuroacanthocytosis Syndromes II*.

1 Introduction

The McLeod blood group phenotype is characterized by absent expression of Kx erythrocyte antigen, weak expression of Kell glycoprotein antigens, and X-linked inheritance [1, 18]. Most carriers of the McLeod blood group phenotype have elevated serum levels of creatine kinase (CK) as a sign of muscle cell abnormality and are prone to develop central nervous system and neuromuscular symptoms with a mean onset ranging between 30 and 40 years [10, 32]. Central nervous system manifestations of McLeod syndrome (MLS) resemble Huntington's disease and comprise a choreatic movement disorder, variable psychiatric manifestations, and cognitive deficits of the frontal-dysexecutive type [5, 12]. In addition, about 40% of McLeod patients were reported to have seizures, which were mostly generalized [5]. Neuromuscular manifestations are present in about half of McLeod patients. Weakness and atrophy are usually mild, and only a minority of patients develops disabling weakness [5, 13, 14]. Rarely, symptomatic rhabdomyolysis has been reported [11].

MLS is caused by mutations in the *XK* gene, a membrane transport protein of as yet unknown function that contains the Kx erythrocyte antigen [8]. The XK protein is linked to the Kell glycoprotein, and the two proteins most probably form a functional complex [26]. However, the precise role of the XK-Kell complex is still not known.

The clinical spectrum of MLS has been extensively reviewed in the first edition of "Neuroacanthocytosis Syndromes" [10]. In this report, an overview of recent clinical, pathological and imaging studies will be given. In addition, novel findings regarding genotype–phenotype correlation, the molecular basis, and the perspective of animal models of MLS [28] will be discussed.

2 McLeod Syndrome Worldwide

About 150 patients with MLS have been reported in the literature. The patients originated from Europe, North America, Australia, New Zealand, Japan, Brazil, and Chile, indicating a widespread occurrence of McLeod syndrome at least in the Eurasian population [5, 7, 12, 20, 29, 30]. Although no African or African-American McLeod patients have been reported to date, this observation has to be interpreted with caution due to the rarity of the disorder.

Several distinct *XK* mutations were present in unrelated families of different geographic origin [20, 31]. A 5 bp deletion in the XK gene (938-942delCTCTA), in particular, was detected in a Chilean family of German descent, a North American patient of Anglo-Saxon descent and a Japanese family [5, 20, 30]. This widespread geographic distribution strongly argues against a significant founder effect. In addition, a broad phenotypic variability was observed between the families [5, 20, 30, 33].

3 Neuroimaging

Computed tomography and magnetic resonance imaging studies demonstrate atrophy of caudate nucleus and putamen, particularly with advanced disease [5, 7, 12]. Exceptionally, there may also be white matter changes [21]. [^{18}F]-fluorodeoxyglucose (FDG)-positron emission tomography (PET) demonstrates impaired striatal glucose metabolism [12, 22], but no impairment of glucose metabolism was found in the cerebral cortex by quantified FDG-PET [12]. Magnetic resonance spectroscopy demonstrates subtle metabolic abnormalities in different extrastriatal brain regions related to the psychiatric and cognitive findings [6]. The neuroradiological findings will be discussed in more detail in the chapter by Leenders et al.

4 Neuropathology

Neuropathological examination of one McLeod patient revealed marked neuronal loss and astrocytic gliosis in caudate nucleus and putamen. In contrast, no alterations were found in cortex, thalamus, subthalamic nucleus, brainstem and cerebellum [2, 7]. Similar findings were observed in the exceptional case of a female *XK* mutation carrier who clinically manifested MLS [7, 9]. Although the prominent psychiatric and cognitive manifestations in McLeod patients indicate a significant and widespread cortical, rather than purely subcortical dysfunction, the cerebral pathological alterations found in prior studies were restricted to the striatum indicating that the neuropsychiatric symptoms in MLS are caused by a neuronal dysfunction due to impaired basal ganglia-cortical circuits. In chapter 17 by Geser et al., however, we present the neuropathology of a McLeod patient with prominent neuropsychiatric alterations who had subtle but clear cortical pathological alterations, most probably corresponding to the extrastriatal metabolic alterations found in magnetic resonance spectroscopy [6].

5 The *XK* Gene

The *XK* gene contains three exons and is located on Xp21.1. It encodes the 444 amino acid residue XK protein, that is predicted to have ten trans-membrane domains and to show structural characteristics of prokaryotic and eukaryotic membrane transport proteins [8]. The XK protein contains the Kx red blood cell (RBC) antigen [15], and shares important homologies with the *ced-8* protein of the nematode *Caenorhabditis elegans*. In this nematode, *ced-8* controls the timing of programmed cell death [28].

The XK protein is linked to the Kell glycoprotein by a single disulfide bond (XKcys347-KellCys72) [18]. The Kell protein is a 93 kDa glycoprotein which is encoded by the *KEL* gene on chromosome 7q34. *KEL* contains 19 exons and shares a consensus sequence with a large family of zinc-dependent endopeptidases [16, 17]. The Kell glycoprotein is a type II red cell membrane protein with a short intracellular amino-terminal, a single transmembrane and a large extracellular domain [6]. It shows substantial homology with the M13 subfamily of mammalian neutral endopeptidases, including endothelin converting enzyme-1 (ECE1; 600423). ECE-1 converts big endothelin-3 into endothelin-3, the bioactive peptide [7]. Its extracellular conformation is stabilized by five intramolecular disulfide bonds preserving endopeptidase activity of Kell protein as well as most of the KEL antigens.

Several studies demonstrate that XK and Kell are co-expressed in erythroid tissue [3, 4, 8, 27]. In the erythroid tissue, Kell and XK most probably form a functional complex, since Kell antigen expression, as well as Kell protein density on the RBC membrane, are dependent on the expression of XK protein [25, 26]. In other tissues, however, the Kell and XK protein have a differential expression pattern. In skeletal muscle, there is no co-localization of Kell and XK [13]. In rodent and human brain, XK is expressed in intracellular compartments of neurons, whereas Kell expression is restricted to red blood cells in cerebral vessels [4]. Another study showed that XK, but not Kell, was significantly expressed in brain, spinal cord, muscle, heart, small intestine, stomach, bladder, and kidney [19]. In brain, XK was predominantly expressed in neuronal cells [19]. Coexpression of Kell and XK in erythroid tissues and the different expressions in non-erythroid tissues suggest that XK may have a complementary hematological function with Kell and a separate role in other tissues.

6 The *XK* Gene Family

Databank analysis demonstrates that the human *XK* gene belongs to a gene family with several different members [3, 24]. Sequence similarity searches identify a family of nine full-length human genes related to *ced-8*, as well as a series of eight Y chromosome-linked partial sequences. Sequence comparisons confirm that previously identified highly conserved motifs in *ced-8* and *XK* are shared with the related genes. Several residues in these motifs are identical between *ced-8* and other family members, but not conserved with *XK* [24]. Recently, two cDNAs, *XPLAC* and *XTES*, have been cloned. *XPLAC*, like *XK*, has three exons and is located on X-chromosome at q22.1, while *XTES* has four exons and is located at 22q11.1. Whereas *XK* is ubiquitously expressed, *XPLAC* is expressed predominantly in placenta and adrenal gland, and *XTES* is exclusively expressed in primate testis [3].

7 Models of McLeod Syndrome

Development of a murine model of McLeod syndrome is in progress (M. Ho, PhD, personal communication). The close homology of *XK* to *ced-8* encourages the search for a *C. elegans* MLS model. The *ced-8* gene was first identified in a screen for delayed clearance of apoptotic cell corpses [28]. The *ced-8* sequence shows a distinct, although significantly divergent, similarity to the *XK* gene. However, it is an open question as to whether these genes share a conserved function [23]. There is ongoing work to analyze *ced-8* mutant *C. elegans*, to identify possibly *ced-8*-interacting genes, to determine the transmembrane topology of CED-8, and to identify the functional mammalian ortholog of *ced-8* by rescue of the *ced-8* mutant phenotype with a human cDNA [23].

8 Conclusion

MLS is an X-linked Huntington-like neurological syndrome with additional neuromuscular features. Although rare, it represents a fascinating model disorder for the examination of molecular processes underlying neurodegeneration. The generation of suitable model systems is on the way. These model systems will hopefully allow the identification of the molecular mechanisms underlying the striatal neurodegeneration in MLS. In addition, these model systems might become a tool for the generation and examination of novel therapeutic strategies in this otherwise relentlessly progressive and fatal disorder.

References

1. Allen FH, Krabbe SMR, Corcoran PA (1961) A new phenotype (McLeod) in the Kell blood-group system. Vox Sang 6:555–560
2. Brin MF, Hays A, Symmans WA, Marsh WL, Rowland LP (1993) Neuropathology of McLeod phenotype is like choreaacanthocytosis (CA). Can J Neurol Sci 20(Suppl):S234
3. Calenda G, Peng J, Redman CM, Sha Q, Wu X, Lee S (2006) Identification of two new members, XPLAC and XTES, of the XK family. Gene 370:6–16
4. Claperon A, Hattab C, Armand V, Trottier S, Bertrand O, Ouimet T (2007) The Kell and XK proteins of the Kell blood group are not co-expressed in the central nervous system. Brain Res. 1147:12–24
5. Danek A, Rubio JP, Rampoldi L, Ho M, Dobson-Stone C, Tison F et al (2001) McLeod neuroacanthocytosis: genotype and phenotype. Ann Neurol 50:755–764
6. Dydak U, Mueller S, Sandor PS, Meier D, Boesiger P, Jung HH (2006) Cerebral metabolic alterations in McLeod syndrome. Eur Neurol 56(1):17–23
7. Hardie RJ, Pullon HW, Harding AE, Owen JS, Pires M, Daniels GL et al (1991) Neuroacanthocytosis. A clinical, haematological and pathological study of 19 cases. Brain 114:13–49
8. Ho M, Chelly J, Carter N, Danek A, Crocker P, Monaco AP (1994) Isolation of the gene for McLeod syndrome that encodes a novel membrane transport protein. Cell 77:869–880
9. Ho MF, Chalmers RM, Davis MB, Harding AE, Monaco AP (1996) A novel point mutation in the McLeod syndrome gene in euroacanthocytosis. Ann Neurol 39:672–675
10. Jung HH (2004) McLeod syndrome: a clinical review. In: Danek A (ed) Neuroacanthocytosis syndromes. Springer, Dordrecht, pp 45–53
11. Jung HH, Brandner S (2002) Malignant McLeod myopathy. Muscle Nerve 26:424–427

12. Jung HH, Hergersberg M, Kneifel S, Alkadhi H, Schiess R, Weigell-Weber M et al (2001) McLeod syndrome: a novel mutation, predominant psychiatric manifestations, and distinct striatal imaging findings. Ann Neurol 49:384–392
13. Jung HH, Russo D, Redman C, Brandner S (2001) Kell and XK immunohistochemistry in McLeod myopathy. Muscle Nerve 24:1346–1351
14. Kawakami T, Takiyama Y, Sakoe K, Ogawa T, Yoshioka T, Nishizawa M et al (1999) A case of McLeod syndrome with unusually severe myopathy. J Neurol Sci 166:36–39
15. Khamlichi S, Bailly P, Blanchard D, Goossens D, Cartron JP, Bertrand O (1995) Purification and partial characterization of the erythrocyte Kx protein deficient in McLeod patients. Eur J Biochem 228:931–934
16. Lee S (1997) Molecular basis of Kell blood group phenotypes. Vox Sang 73:1 11
17. Lee S, Lin M, Mele A, Cao Y, Farmar J, Russo D et al. (1999) Proteolytic processing of big endothelin-3 by the Kell blood group protein. Blood 94:1440–1450
18. Lee S, Russo D, Redman CM (2000) Kell blood group system: Kell and XK membrane proteins. Semin Hematol 37:113–121
19. Lee S, Sha Q, Wu X, Calenda G, Peng J (2007) Expression profiles of mouse Kell, XK, and XPLAC mRNA. J Histochem Cytochem 55:365–374
20. Miranda M, Castiglioni C, Frey BM, Hergersberg M, Danek A, Jung HH (2007) Phenotypic variability of a distinct deletion in McLeod syndrome. Mov Disord. 22:1358–1361
21. Nicholl DJ, Sutton I, Dotti MT, Supple SG, Danek A, Lawden M (2004) White matter abnormalities on MRI in neuroacanthocytosis. J Neurol Neurosurg Psychiatry 75:1200–1201
22. Oechsner M, Buchert R, Beyer W, Danek A (2001) Reduction of striatal glucose metabolism in McLeod choreoacanthocytosis. J Neurol Neurosurg Psychiatry 70:517–520
23. Phelan JK, Pinto SM, Falquet L, Jung HH, Hengartner MO (2006). Characterization of *ced-8* and *ced-8* interacting genes. European Worm Meeting 2006, Hersonissos, Crete, Greece, April 29–May 3
24. Phelan JK, Wong K, Jung HH, Hengartner MO (2005). Sequence analysis identifies a family of human genes related to *ced-8*. 15th Biennial International *C. elegans* Conference, University of California, Los Angeles, CA, USA, June 25–29
25. Russo D, Lee S, Redman C (1999) Intracellular assembly of Kell and XK blood group proteins. Biochim Biophys Acta 1461:10–18
26. Russo D, Redman C, Lee S (1998) Association of XK and Kell blood group proteins. J Biol Chem 273:13950–13956
27. Russo D, Wu X, Redman CM, Lee S (2000) Expression of Kell blood group protein in nonerythroid tissues. Blood 96:340–346
28. Stanfield GM, Horvitz HR (2000) The *ced*-8 gene controls the timing of programmed cell deaths in *C. elegans*. Mol Cell 5:423–433
29. Starling A, Schlesinger D, Kok F, Passos-Bueno MR, Vainzof M, Zatz M (2005) A family with McLeod syndrome and calpainopathy with clinically overlapping diseases. Neurology 65:1832–1833
30. Wada M, Kimura M, Daimon M, Kurita K, Kato T, Johmura Y et al (2003) An unusual phenotype of McLeod syndrome with late onset axonal neuropathy. J Neurol Neurosurg Psychiatry 74:1697–1698
31. Walker RH, Danek A, Uttner I, Offner R, Reid M, Lee S (2007) McLeod phenotype without the McLeod syndrome. Transfusion 47:299–305
32. Walker RH, Jung HH, Dobson-Stone C, Rampoldi L, Sano A, Tison F et al (2007) Neurologic phenotypes associated with acanthocytosis. Neurology 68:92–98
33. Walker RH, Jung HH, Tison F, Lee S, Danek A (2007) Phenotypic variation among brothers with the McLeod neuroacanthocytosis syndrome. Mov Disord 22:244–248

Huntington's Disease-Like 2

R.L. Margolis(✉) and D.D. Rudnicki

Abstract Huntington's disease like-2 (HDL2) is an autosomal dominant disorder that clinically and pathologically closely resembles Huntington's disease (HD). Like HD, HDL2 is characterized by mid-life onset, abnormalities of voluntary and involuntary movement, psychiatric syndromes, and dementia, with a relentless progress to death. The disease is rare, and thus far has only been detected in individuals of African ancestry. Striatal and cortical atrophy is prominent, as are intranuclear protein aggregates. Some, but not all, affected individuals have acanthocytosis. HDL2 is caused by a CAG/CTG expansion mutation on chromosome 16q24.3, in an alternatively spliced exon of junctophilin-3. The mechanism of HDL2 pathogenesis is uncertain, but may involve the toxic properties of mutant transcripts containing expanded CUG repeats.

R.L. Margolis
Laboratory of Genetic Neurobiology, Division of Neurobiology, Department of Psychiatry and Behavioral Sciences, Johns Hopkins University School of Medicine, CMSC 8-121, 600 N. Wolfe Street, Baltimore, MD 21287, USA
rmargoli@jhmi.edu

1 Introduction

The 1993 discovery of the genetic mutation that causes Huntington's disease (HD) confirmed that almost all cases of clinically diagnosed HD, regardless of age of onset, course, or clinical presentation, were caused by a CAG repeat expansion on chromosome 4p, in a gene eventually termed *huntingtin*. The nature of the repeat expansion explained at least some of the phenotypic variation, as longer repeats tend to lead to an earlier age of onset and repeat length tends to expand during paternal transmission. HD research has subsequently focused on pathogenic mechanisms, with much of the emphasis on how the long tracts of polyglutamine, encoded by the CAG repeat expansion, lead to neurotoxicity.

Careful inspection of clinical populations revealed a small number of suspected HD cases that did not have the HD mutation [18]. Some of these individuals were eventually diagnosed with another neurological disorder, such as dentatorubro-pallidoluysian atrophy (DRPLA) or prion disease, or a somatoform disorder (i.e., symptoms derived from psychological factors). A few, however, clearly had a personal and family history of a disease that could not be distinguished from HD, and for which no genetic cause could be found. In 2001, we identified a mutation on chromosome 16q segregating with illness in one such family, and termed their disease Huntington's disease-like 2 (HDL2). Subsequent work in our clinic and lab, and by others, has led to a preliminary understanding of the phenotype, epidemiology, and pathology of HDL2, and some intriguing leads into disease pathogenesis.

2 The HDL2 Phenotype

2.1 HDL2 Clinical Presentation

The phenotype of HDL2 was initially described in eight members of the index family (Fig. 1) [11]. Segregation was consistent with autosomal dominant inheritance. The age of disease onset ranged from 26 to 48, with a non-significant trend toward younger onset in the later of the two generations evaluated. Dysarthria, rigidity, hyperreflexia, chorea or dystonia, weight loss, dementia, and psychiatric symptoms were detected in every case examined. Almost all cases had an action tremor, abnormal gait, and bradykinesia. Eye movements were relatively spared. Disease course was quite stereotyped, beginning with weight loss and a decline in coordination, with gradually progressive cognitive, psychiatric, and motor abnormalities. After about 10–15 years, affected individuals were profoundly demented, rigid, and essentially bed bound. As in other neurodegenerative diseases, death then followed from a combination of inanition and infection.

Walker and colleagues [26] described a similar family that only later was shown to have the HDL2 mutation. The proband, described late in his disease course, had

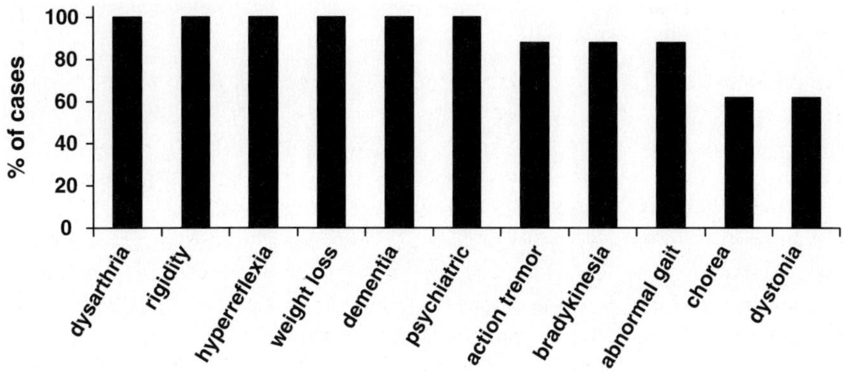

Fig. 1 Relative frequency of signs and symptoms of HDL2 in the index family. (Derived from Margolis et al. [11].)

mild chorea, was unable to speak or stand, and had lower extremity hyperreflexia (but overall decreased tone). One other family member had a similar phenotype, and another showed marked parkinsonism. In contrast, a member of a third pedigree was noted to have prominent chorea, abnormal eye movements, mild cognitive impairment and bradykinesia, and normal tone and reflexes [28]. A second case with this more classical Huntington's disease presentation has also been reported [20]. Overall, the clinical presentation of HD and HDL2 cannot be distinguished in a given individual, though a presentation characterized by prominent parkinsonism appears to be more common in HDL2 than in HD. Video clips of selected HDL2 cases have been published [27]. Unlike the autosomal recessive disorder chorea-acanthocytosis, muscle weakness, lip and tongue biting, and seizures are not part of the typical HDL2 clinical presentation [2]. There is also no evidence of the cardiac manifestations common in the X-linked McLeod syndrome, or the muscle and liver enzyme elevations common in either of these diseases.

2.2 *Neuropathology*

Neuroimaging studies have consistently shown cortical and basal ganglia atrophy in individuals with HDL2, even in relatively early cases [11, 20, 26]. Qualitatively, MRI images cannot be distinguished from those of HD patients (Fig. 2). While quantitative studies have not been performed, there is no consistent qualitative evidence of atrophy outside of the cortex and basal ganglia.

The neuropathology of 4 HDL2 cases has been published. The case from the index HDL2 family came to autopsy 20 years after disease onset. Gross examination showed mild frontal, temporal, and mesial parietal and occipital atrophy, with severe atrophy of the caudate and putamen. White matter appeared normal. Microscopic examination showed marked neuronal loss with relative sparing of

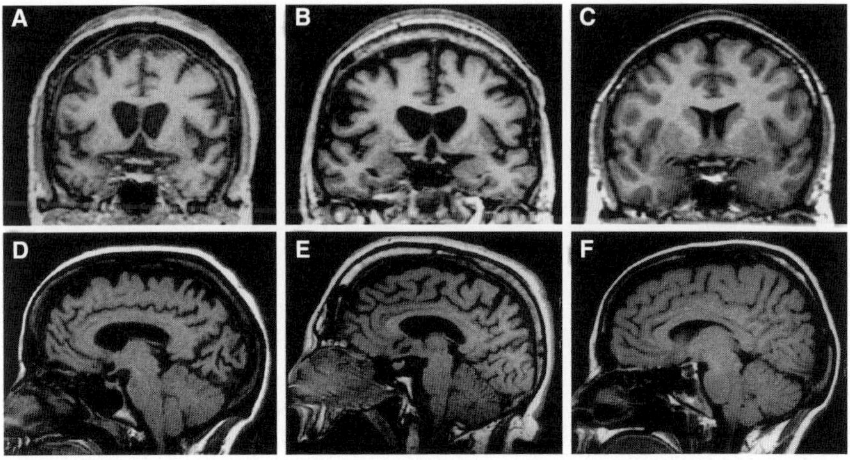

Fig. 2 Striatal and cerebral cortex atrophy in HDL2 resembles HD. (**a, d**) HDL2 case, age 36, 10 year disease duration. (**b, e**) HD case, age 48 years, 12 year, disease duration. (**c, f**) Normal control, age 43. (Reprinted, with permission, from Margolis et al. [11].)

larger neurons, astrocytosis, and neuropil vacuolation of the striatum (caudate worse than putamen), in a dorsal to ventral gradient. The globus pallidus was affected to a lesser extent, and neurodegeneration without Lewy bodies was observed in the substantia nigra. There was no evidence of neurofibrillary tangles or amyloid plaques. Intranuclear inclusion bodies are typically found in various regions, discussed in more detail below.

An unrelated individual [26], autopsied about 23 years after disease onset and retrospectively determined to have the HDL2 mutation, displayed similar, but not identical, findings. Cortical atrophy was diffuse, except for sparing of the hippocampus and medial temporal lobe. The striatum was severely atrophic, with much less atrophy of the globus pallidum. Pigmentation of the locus ceruleus and the substantia nigra was reduced. Substantial gliosis and neuronal loss, sparing large somatostatin-positive inter-neurons, was apparent in the striatum, with milder neuronal loss detected in multiple other brain regions, including substantia nigra, locus ceruleus, and amygdala. The frequency of scattered neurofibrillary tangles was considered to be within the normal range for age.

Two cases recently reported showed similar findings with a few interesting differences [3]. In the first case, mild atrophy was detected in the acumbens, and isolated regions of parietal and occipital cortex showed loss of large neurons in layers V and VI. There was evidence of mild neurodegeneration in the substantia nigra and the locus ceruleus, and mild AD-type findings in paralimbic and allocortical regions. The second case was notable, in addition to marked striatal neurodegeneration, for very prominent occipital lobe involvement, particularly in the primary visual cortex. Substantia nigra was mildly affected.

One conclusion from these preliminary findings is that HDL2 and HD cannot be distinguished on pathological grounds in any given case, but there may be more occipital lobe involvement, and perhaps substantia nigra involvement, in HDL2.

2.3 Protein Aggregates

In all four cases reported above, and in four other cases examined by our group, there were intranuclear aggregates that stained with anti-ubiquitin antibodies and an antibody (1C2) with partial specificity for long polyglutamine tracts [25] (Fig. 3). In the index case, aggregates were relatively infrequent, and were more common in striatum then cortex. They did not stain with anti-huntingtin antibodies. In the second case [26], aggregates were most common in the insula of the cortex, and were not detected in the basal ganglia. Interestingly, the inclusions stained for torsinA, but not for tau, alpha-synuclein, or p53. In the third case, scattered aggregates were found in multiple brain regions, and were generally most frequent near, but not in, those regions with the greatest neurodegeneration. In the fourth case, aggregates stained with anti-ubiquitin antibodies were found in the nuclei of neurons in multiple brain regions. Aggregates ranged in size from punctate to 5μm, and aggregate frequency did not appear to correlate with the extent of neuronal loss. In all cases,

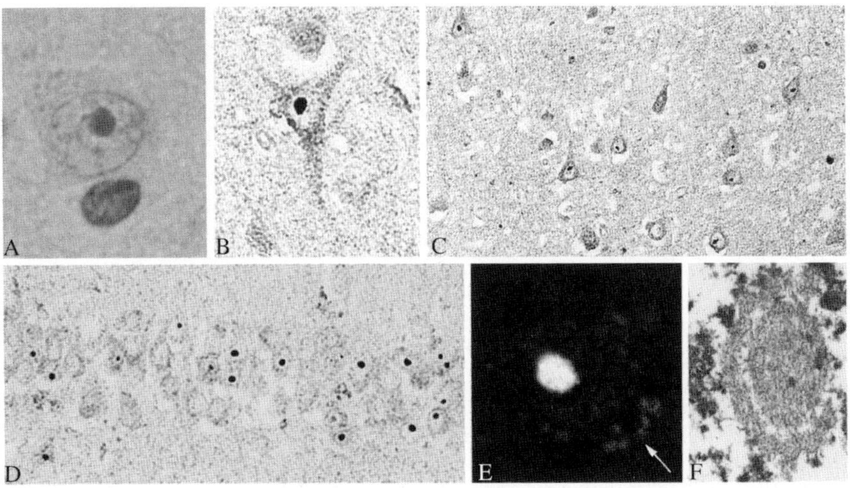

Fig. 3 Protein aggregates in HDL2. (**a**) Aggregates in cortical neurons detected by anti-ubiquitin antibodies. (**b**, **c**) Aggregates in cortex detected with anti-torsin A antibodies at higher power (×60) and lower power (×20). (**d**) Anti-torsin A staining of hippocampal granule cells. (**e**) 1C2 staining of aggregate, as detected by confocal microscoping using fluorochrome-labeled immunoreactivity (*arrow* indicates cytoplasmic lipofuscin). (**f**) Electron micrograph of protein aggregates (×20,000). (Figure and portions of the figure legend, are reprinted with permission from Walker et al. [26].)

the aggregates were round to oval in shape, and resembled those seen in HD. Aggregates outside of the nucleus, as seen in HD and other polyglutamine disorders, have not yet been reported in HDL2.

2.4 Genetic Features of HDL2

The HDL2 locus on 16q24.3 is a CTG repeat that is highly polymorphic in length in the general population, with a range of 6–28 and a modal length of 13. Repeat length associated with HDL2 varies from 40 to 59, with potential incomplete penetrance at the lower end of this range. The potential of repeat lengths between 29 and 39 triplets to contribute to disease is uncertain, though repeat length instability may be present. For instance, a mother with an HDL2 repeat of 33 triplets developed a nonprogressive cerebellar disorder after hospitalization for hyperglycemia. Her son, who inherited a slight expansion of this allele to 35 triplets, developed Cogan's syndrome, an incompletely characterized disorder of uncertain etiology involving interstitial keratitis with prominent optic and audiovestibular findings.

The length of the HDL2 repeat is clearly associated with onset age (Fig. 4) [12]. This provides strong evidence that the repeat itself, rather than another mutation in linkage disequilibrium with the repeat, is causative. The repeat length is not stable in vertical transmission, as evident by the small variability in HDL2 repeat lengths found in the sibships of the index HDL2 family [6]. A trend toward longer repeat lengths in subsequent generations has been detected, but will require more transmissions to confirm.

The relationship between genotype and phenotype is remarkably similar in HDL2 and HD, including the polymorphism of the repeat in the normal population, the length at which repeats are associated with disease, the instability of long repeats below the disease threshold, the incomplete penetrance of repeat lengths at the low end of the disease range, and the correlation of longer repeats with early onset age. The primary difference appears to be that the average length of the HDL2 disease allele is somewhat longer than the HD disease allele.

2.5 HDL2 is Associated with African Ethnicity

HDL2 has thus far been identified in about 1% of individuals with an HD-like disorder who do not have HD. The actual percentage, at least in the U.S., is probably somewhat higher if analysis is restricted to individuals with a family history and a clear HD-like presentation, with exclusion of individuals who likely have tardive dyskinesia or a tic disorder. So far, all individuals with HDL2 have either definite or probable African ancestry. In South Africa, preliminary analysis suggests that HDL2 is almost as common as HD in individuals of African descent [12]. It therefore appears that the HDL2 expansion has an African origin. The distribution of HDL2, and the length of the normal allele in various African populations, remains to be determined.

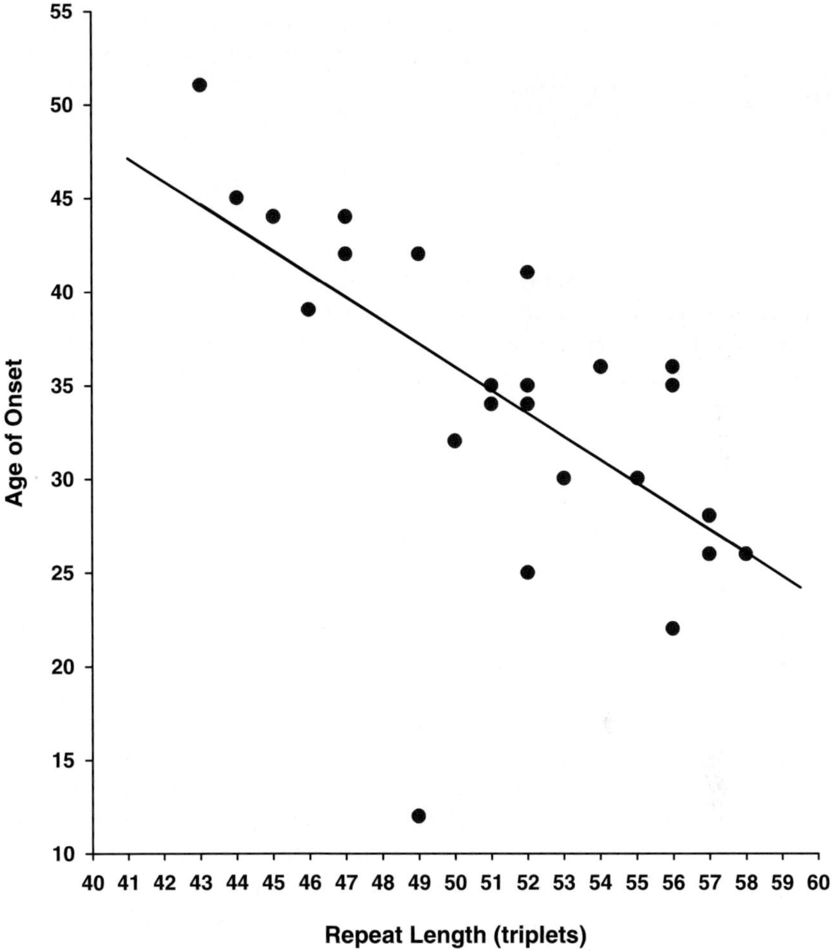

Fig. 4 HDL2 repeat length is correlated with an earlier onset age. $N = 24$, $R = 0.62$, $r^2 = 0.39$, $p = 0.0011$. (Reprinted, with permission, from Margolis et al. [12].)

3 HDL2 and Acanthocytosis

The possibility that HDL2 is a form of neuroacanthocytosis emerged in a family studied by Walker et al. prior to the discovery of the HDL2 mutation. All three individuals in the family, clinically described above, had 30–35% acanthocytes as measured both by peripheral blood smears and scanning EM (Fig. 5). RBC membrane extract showed a prominent band not present in a normal control, and consistent with a band 3 breakdown product. Subsequently, peripheral blood smears from three members of a pedigree from Mexico, two members of the index pedigree, and one member of a fourth pedigree were examined [28]. Only one of the individuals from the Mexican family had acanthocytosis (defined as >30% acanthocytes on a

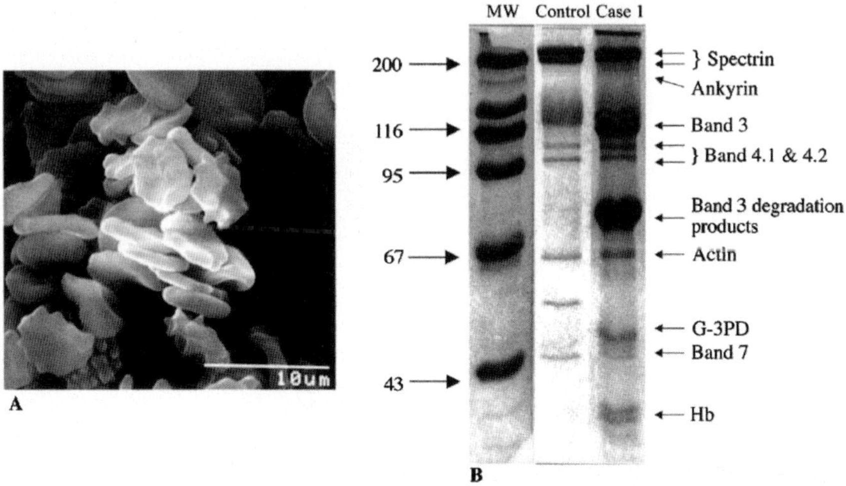

Fig. 5 Acanthocytes in HDL2. (**a**) Scanning EM of peripheral blood. (**b**) Band 3 abnormalities in peripheral RBC membranes. MW = molecular weight markers. (Figure and portions of the figure legend, are reprinted with permission from Walker et al. [26].)

peripheral smear). The presence of acanthocytes in two unrelated pedigrees with HDL2 is unlikely to be coincidental. It therefore appears that acanthocytosis is a variably penetrant feature of HDL2, and that mutation of JPH3 may lead to disruption of RBC membranes.

4 The Structure and Function of *JPH3* and Its Encoded Protein

JPH3 contains at least six exons (Fig. 6). The HDL2 repeat is located in a previously unidentified alternative 3′ exon of the gene. Surprisingly, the repeat is oriented in the CTG direction, and various splice acceptor sites place the repeat in-frame to encode polyalanine or polyleucine, or in 3′ UTR. *JPH3* is one of a four-member gene family encoding proteins thought to be involved in physically bridging the gap between plasma membrane and sarcoplasmic reticulum (SR, muscle) or endoplasmic reticulum (ER, neurons) [16, 22]. Each protein in the family has a terminal region that serves to anchor the protein into plasma membrane, and a C-terminal domain that functions to attach the protein to the SR or ER. Myocytes of mice in which junctophilin-1 is not expressed show abnormal ultrastructure of the junctions between SR and plasma membrane, and the muscle develops an abnormally low contractile force after low frequency electrical stimulation [8].

Knock out of *JP3* (the mouse orthologue of *JPH3*) resulted in only a mild motor phenotype and no clear electrophysiological or pathological abnormalities, though

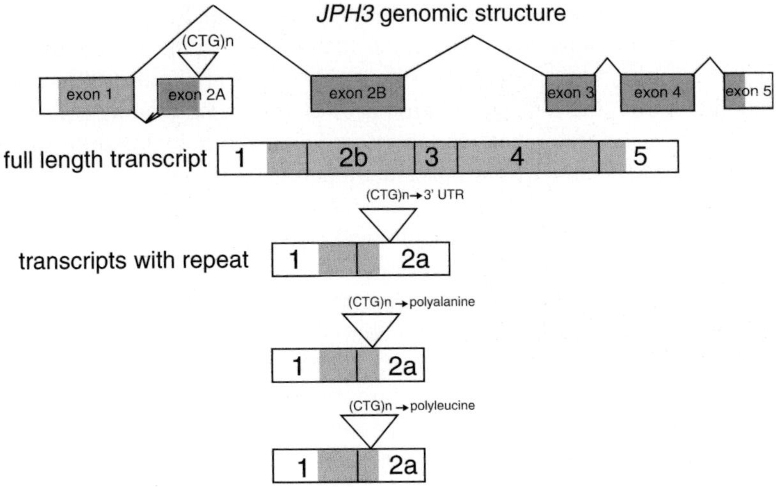

Fig. 6 *JPH3* genomic organization. The repeat is located in alternatively spliced exon 2A. Multiple reading frames have been detected in exon 2A, determined by the use of different splice acceptor sites. (Derived in part from Holmes et al. [6].)

analysis has been limited to relatively young animals [17]. A mouse double knockout, missing the two junctophilins expressed in brain (types 3 and 4), showed no evidence of neuropathology or, at least in the hippocampus, ultrastructural abnormalities of the junctional complex [14]. However, the mice did have cognitive and motor abnormalities, an abnormal response of small conductance calcium activated potassium (SK) channels to electrical stimulation, impaired long-term potentiation, and hyperactive calcium/calmodulin-dependent protein kinase II. Together, this led to the hypothesis that loss of junctophilins disconnected the "functional communications" among NMDA receptors (glutamate-activated plasma membrane channels), ryanodine receptors (ER membrane channels permitting release of ER calcium stores), and SK channels, with consequences for hippocampal plasticity and electrophysiologic function, explaining the motor and cognitive deficits of these animals.

5 HDL2 Pathogenesis: Three Hypotheses

5.1 Poly-Amino Acid Toxicity

HDL2 strongly resembles the diseases caused by polyglutamine expansion; midlife onset of progressive neurodegeneration, repeat expansion with a disease threshold at about 40 triplets, repeat length correlating with onset age, and the presence of intranuclear protein aggregates that stain with 1C2 and anti-ubiquitin antibodies. The obvious hypothesis was that HDL2 would be another polyglutamine disorder, our original prediction [11]. We were surprised to find that the mutation was in the CTG orientation in *JPH3*. An open reading frame on the reverse strand, where the repeat was in frame to encode polyglutamine, made us optimistic that we could find a polyglutamine encoding gene. However, there was little database or bioinformatic evidence that this open reading frame was actually expressed, and our Northern blot and cDNA library screens to search for a CAG repeat-containing transcript similarly were negative. Given the recent finding of transcript expression on both strands of the SCA8 expansion [15], it remains possible that CAG-repeat containing transcripts may nonetheless be expressed at very low levels. However, we have not yet found evidence for expression of a polyglutamine protein from the HDL2 locus other than the 1C2 staining of aggregates. This leaves two possible explanations for the presence of 1C2 positive aggregates: either the antibody is recognizing epitopes other than expanded polyglutamine in the aggregates [7, 21], or a polyglutamine-containing peptide below our threshold for detection is expressed from the HDL2 locus. We therefore hypothesize that polyglutamine expression at best plays a contributing role in HDL2 pathogenesis, and is unlikely in itself to fully explain HDL2 neurotoxicity.

Alternatively, HDL2 neurotoxicity could arise from expression of long tracts of polyalanine or polyleucine. Long tracts of these amino acids are toxic to cells in culture, and at least one neurodegenerative disease results from an alanine expansion [1]. However, while exon 2A from alleles with normal length repeats is expressed in human brain, to date we have not found evidence for expression of the expanded repeat (Rudnicki, unpublished observations).

5.2 JPH3 *Loss of Function*

A second possibility is that the expansion mutation leads to a loss of *JPH3* expression and subsequently to neuropathology. Direct assay of this possibility in patient brains suggests at least a partial loss of expression of the *JPH3* transcript and protein, but the evidence is difficult to interpret because of substantial variability among the available brains. Mice with a hemizygous loss of JP3 expression, presumably the closest model of the loss of function hypothesis in HDL2, show some deficits, but are not profoundly affected [17]. More striking is the lack of pathology in mice missing both JP3 and JP4 [14]. However, as noted above, these mice have demonstrable cognitive and motor deficits. We therefore have tentatively concluded that loss of *JPH3* function is unlikely to fully explain HDL2 pathogenesis, but could contribute to neurotoxicity, perhaps by increasing the vulnerability to other insults secondary to dysregulated calcium flux (Fig. 7).

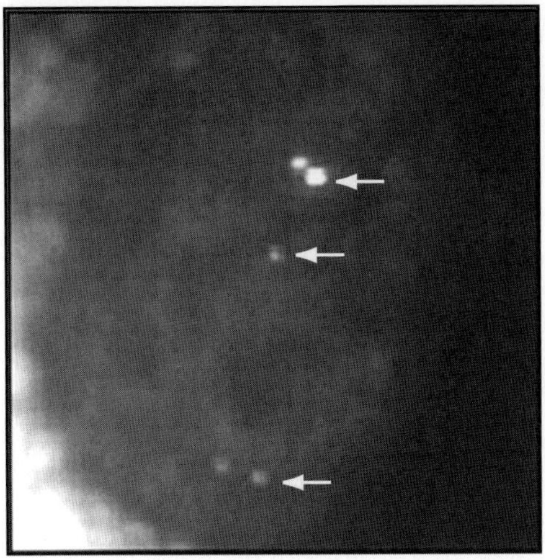

Fig. 7 RNA foci in HDL2 cortical neuron. *Arrows* indicated multiple foci detected by a fluorescently tagged CAG 20-mer riboprobe

5.3 *RNA Gain of Function*

Myotonic dystrophy type 1 (DM1) is caused by a 3′ UTR CTG repeat expansion ranging from 60 to >2,000 triplets in the gene *DMPK* [10]. A series of investigations has demonstrated that the transcript with the CUG expansion is toxic to cells. The mechanism of action appears to be a change in the ratio of the splicing modulators CUG-binding protein 1 (CUGBP1) and muscleblind-like protein 1 (MBNL1). The former is increased by an unknown process while the latter is decreased, perhaps by sequestration or abnormal localization induced by interaction with the CUG-repeat containing transcript [4, 5, 9, 13, 24]. The net effect is an abnormal increase in the ratio of pre-adult to adult splice variants of many genes, including several genes with direct links to the pathophysiology of DM1. DM1 tissue, including brain, contains RNA foci that can be detected by a (CAG)n riboprobe [9, 23]. The foci also contain MBNL1. While it is unclear if the RNA foci themselves are essential to disease pathogenesis, they serve as markers for the presence of potentially toxic RNA transcripts, and have been detected in brain from fragile X tremor ataxia syndrome (CGG repeats) and myotonic dystrophy type 2 (CCUG repeat) patients.

The findings in DM1 led us to consider the hypothesis that the transcript with CUG repeats might also play a role in HDL2 pathogenesis [19]. We detected RNA foci in HDL2 brain that strongly resembled DM1 foci (Fig. 7). These foci, as with those in DM1, co-stain with antibodies against MBNL1, and the amount of free

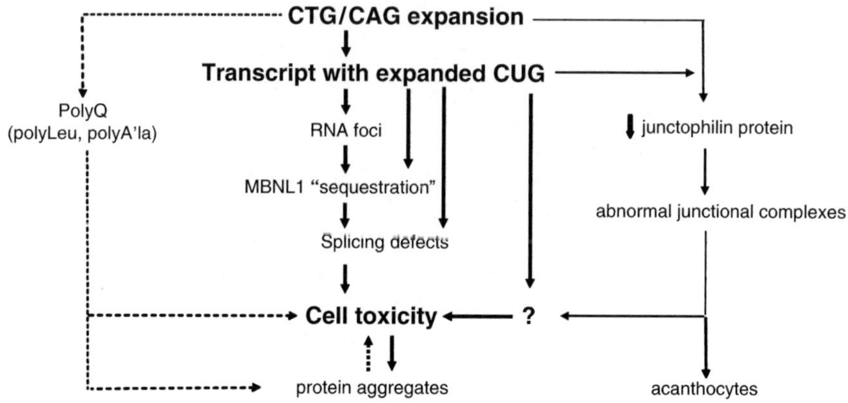

Fig. 8 Model of HDL2 pathogenic pathways

MBNL1 is diminished in HDL2 cortical neurons. Both intronic and exonic *JPH3* was detected in the foci, suggesting that at least a portion of the *JPH3* transcript may be sequestered within the foci prior to splicing. Splicing of *APP* and *MAPT* is abnormal in HDL2 brain, consistent with a loss of MBNL1 expression. Overexpression of a fragment of the *JPH3* transcript containing an expanded repeat and engineered to prevent translation was toxic to neuronal and non-neuronal cells in culture. Together, these results provide evidence that at least a portion of the neuronal dysfunction and death in HDL2 may derive from toxicity of the untranslated expanded CUG repeat.

6 Summary

For the first time, it is reasonable to speculate on the interacting pathways contributing to HDL2 pathogenesis. Data that need to be accounted for include the phenotypic and genetic similarity to HD and other polyglutamine disorders (rather than to DM1), the presence of protein inclusions staining with 1C2 antibodies, and the mild phenotype of JP3 knockout mice. We propose the tentative model shown in Fig. 8, recognizing that some aspects of the model are almost certainly wrong. The left-most pathway depicts the possible role of poly-amino acid tracts, either polyalanine or polyleucine encoded from *JPH3*, or polyglutamine cryptically encoded from the reverse strand. The dotted lines indicate that this pathway is not well supported by the available data. However, this pathway is the most obvious explanation for the protein aggregates observed in HDL2. Perhaps a small amount of protein with an expanded polyglutamine tract, undetectable by conventional assays, is sufficient to seed aggregates which are ultimately composed of other proteins. Toxicity would stem from one or more of the mechanisms postulated to contribute to other polyglutamine diseases.

The righthand pathway depicts the possible role of loss of *JPH3* expression. Assays of *JPH3* transcript and protein in HDL2 brain provide modest support for loss of expression. Decreased expression could arise from a direct effect of the repeat on transcription or splicing, or by sequestration of the transcript into RNA foci. Loss of expression might lead to incomplete construction of junctional complexes, with disruptions in calcium flux and increased vulnerability to factors such as excitotoxity and metabolic stress. Loss of JPH3, a membrane protein, might contribute to the formation of acanthocytes, but perhaps only in predisposed individuals.

The largest arrows depict a pathway of RNA toxicity. The CUG repeat leads to RNA foci formation. Whether these foci themselves contribute to toxicity is unclear, but the net result is a decrease in functional MBNL1, with subsequent effects on splicing of many genes, destabilizing neurons. CUG expansions may induce toxicity through mechanisms other than MBNL1 and splicing dysregulation. We predict that cells subject to CUG repeat toxicity may aggregate protein, partially accounting for HDL2 protein aggregates.

This model is speculative, but provides a basis for further exploration of HDL2 pathogenesis.

Acknowledgements The authors would like to thank Drs. Susan E. Holmes, Christopher A. Ross, Ruth H. Walker, Amanda Krause, Charles Thornton, Olga Pletnikova, Juan Troncoso, Adam Rosenblatt, Nancy Sachs, and Elizabeth O'Hearn for their insights into HDL2. We also thank the HDL2 families who have so willing cooperated with our investigations. This work was supported by the Hereditary Disease Foundation and NIH NS16375.

References

1. Brais B, Bouchard JP, Xie YG, Rochefort DL, Chretien N, Tome FM, Lafreniere G, Rommens JM, Uyama E, Nohira O, Blumen S, Korczyn AD, Heutink P, Mathieu J, Duranceau A, Codere F, Fardeau M, Rouleau GA (1998) Short GCG expansions in the PABP2 gene cause oculopharyngeal muscular dystrophy. Nat Genet 18:164–167
2. Danek A, Walker RH (2005) Neuroacanthocytosis. Curr Opin Neurol 18:386–392
3. Greenstein PE, Vonsattel JP, Margolis RL, Joseph JT (2007) Huntington's disease like-2 (HDL2). Neuropathol Mov Disord 22:1416–1423
4. Ho TH, Charlet BN, Poulos MG, Singh G, Swanson MS, Cooper TA (2004) Muscleblind proteins regulate alternative splicing. EMBO J 23:3103–3112
5. Ho T, Bundman D, Armstrong DL, Cooper TA (2005) Transgenic mice expressing CUG-BP1 reproduce splicing mis-regulation observed in myotonic dystrophy. Hum Mol Genet 14:1539–1547
6. Holmes SE, O'Hearn E, Rosenblatt A, Callahan C, Hwang HS, Ingersoll-Ashworth RG, Fleisher A, Stevanin G, Brice A, Potter NT, Ross CA, Margolis RL (2001) A repeat expansion in the gene encoding junctophilin-3 is associated with Huntington disease-like 2. Nat Genet 29:377–378
7. Ishikawa K, Owada K, Ishida K, Fujigasaki H, Shun Li M, Tsunemi T, Ohkoshi N, Toru S, Mizutani T, Hayashi M, Arai N, Hasegawa K, Kawanami T, Kato T, Makifuchi T, Shoji S, Tanabe T, Mizusawa H (2001) Cytoplasmic and nuclear polyglutamine aggregates in SCA6 Purkinje cells. Neurology 56:1753–1756
8. Ito K, Komazaki S, Sasamoto K, Yoshida M, Nishi M, Kitamura K, Takeshima H (2001) Deficiency of triad junction and contraction in mutant skeletal muscle lacking junctophilin type 1. J Cell Biol 154:1059–1067

9. Jiang H, Mankodi A, Swanson MS, Moxley RT, Thornton CA (2004) Myotonic dystrophy type 1 is associated with nuclear foci of mutant RNA, sequestration of muscleblind proteins and deregulated alternative splicing in neurons. Hum Mol Genet 13:3079–3088
10. Mahadevan M, Tsilfidis C, Sabourin L, Shutler G, Amemiya C, Jansen G, Neville C, Narang M, Barcelo J, O'Hoy K et al. (1992) Myotonic dystrophy mutation. an unstable CTG repeat in the 3′ untranslated region of the gene. Science 255:1253–1255
11. Margolis RL, O'Hearn E, Rosenblatt A, Willour V, Holmes SE, Franz ML, Callahan C, Hwang HS, Troncoso JC, Ross CA (2001) A disorder similar to Huntington's disease is associated with a novel CAG repeat expansion. Ann Neurol 50:373–380
12. Margolis RL, Holmes SE, Rosenblatt A, Gourley L, O'Hearn E, Ross CA, Seltzer WK, Walker RH, Ashizawa T, Rasmussen A, Hayden M, Almqvist EW, Harris J, Fahn S, MacDonald ME, Mysore J, Shimohata T, Tsuji S, Potter N, Nakaso K, Adachi Y, Nakashima K, Bird T, Krause A, Greenstein P (2004) Huntington's disease-like 2 (HDL2) in North America and Japan. Ann Neurol 56:670–674
13. Miller JW, Urbinati CR, Teng-Umnuay P, Stenberg MG, ByrneBJ, Thornton CA, Swanson MS (2000) Recruitment of human muscleblind proteins to (CUG)(n) expansions associated with myotonic dystrophy. EMBO J 19:4439–4448
14. Moriguchi S, Nishi M, Komazaki S, Sakagami H, Miyazaki T, Masumiya H, Saito SY, Watanabe M, Kondo H, Yawo H, Fukunaga K, Takeshima H (2006) Functional uncoupling between Ca^{2+} release and after hyperpolarization in mutant hippocampal neurons lacking junctophilins. Proc Natl Acad Sci U S A 103:10811–10816
15. Moseley ML, Zu T, Ikeda Y, Gao W, Mosemiller AK, Daughters RS, Chen G, Weatherspoon MR, Clark HB, Ebner TJ Day JW, Ranum LP (2006) Bidirectional expression of CUG and CAG expansion transcripts and intranuclear polyglutamine inclusions in spinocerebellar ataxia type 8. Nat Genet 38:758–769
16. Nishi M, Mizushima A, Nakagawara K, Takeshima H (2000) Characterization of human junctophilin subtype genes. Biochem Biophys Res Commun 273:920–927
17. Nishi M, Hashimoto K, Kuriyama K, Komazaki S, Kano M, Shibata S, Takeshima H (2002) Motor discoordination in mutant mice lacking junctophilin type 3. Biochem Biophys Res Commun 292:318–324
18. Rosenblatt A, Ranen NG, Rubinsztein DC, Stine OC, Margolis RL, Wagster MV, Becher MW, Rosser AE, Leggo J, Hodges JR, French-Constant CK, Sherr M, Franz ML, Abbott MH, Ross CA (1998) Patients with features similar to Huntington's disease, without CAG expansion in huntingtin. Neurology 51:215–220
19. Rudnicki DD, Holmes SE, Lin MW, Thornton CA, Ross CA, Margolis RL (2007) Huntington's disease-like 2 is associated with CUG repeat-containing RNA foci. Ann Neurol 61:272–282
20. Stevanin G, Camuzat A, Holmes SE, Julien C, Sahloul R, Dode C, Hahn-Barma V, Ross CA, Margolis RL, Durr A, Brice A (2002) CAG/CTG repeat expansions at the Huntington's disease-like 2 locus are rare in Huntington's disease patients. Neurology 58:965–967
21. Takahashi J, Fukuda T, Tanaka J, Minamitani M, Fujigasaki H, Uchihara T (2000) Neuronal intranuclear hyaline inclusion disease with polyglutamine-immunoreactive inclusions. Acta Neuropathol (Berl) 99:589–594
22. Takeshima H, Komazaki S, Nishi M, Iino M, Kangawa K (2000) Junctophilins: a novel family of junctional membrane complex proteins. Mol Cell 6:11–22
23. Taneja KL, McCurrach M, Schalling M, Housman D, Singer RH (1995) Foci of trinucleotide repeat transcripts in nuclei of myotonic dystrophy cells and tissues. J Cell Biol 128:995–1002
24. Timchenko NA, Patel R, Iakova P, Cai ZJ, Quan L, Timchenko LT (2004) Overexpression of CUG triplet repeat-binding protein, CUGBP1, in mice inhibits myogenesis. J Biol Chem 279:13129–13139
25. Trottier Y, Lutz Y, Stevanin G, Imbert G, Devys D, Cancel G, Saudou F, Weber C, David G, Tora L et al. (1995) Polyglutamine expansion as a pathological epitope in Huntington's disease and four dominant cerebellar ataxias. Nature 378:403–406

26. Walker RH, Morgello S, Davidoff-Feldman B, Melnick A, Walsh MJ, Shashidharan P, Brin MF (2002) Autosomal dominant chorea-acanthocytosis with polyglutamine-containing neuronal inclusions. Neurology 58:1031–1037
27. Walker RH, Jankovic J, O'Hearn E, Margolis RL (2003) Phenotypic features of Huntington's disease-like 2. Mov Disord 18:1527–1530
28. Walker RH, Rasmussen A, Rudnicki D, Holmes SE, Alonso E, Matsuura T, Ashizawa T, Davidoff-Feldman B, Margolis RL (2003) Huntington's disease-like 2 can present as chorea-acanthocytosis. Neurology 61:1002–1004

Neuroacanthocytosis in Japan – Review of the Literature and Cases

G. Hirose

Abstract Since the first case report of this disease in 1974, a total of 71 cases of probable chorea-acanthocytosis (ChAc) were collected in Japan up to the end of 2006. These reports were reviewed for their clinical features and to document research achievements in Japan in this field. Whilst the clinical phenotype of these patients was typical of ChAc, most of these cases were diagnosed clinically without molecular diagnosis, so the diagnosis of McLeod syndrome cannot be completely excluded.

The mean age of onset was 30.5 (range 18–42) years and the male:female ratio was 18:7. Involuntary movements consisting of oro-lingual-facial dyskinesias and choreiform limb movements were seen in over 90% of cases. Self-mutilation of the lower lip was also seen with the same incidence. Depression or absence of deep tendon reflexes was noted in almost all cases. Cognitive impairment with or without psychiatric symptoms was noted in 40% of cases. The degree of acanthocytosis of peripheral red blood cells varied from 6 to 80% (mean value 24%). Serum creatine phosphokinase activity was increased in 86%. Computed tomography of the brain revealed symmetrical atrophy of the caudate nuclei in almost all cases examined. Forty percent of patients had seizures. The mode of transmission was predominantly autosomal recessive, but four families have been reported with apparent dominant inheritance. Sural nerve biopsy showed evidence of chronic denervation with axonopathy. Grouped atrophy of muscle fibers was also reported, but recent studies suggest a primary disorder of the muscle membrane or muscle fibers as a

G. Hirose
Department of Neurology, Neurological Center, Asanogawa General Hospital, Japan
Mailing address: Naka 83, Kosaka-cho, Kanazawa City, Ishikawa Prefecture, 920-8621, Japan
ghirose@asanogawa-gh.or.jp

R.H. Walker et al. (eds.), *Neuroacanthocytosis Syndromes II*.

cause of elevated creatine kinase. *Post mortem* examination of the brain revealed marked neuronal loss and gliosis affecting the caudate nucleus and pallidum. The cerebral cortex and substantia nigra seemed to be spared. 15 cases of McLeod syndrome were identified in Japan between 1994 and 2006.

Many scientific advances were made in Japan with respect to ChAc. The red cell membrane pathology was studied morphologically as well as biochemically. Abnormal conformation of the red cell membrane was found, with an increase of palmitic and docosahexaenoic acids and a decrease of stearic acid. A mutation in the gene coding for a protein designated "chorein" was found and reported in a Japanese family with autosomal recessive inheritance, and a new single heterozygous frame shift mutation was also found from a family with apparent autosomal dominant inheritance. In addition, a gene-targeted mouse model of ChAc was reported from our country. The future of research in the area of ChAc in Japan is very promising.

1 Introduction and Historical Review

Levine and his colleagues [19] and Critchley and his associates [5] independently published a new syndrome of hereditary neurological disease with acanthocytosis in two different families, the Goode family from New England and the Stevens family from Eastern Kentucky in 1968. Thereafter another three families were reported, two from England and one family from the United States. However, recognition of the disease in Japan was delayed for about 10 years.

The first case of probable ChAc was reported in a local medical journal of Hiroshima Prefecture, the Journal of the Hiroshima Medical Association, in 1978 by Professor Kito and colleagues [11] of Hiroshima University Medical School. Clinical studies were reported, in addition to scanning microscopic findings of acanthocytes, neurogenic atrophy of muscle fibers, and caudate nucleus atrophy, in a 50-year-old man with a family history of a similar neurological disease in his younger brother and paternal uncle. The authors published a detailed family pedigree with apparent autosomal dominant inheritance in 1980 [16]. The authors claimed that their family was the first with this disorder in Japan and the sixth family in the world. However, looking back at previous reports of this disorder in our country, there is an earlier case report by Shimizu et al. [32] in 1974. These authors described an adult patient with self-mutilation, choreoathetosis, hypotonia, areflexia and normouricemia. They considered this patient to have a new disease, distinct from Lesch–Nyhan syndrome, without recognising acanthocytosis. In 1978 [33], however, they found acanthocytes in the patient's younger brother in addition to the patient. The authors subsequently reported the family in more clinical detail, and also found a reduced ratio of C24:1/C24:0 fatty acids in sphingomyelin from red cell membranes [14].

During the following 10 years, 25 well-studied cases of probable ChAc were rapidly accumulated with salient clinical features [1, 2, 8, 11, 15, 17, 18, 21, 22, 28, 32–34, 41], including our familial cases. The clinical characteristics, family history, neuroimaging and hematological results, of these earlier fully reported 25 cases, are summarized in Table 1.

Japanese cases of probable ChAc were further identified by search of the Japana Centro Revuo Medicina (JCRV), the Japanese version of Index Medicus, a collection of abstract-form reports from about 4700 Japanese journals and periodicals, established in 1903 by Dr. Shiro Amako. By the end of 2006, a total of at least 71 cases were collected. In addition to the 25 cases from 1974 to 1983, reported above, 25 cases were reported from 1984 to 1990, 10 cases from 1991 to 2000, and 11 cases from 2001 to 2006. This abundance of reports has facilitated the progress of molecular biological studies of these disorders.

Cases of MLS were also identified from the same source (JCRV) and PubMed. The first case of MLS in Japan was reported in 1994 by Takashima et al. with molecular genetic diagnosis [35]. Since then, 12 cases of molecularly-diagnosed MLS were published, with three cases of chronic granulomatous disease associated with acanthocytosis. In addition to these 13 cases, two cases of Japanese patients were reported in the Western literature [40]. No cases of Huntington's disease-like 2 have yet been reported in our country among patients referred for Huntington's disease testing [20].

2 Epidemiology and Genetic Inheritance of Probable Chorea-Acanthocytosis

From 1974 to 2005, at least 71 cases of probable ChAc were reported in Japan, according to this search of the Japanese literature. Most cases were diagnosed clinically without any molecular studies, except for the most recent five cases of ChAc. Without definitive molecular testing, McLeod syndrome or other diseases with acanthocytosis and choreiform involuntary movments cannot be excluded. The patients were widely distributed throughout Japan, including Hokkaido, Honshu, Sikoku and Kyushu islands, with some preponderance of the western part of Japan, but without any endemic areas (Fig. 1).

The mode of transmission seems to be predominantly autosomal recessive. Among the early well-studied 25 cases, 20 cases occurred in 14 families. There was autosomal recessive inheritance in 12 families and autosomal dominant (AD) inheritance in 2 families. In these two AD families, a paternal uncle and paternal grandfather suffered a similar neurodegenerative disease, excluding X-linked inheritance. No X-linked pedigree was reported in these cases. Five cases were apparently sporadic in this series. Recently two more families of molecularly confirmed ChAc were reported with an inheritance pattern strongly suggestive of AD inheritance [10, 26].

Table 1 Summary of clinical features of 25 early Japanese cases of probable chorea-acanthocytosis

References	Age	Sex	Age of onset (year)	IVM	Self-mutil.	DTR	Sz	IQ < 80	Acanthocytes (%)	Caudate atrophy	CPK	FH Inherit
[32]	22	M	18	+	+	hypo	–	98	6–7	+	high	+AR
[33]	34	F	20	+	+	hypo	GM	72	6–7	+	NL	–
[11]	50	M	36	+	ND	hypo	–	104	+	+	ND	+AD
[22]	38	F	29	+	+	NL	GM	90	30	+	high	+AR
[15]	44	M	34	+	+	hypo	–	NL	8	–	high	+AR
	47	F	25	+	+	hypo	+	NL	6–7	+	high	+AR
	45	F	31	+	–	ND	+	ND	bizarre E	ND	ND	+AR
[41]	41	M	34	+	+	hypo	–	low	30–35	+	high	+AR
	40	F	34	+	+	hypo	+	low	6–7	+	high	+AR
[18]	36	M	32	+	+	hypo	+	76	40–50	+	high	+AR
	37	M	34	+	+	hypo	GM	90	10–20	+	high	+AR
[8]	50	M	35	+	+	hypo	GM	73	20	+	NL	+AR
[21]	35	F	31	+	+	ND	GM	ND	10	ND	ND	+AR
[2]	38	M	24	+	+	hypo	–	NL	10–20	+	high	+AD
[29]	31	M	28	+	+	hypo	–	NL	30–80	+	high	+AR
	47	M	27	+	+	hypo	–	NL	50–80	+	high	+AR
	37	M	34	+	+	hypo	–	90	20–30	+	high	–
[17]	37	M	32	–	+	hypo	–	low	45	+	high	+AR
	30	F	27	+	+	hypo	–	81	26	+	high	+AR
	46	M	42	+	+	hypo	GM	88	26	+	high	–
	47	M	40	–	+	hypo	–	97	15	+	high	–
	28	M	28	+	+	hypo	–	64	57	+	high	+AR
	22	M	ND	+	–	hypo	–	90	58	+	high	+AR
[34]	26	M	21	+	+	hypo	–	72	6	+	high	+AR
[1]	46	F	37	+	+		–	NL	20	+	NL	–
Total 25 cases	–	18:7	30.5	23/25	22/24	22/23	10/25	10/25	6–80%	23/23	19/22	AR:AD=18:2
Mean value	38.2	–	–	–	–	–	–	–	24%	–	–	–

AD autosomal dominant, *AR* autosomal recessive, *bizarre E* bizarre erythrocytes, *CPK* creatine phophokinase, *DTR* deep tendon reflexes, *F* female, *FH inherit* family history and inheritance pattern, *GM* grand mal type of seizure, *hypo* hypoactive, *IVM* involuntary movements, *M* male, *ND* not described, *NL* normal, trait, *Sz* seizure, *Self mutil.* self-mutilation

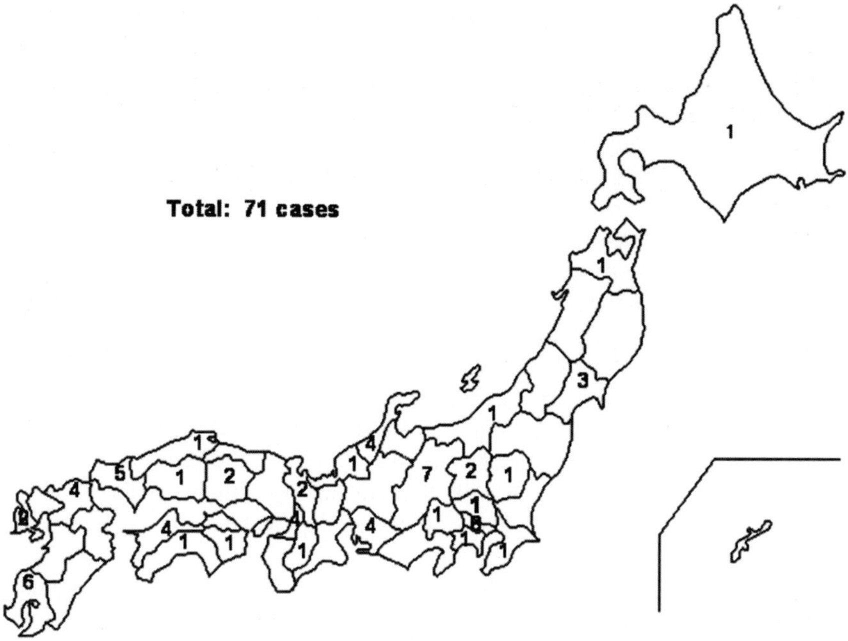

Fig. 1 Distribution of patients with probable chorea-acanthocytosis in Japan

3 Clinical Characteristics of Probable Chorea-Acanthocytosis from 25 Early Cases

The disease onset varied from 18 to 42 years old with the mean age of onset at 30.5 years. The male:female ratio was 18:7. No X-linked family pedigree of disease inheritance was reported in these male patients. Choreiform involuntary movements were noted in 23 of 25 cases, with mainly oro-lingual-buccal dyskinesia and choreiform limb movements. The knee-buckling gait (flamingo-like walk), often seen in patients with Huntington's disease (HD), was noted in four cases among 25 [22, 33, 42]. Twenty-two patients of 23 had decreased deep tendon reflexes suggesting polyneuropathy. Self-mutilation was noted in 22 patients of 24. Among 14 cases of MLS reported in Japan, self-mutilation was not noted except for one patient with a tongue scar due to a bite [35]. Severe oro-facial dyskinesia with lip-biting was considered to most likely indicate a diagnosis of ChAc. WAIS-IQ less than 80 was reported in ten cases of 25. Seizures were seen in ten, and of those, generalized seizures of the grand mal type were reported in six, complicated in one by additional psychomotor seizures.

The clinical characteristics of the involuntary movements (IVM) in probable ChAc included orolingual choreic movements associated with vocalization, biting of the lower lip and tongue, and facial grimacing, dysphagia with bird-like drinking, choreiform limb and trunk movements, and postural lapse with abrupt buckling of the knees occasionally seen in patients with HD [42]. When 14 molecularly diagnosed MLS cases were compared to patients with ChAc, choreiform limb movements were milder, and no lip self-mutilations, facial tics or dysphagia were noted. In general, these differences appear to clinically differentiate ChAc from MLS in most cases of neuroacanthocytosis (NA).

Polygraphic and jerk-locked averaging techniques revealed different characteristics of IVM between patients with probable ChAc and HD [31]. Slow negativity before IVM, similar to the *Bereitschaftpotential* was seen only in the patients with probable ChAc, but not in those with HD.

Neuroimaging studies in these disorders were usually performed by computed tomography (CT) of the brain, but single photon emission computed tomography (SPECT) and positron emission tomography (PET) were also used on rare occasions. Brain CT was reported in 23 cases and atrophy of the caudate nucleus was noted in all. ^{18}Fluoro-deoxyglucose-PET study in five patients with choreiform syndromes, including probable ChAc, revealed hypometabolism in the striatum bilaterally, similar to that seen in HD [9].

Laboratory examinations including routine blood and urine analyses were unremarkable. The most important hematological finding was of acanthocytes in the peripheral wet smear preparation, with a range of 6–80% of red blood cells and a mean value of 24%. Markedly raised serum creatine phosphokinase (CPK) was noted in 19 cases out of 22. These characteristics were quite similar to the clinical data described among cases from England by Hardie and associates [7].

4 Further Neurobiologic Studies

Based on the abundant reports of probable ChAc in Japan, basic scientific studies were performed in a number of institutions. A number of erythrocyte membrane studies were carried out [3, 24, 38]. A study using freeze-fracture electron microscopy showed a significant increase of the intramembranous particle free areas in both P and F faces [38]. The fluidity deep inside the red cell membrane in this disorder was studied using a spin labeling technique, and low fluidity was found when compared to that of normal red blood cells [24]. The same group also studied the capacity of self-digestion of red blood cells and reported easily self-digestible conformation of the red cell membrane in patients with probable ChAc [3]. Sakai et al. reported the abnormal membrane property of acanthocytes in patients with probable ChAc [29]. Analysis of covalently (tightly) bound fatty acids in erythrocyte membrane proteins after alkaline hydrolysis disclosed an increase of palmitic and docosahexaenoic (C22:6) acids and a decrease of stearic acid in this disease.

Muscle biopsy examination in the patients with distal amyotrophy due to probable ChAc revealed a typical neurogenic grouped atrophy with fibre-type grouping, suggestive of chronic polyneuropathy [11, 17]. Histological studies of the sural nerve biopsy showed axonal degeneration associated with secondarily demyelinating changes. The density of total myelinated fibers was slightly decreased with more involvement of large myelinated fibers [22, 23].

In addition to neurogenic changes in muscle biopsy, evidence now supports a primary muscle fiber or membrane disorder to explain elevated CPK in this disease. Nemaline rods were noted in the subsarcolemmal and paranuclear locations of the muscle biopsy of probable ChAc [36]. This strongly suggests a primary disorder in the muscle membranous structure in this disease. Dilated cardiomyopathy with skeletal myopathy was also reported in a case of ChAc in which MLS was definitively excluded [13]. We also have an evidence of abnormally disrupted ChAc expression in the muscle membrane, stained histochemically with the anti-chorein antibody (in preparation).

Post mortem neuropathological studies of probable ChAc were reported by Iwata et al. [12] and Sato et al. [30]. These authors found atrophy and gliosis of the caudate nuclei and putamen, with no neuronal loss in the cerebral cortex or other parts of the brain, including the substantia nigra. Degeneration and gliosis was the most marked in the head of the caudate, followed in the body and the tail. The numbers of small neurons in the caudate nucleus and putamen were greatly reduced to 1% and 20% of each healthy control respectively. On the other hand, large neurons of the caudate decreased in diameter but not in number. The ventral portion of the putamen was also moderately degenerated. These findings are quite similar to those reported by Bird et al. [4] and Hardie et al. [7]. Sato's group also studied biochemical changes of the striatum in their autopsy case. They found marked a decrease of substance P (SP) level without any changes of choline acetyltransferase or glutamic acid decarboxylase (GAD) in both the caudate nucleus and putamen [30]. However, GAD and SP activity were decreased in the substantia nigra where there were no histopathological abnormalities.

Neurosurgical management of probable ChAc with posteroventral pallidotomy (PVP) was reported in one case with severe intractable involuntary movements [6] and see chapter by Yokochi and Burbaud. A 41-year-old man suffered from marked oro-lingual dyskinesia which gave him marked difficulty with eating and swallowing. A left PVP was done initially with a marked reduction of oro-lingual dyskinesia and chorea of the right limbs. Subsequently he had a PVP of the right side with reduction of the left-sided choreo-ballistic movements. The same authors have performed a bilateral PVP for another patient with ChAc with a complete remission of involuntary movements (see chapter by Yokochi and Burbaud). The exact mechanism of the effect by this surgical procedure has not yet been clearly explained.

5 Contemporary Neurobiological Studies of Neuroacanthocytosis

The heterogeneous group of NA disorders can now be clearly classified on the basis of molecular genetic studies. The principal NA syndromes are autosomal recessive ChAc and X-linked McLeod syndrome, but now Huntington's disease-like 2 and pantothenate kinase-associated neurodegeneration can be included. All of these diseases share common neurological manifestations with involuntary movements, progressive deterioration of the higher cortical function, and can be diagnosed by clinical, laboratory and imaging techniques as well as by detecting gene mutations.

A number of highly significant advances in autosomal recessive ChAc have recently been reported from Japan. The disease has been linked in most families to chromosome 9q21, where Ueno and colleagues found a mutation in the gene encoding a large (3100 amino acids) protein designated *"chorein"* in 2001 [39]. A deletion was found in the coding region of the cDNA leading to a frame shift, resulting in the production of a truncated protein in both alleles of the patients and single alleles of the obligate carriers. This protein is thought to be an evolutionarily conserved protein that is probably involved in the cellular protein sorting and trafficking [25]. We have also reported a family with apparent autosomal dominant inheritance. From this family, our group found a novel single heterozygous frame shift mutation in the last nucleotide of exon 57 of the ChAc gene in 2003 [26]. In order to rule out Huntington's disease-like 2, in which autosomal dominant inheritance is seen, expansion of the CTG/CAG repeat within junctophilin-3 gene was excluded [27]. Our findings confirm that clinical features in patients with ChAc with apparent autosomal dominant inheritance does not differ from those in a recessive form. It remains unclear why one mutation in the ChAc gene causes recessive inheritance in one family and another mutation within the same gene causes an apparent autosomal dominant inheritance of ChAc.

Tomemori and colleagues produced a gene-targeted mouse model for ChAc in 2005 [37]. They identified the mouse ChAc cDNA sequence and the exon–intron structures of the gene, and produced a ChAc model mouse by introducing a deletion of 60–61 exons, using a gene-targeting technique. Hematological study of this model revealed typical acanthocytes in the peripheral blood with a marked increase of the osmotic fragility of red blood cells. Motor evaluation of these animals during late adult stages showed statistically significant changes with a shorter stride length, poorly coordinated balance in the Rotarod test, and decreased locomotor activity as compared to control animals. No involuntary movements were observed in this mouse model. Histopathological study also confirmed striatal degeneration with gliosis, consistent with the findings in humans with ChAc.

As a number of important neuroscientific advances in ChAc have come from Japan in recent years, we anticipate that this is a most promising arena for further developments in the study of NA syndromes, in particular ChAc.

References

1. Araki J, Tatsumi Y, Sannomiya Y et al. (1983) A case of chorea-acanthocytosis. Jpn J Clin Hematol 24:1055–1059
2. Arima S, Mori R, Kato N (1980) A case of the familial chorea with acanthocytosis. Clin Neurol (Rinsho Shinkeigaku) 20:1069–1070

3. Asano K, Osawa Y, Yanagisawa N et al. (1985) Erythrocyte membrane abnormalities in patients with amyotrophic chorea with acanthocytosis. Part 2. Abnormal degradation of membrane protein. J Neurol Sci 68:161–173
4. Bird TD, Cederbaum S, Valpey RW et al. (1978) Familial degeneration of the basal ganglia with acanthocytosis: a clinical, neuropathological, and neurochemical study. Ann Neurol 3:253–258
5. Critchley EMR, Clark DB, Wikler A (1968) Acanthocytosis and neurological disorder without abetalipoproteinemia. Arch Neurol 18:134–140
6. Fujimoto Y, Isozaki E, Yokochi F et al. (1997) A case of chorea-acanthocytosis successfully treated with posteroventral pallidotomy. Clin Neurol (Rinsho Shinkeigaku) 37:891–894
7. Hardie RJ, Pullon HW, Harding AE et al. (1991) Neuroacanthocytosis. A clinical, haematological and pathological study of 19 cases. Brain 114:13–49
8. Hirayama K, Yamada T (1980) Familial neuro-acanthocytosis in adult. Special consideration on symptomatology of involuntary movements. Clin Neurol (Rinsho Shinkeigaku) 20:1064–1066
9. Hosokawa S, Ichiya Y, Kuwabara Y et al. (1987) Positron emission tomography in cases of chorea with different underlying diseases. J Neurol Neurosurg Psychiatry 50:1284–1287
10. Ichiba M, Muroya S, Mizuno E et al. (2005) A CHAC-pedigree of with semi-dominant inheritance. Ann Rep Mitsubishi Pharma Res Found 37:35–41
11. Itoga E, Kito S, Tsubota W et al. (1978) A case of amyotrophic chorea with acanthocytosis. J Hiroshima Med Ass 31:772–775
12. Iwata M, Toyokura Y, Sakuta M et al. (1981) Neuropathology of chorea-acanthocytosis (Levine–Critchley syndrome). Neurol Med 15:132–145
13. Kageyama Y, Kodama Y, Tadano M et al. (2000) A case of chorea-acanthocytosis with dilated cardiomyopathy and myopathy. Clin Neurol (Rinsho Shinkeigaku) 40:816–820
14. Kamakura K, Shimizu T, Toyokura Y et al. (1981) Chorea-acanthocytosis (Levine–Critchley syndrome). Report of a family with biochemical study. Neurol Med 15:1–7
15. Kawazawa S, Kashiwamura K, Takamatsu S et al. (1979) Self-mutilation, choreiform movement, areflexia and acanthocytosis. Report of three cases. Neurol Med 10:486–488
16. Kito S, Itoga E, Hiroshige Y et al. (1980) A pedigree of amyotrophic chorea with acanthocytosis. Arch Neurol 37:514–517
17. Kondo K, Yanagisawa N (1981) Familial chorea with neuropathy and acanthocytosis. Report of six cases with special reference to its clinical variety. Neurol Med 15:118–127
18. Kooriyama T, Yoshida Y, Nakamura N et al. (1980) Pathogenesis of chorea with acanthocytosis. Clin Neurol (Rinsho Shinkeigaku) 20:1076–1078
19. Levine IM, Estes JW, Looney JM (1968) Hereditary neurological disease with acanthocytosis. Arch Neurol 19:403–409
20. Margolis RL, Holmes SE, Rosenblatt A et al. (2004) Huntington's disease-like 2 in North America and Japan. Ann Neurol 56:670–674
21. Mori K, Ide Y, Haku R et al. (1980) A case with choreic involuntary movement and acanthocytosis, who responded to steroid treatment. Clin Neurol (Rinsho Shinkeigaku) 20:1071–1072
22. Nagashima T, Iwashita H, Kuroiwa Y et al. (1979) Chorea-acanthocytosis. A report of a family. Clin Neurol (Rinsho Shinkeigaku) 19:609–615
23. Ohnishi A, Sato Y, Nagara H et al. (1981) Neurogenic muscular atrophy and low density of large myelinated fibers of sural nerve in chorea-acanthocytosis. J Neurol Neurosurg Psychiatry 44:645–648
24. Oshima M, Osawa Y, Asano K et al. (1985) Erythrocyte membrane abnormalities in patients with amyotrophic chorea with acanthocytosis. Part 1. Spin labeling studies and lipid analysis. J Neurol Sci 68:147–160
25. Rampoldi L, Dobson-Stone C, Rubio JP et al. (2001) A conserved sorting-associated protein is mutant in chorea-acanthocytosis. Nat Genet 28:119–120
26. Saiki S, Sakai K, Kitagawa Y et al. (2003) Mutation in the CHAC gene in a family of autosomal dominant chorea-acanthocytosis. Neurology 61:1614–1616

27. Saiki S, Sakai K, Saiki M et al. (2004) Huntington's disease-like 2 can present as chorea-acanthocytosis. Neurology 63:939–940
28. Sakai T, Iwashita H, Sato Y et al. (1980) Chorea-acanthocytosis: familial and sporadic case. Clin Neurol (Rinshou Shinkeigaku) 20:1062–1063
29. Sakai T, Antoku Y, Iwashita H et al. (1991) Chorea-acanthocytosis: abnormal composition of covalently bound fatty acids of erythrocyte membrane proteins. Ann Neurol 29:664–669
30. Sato Y, Ohnishi A, Tateishi J et al. (1984) An autopsy case of chorea-acanthocytosis. Special reference to the histopathological and biochemical findings of basal ganglia. Brain Nerve 36:105–111
31. Shibasaki H, Sakai T, Nishimura H et al. (1982) Involuntary movements in chorea-acanthocytosis: a comparison with Huntington's chorea. Ann Neurol 12:311–314
32. Shimizu T, Inoue K, Sugita H et al. (1974) Self-mutilation, choreoathetosis, muscular hypotonia, absence of deep tendon reflexes and normouricemia. Report of adult case. Neurol Med 1:135–136
33. Shimizu T, Kamakura K (1978) A new acanthocytosis syndrome associated with self-biting, choreoathetosis, epilepsy and peripheral neuropathy. Neurol Med 9:206
34. Takahashi Y, Kojima T, Atsumi Y et al. (1983) A case of chorea-acanthocytosis with various psychotic symptoms. Psychiat Neurol Japonicum 85:457–472
35. Takashima H, Sakai T, Iwashita H et al. (1994) A family of McLeod syndrome, masquerading as chorea acanthocytosis. J Neurol Sci 124:56–60
36. Tamura Y, Matsui K, Yaguchi H et al. (2005) Nemaline rods in chorea acanthocytosis. Muscle Nerve 31:516–519
37. Tomemori Y, Ichiba M, Kusumoto A et al. (2005) A gene-targeted mouse model for chorea-acanthocytosis. J Neurochem 92:759–766
38. Ueno E, Oguchi K, Yanagisawa N (1982) Morphological abnormalities of erythrocyte membrane in the hereditary neurological disease with chorea, areflexia and acanthocytosis. A study with freeze-fracture electron microscopy. J Neurol Sci 56:89–97
39. Ueno S, Maruki Y, Nakamura M et al. (2001) The gene encoding a newly discovered protein, chorein, is mutated in chorea-acanthocytosis. Nat Genet 28:121–122
40. Wada M, Kimura M, Daimon M et al. (2003) An unusual phenotype of McLeod syndrome with late onset axonal neuropathy. J Neurol Neurosurg Psychiatry 74:1697–1699
41. Yamamoto T, Hirose G, Takado M et al. (1980) A familial neurological disorder with acanthocytosis. Special reference to movement disorders. Clin Neurol (Rinsho Shinkeigaku) 20:1067–1068
42. Yamamoto T, Hirose G, Shimazaki K et al. (1982) Movement disorders of familial neuroacanthocytosis syndrome. Arch Neurol 39:298–301

Part II
Basic Research - Proteins and Erythrocytes

The Function of Chorein

A. Velayos-Baeza(✉), C. Lévecque, C. Dobson-Stone, and A.P. Monaco

Abstract Chorein is the protein encoded by gene *VPS13A* which is altered in chorea-acanthocytosis (ChAc). It belongs to the VPS13 protein family which, in mammals, has three other members: VPS13B, VPS13C and VPS13D. These proteins are similar to Vps13p, a yeast protein shown to be involved in intra-cellular trafficking of a number of transmembrane proteins. Chorein and its homologous human proteins lack domains or motifs of known function. This, together with their large size, makes the functional characterisation of these proteins a difficult task. Nevertheless, we have undertaken this task following a molecular and cellular biology approach. We have cloned the cDNA for the human *VPS13* genes and used them for transfection of mammalian cell lines. We present here an overview of the

A. Velayos-Baeza
The Wellcome Trust Centre for Human Genetics, University of Oxford, Roosevelt Drive, Headington, Oxford, OX3 7BN, UK
avelayos@well.ox.ac.uk

R.H. Walker et al. (eds.), *Neuroacanthocytosis Syndromes II.*

results obtained. We also analyse the data available for similar proteins and the information provided by mutational screening in ChAc patients in the context of the implications in protein function.

1 Introduction

McLeod syndrome and chorea-acanthocytosis (ChAc) constitute the classical syndromes known as "neuroacanthocytosis" (for recent reviews, see [8, 9, 42, 44]). While the genes altered in these diseases (*XK* and *VPS13A*, respectively) were identified several years ago [20, 26, 40], knowledge regarding the functions of the proteins encoded by these genes has proven to be slow and difficult to acquire. A generally accepted idea is that whatever the actual function of each of these proteins, a connection between them must exist in order to explain the similarity at the phenotypic level between both diseases. The neurological findings also resemble those of Huntington's disease, and a shared vulnerability of basal ganglia neurons due to mutations leading to these different syndromes has been proposed [9].

The function of XK is still unknown, although its structural features suggest that it is a membrane transport protein [20]. Its close interaction with Kell, a transmembrane protein with endothelin-3 converting activity, or, rather, the alteration of this interaction in McLeod syndrome, is a main focus of the basic research on this disease (see [28, 29] for a review). Here, we will address the possible function(s) of chorein, the protein encoded by the *VPS13A* gene that is altered in ChAc, and other similar proteins, in light of the data available so far.

2 Chorein and the VPS13 Protein Family

The human *VPS13A* gene (formerly known as *CHAC*) was reported as the gene altered in ChAc patients in 2001 [26, 40], and the 3,174-amino acid (aa) deduced protein is called chorein. The only functional data available about similar proteins involved the yeast Vps13p [2] and the *Dictyostelium discoideum* TipC proteins [35], and the situation has changed little since. In this section we will try to summarise the data from these lower organisms and from the analysis of other similar proteins.

2.1 The Yeast Vps13p Protein

Vps13p (3,144 aa, 358kDa) is the yeast protein homologous to chorein and is encoded by the *VPS13* gene, also known as *SOI1*, cloned by complementation of a sporulation defect [2]. For TipC, the only available information is its involvement in cell type differentiation and tip formation in early *Dictyostelium* development [35]. In contrast, the powerful tools available in yeast allowed Brickner and Fuller [2] to establish a number of functional features of this new large protein. Vps13p appears to be peripherally associated with membranes forming a high molecular weight hetero- or homo-oligomeric complex. It was shown to be involved in the efficient *trans*-Golgi network (TGN) localisation of three transmembrane proteins, Kex2p, Ste13p and Vps10p. No effect was detected on the transport of proteins from the endoplasmic reticulum (ER) to the Golgi or through the Golgi when *VPS13* was mutated. Vps13p is not required for transport of proteins between the TGN and the prevacuolar compartment (PVC; equivalent to the mammalian late endosome) *per se* but it seems to regulate this transport. The authors hypothesise a double mechanism for this regulation, through interaction with localisation signals present in the cytoplasmic domain of the transmembrane proteins subject to the inter-compartmental transport. These localisation signals, called TLS1 and TLS2 for Kex2p, would promote the retrieval of the protein from the PVC to the TGN (TLS1) and retain the protein at the TGN (TLS2). Vps13p would antagonise the TGN-retention function of TLS2, facilitating the transport from TGN to PVC, and would be required for the full function of TLS1, which would promote the trafficking in the opposite direction. Thus, the unifying hypothesis for the data collected in these analyses is that Vps13p is involved in the recruitment of TGN membrane proteins into transport vesicles leaving both the TGN and the PVC. There is no evidence, however, about the nature of the interaction between Vps13p and the TLS1/TLS2 signals, and either a direct physical interaction or an indirect effect through other cytosolic factors is possible.

2.2 *Homologous Proteins*

One of the difficulties found in the characterisation of chorein is the absence of any known domains or motifs that would indicate the biochemical or cellular pathways in which this protein might be involved. This does not mean that functional domains or motifs are not present, but rather that new, not yet described ones might appear in this protein. Intra- and inter-species sequence comparisons of similar proteins can help in the detection of such putative regions.

2.2.1 Human Homologues

In the original report of *VPS13A* (*CHAC*) in chromosome 9q21 as the gene altered in ChAc, Rampoldi *et al.* reported that the encoded protein showed a high degree of similarity with another putative human protein (see supplementary Fig. C in [26]), which would be encoded by a gene on chromosome 15. We showed that, in fact, there are four proteins in humans, including chorein, that are most similar to Vps13p protein in yeast and, therefore, their encoding genes were named as

VPS13A to *D* [41]. One of these genes, *VPS13B*, was reported as *COH1* and is altered in Cohen syndrome patients [21]. Data about the features of these genes and their encoded proteins are summarised in Table 1. As already reported for *VPS13A* [26, 40], all the human *VPS13* genes are widely expressed and several alternative splicing variants have been detected [41]: variant 1/variant 2 are defined by the presence of either exon 28 or exon 28b in *VPS13B*, the absence/presence of exons 6 + 7 in *VPS13C*, and the presence/absence of exon 40 in *VPS13D* (see Table 1). Interestingly, the expression of these three genes is different in brain compared to most other tissues, with *VPS13B* variant 1, *VPS13C* variant 2 and *VPS13D* variant 1 being the main forms in brain, the opposite situation to other tissues. These differences might indicate that the changes introduced by exons 28, 6 + 7, and 40 in the proteins VPS13B, C and D, respectively, are especially important for at least some of the functions of these proteins in the brain.

A number of other alternative splicing variants were found in the human *VPS13* genes. Most of these variants lead to shorter protein versions due to alternative 3′ end exons or to the appearance of stop codons (derived from frameshift or from the use of alternative exons containing such codons). In most cases, the introduction of stop codons in these mRNAs, whether they are splicing errors or "functional" variants, would trigger their degradation by the nonsense-mediated mRNA decay (NMD) response [23, 34].

2.2.2 The VPS13 Proteins Across Species

Vps13p is the only protein of the VPS13 family present in the yeast *Saccharomyces cerevisiae*, but this is not the situation in most organisms. Orthologous genes, encoding proteins equivalent to the human VPS13 proteins, can be found in the animal model organisms *Caenorhabditis elegans* (nematode), *Drosophila melanogaster* (fruit fly), and *Fugu rubripes* (pufferfish), as well as in closer species such as mammals. However, not all the human *VPS13* genes have a counterpart in all

Table 1 Features of the human *VPS13* genes and their encoded proteins

Gene	Chromosome	gDNA[a]	Variant	Exons	RNA[b]	Protein[c]	MW[d]
VPS13A	9q21	240	1A	1–68;70–73	9525	3174	360
–	–	–	1B	1–69	9288	3095	351
VPS13B	8q22	864	1A	1–62	12069	4022	449
–	–	–	2A	1–27;28b;29–62	11994	3997	446
VPS13C	15q21	208	1A	1–5;8–85	11133	3710	417
–	–	–	2A	1–85	11262	3753	422
VPS13D	1p36	252	1A	1–70	13167	4388	492
–	–	–	2A	1–39;41–70	13092	4363	489

[a]Genomic DNA, in kb
[b]Size of coding mRNA from start to stop codons, in bp
[c]Number of amino acids in the encoded protein
[d]Molecular weight of the protein, in kDa

these organisms. Thus, *C. elegans* only presents orthologous genes for *VPS13A* and *VPS13B*, while *D. melanogaster* lacks an orthologue for *VPS13C*. The *VPS13C* gene cannot be found either in the sea squirt *Ciona intestinalis* (http://genome.jgi-psf.org/Cioin2/Cioin2.home.html, release v2.0), although this is still an ongoing sequencing project. The phylogenetically most distant species with four *VPS13* genes is *F. rubripes*, meaning that this gene family can be traced back at least to the group of bony vertebrates or Euteleostomi. The phylogenetic data show that the *VPS13* gene family has been evolving from a single gene (yeast) to a group of four, and suggest that there has been a process of specialisation in the function performed by the VPS13 proteins. The fact that alteration of VPS13A or VPS13B proteins lead to ChAc or Cohen syndrome indicates that the other, non-altered, VPS13 proteins cannot compensate for such defects and, therefore, it is not an example of functional redundancy.

2.3 Sequence Analyses

The availability of a family of four proteins in humans allowed us to perform sequence comparisons between them and also with the proteins present in other organisms [41]. The comparative analyses showed that, among the human proteins, VPS13A is the most similar to Vps13p from yeast. VPS13C is very similar to VPS13A, and its origin seems to be from a very recent duplication event. The other two human proteins are less conserved. VPS13D shows similarities with VPS13A throughout its sequence but this has become much more localised in VPS13B. These results agree with the phylogenetic data in a context in which VPS13A keeps a high similarity with the yeast homologue. This suggests that some selective pressure might be acting, and that this human protein has probably retained more functions than the rest of human VPS13 proteins from its yeast counterpart. The other members of the family have a faster rate of divergence from the original sequence and their function may have also become more different or specialised. VPS13B is probably the result of an ancient duplication, and it has lost most of the similarity with other proteins from the family except for some regions.

Figure 1 shows the features found in the human VPS13 proteins. The most conserved regions are the N- and C-termini but another region, named C2 and present in the four proteins, also shows a good degree of conservation (for alignment data see supplementary Fig. B in [26] and Fig. 3 in [41]). In proteins VPS13A, C and D, this sequence partially overlaps with a region called DUF1162, a domain deduced from comparisons among proteins belonging to the VPS13 family but which is not detected in VPS13B. These three regions (N, C, and C2) are probably involved in functions common to all the four VPS13 proteins as they have remained conserved through their evolution.

Analysis of the VPS13C sequence showed an internal duplication of a 494aa-region, caused by the duplication of 11 exons, responsible for the larger size of this protein when compared with VPS13A [41]. From this finding, it was possible to detect that this internal duplication in VPS13C was not a single event, but that the same process had happened in several occasions during the evolution of this protein family. An extra region similar to the two described above was found in VPS13C, two are present in VPS13A and three in VPS13D. The conservation between all these repeated regions (called R1, R2 and R2b, see Fig. 1) varies, but a common core element around 45 residues long, containing the sequence $P-X_4-P-X_{13-17}-G$, was found in all of them and is also present in other, non-human, members of this family [41]. The fact that this smaller element is conserved in all R regions suggests that it may be important for the function and/or the structure of the VPS13 proteins. The exception is VPS13B, where these regions cannot be detected. However, it is possible that this protein has also undergone internal duplication events (which would explain its size) and that the similarities have been lost during its evolution.

As with chorein, database comparisons with VPS13B, -C and -D protein sequences did not predict any known domains or motifs with a high degree of confidence. The two motifs with the highest probability are UBA (Ubiquitin-associated domain) and Ricin-B-lectin (lectin domain of ricin B chain profile), both in

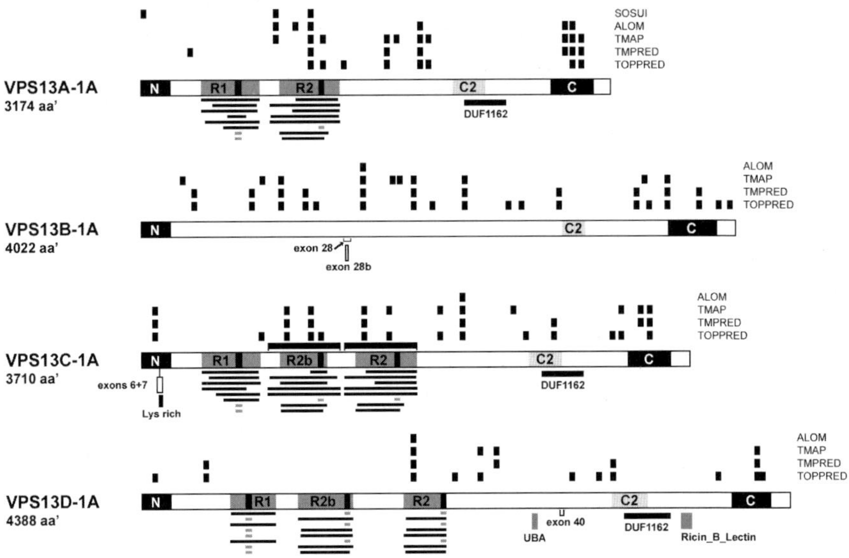

Fig. 1 Features of human VPS13 proteins (variant 1A). Areas in common are indicated in the main bar: *N*, N-terminal region; *C*, C-terminal region; *C2*, C-terminal region 2; *R (1, 2, 2b)*, repeated region (black area shows the 45-aa core element, see Sect. 2.3). *Small black boxes* above the bar indicate TM regions predicted by programs shown on the right; HMMTOP or TMHMM did not predict any TM region. Differences with variant 2 appear below for VPS13B, C, and D (see Table 1). Regions showing an above-threshold score with described motifs appear below the protein bar. *Long black boxes* above VPS13C protein show the duplicated regions 859–1350 and 1373–1864. *Horizontal bars* below R regions indicate the fragment of that R region that shows similarity in pairwise comparison with regions (VPS13)A-R1, A-R2, C-R1, C-R2b, C-R2, D-R1, D-R2b and D-R2, respectively. For more details, please see [41]. (Reprinted from [41] with permission from Elsevier.)

VPS13D [41]. In particular, the domains originally described for VPS13B/COH1 [21] were either undetectable or below the default confidence threshold for the software we used. Another question that has not been resolved since chorein was first described is whether this is a family of soluble or membrane proteins. The VPS13B/COH1 protein was described as a multiple transmembrane protein, with ten transmembrane (TM) domains [21]. However, TM prediction showed an inconsistent pattern, detecting from none to 18 TM domains in the same VPS13 protein depending on the program used (see Fig. 1). The lack of a signal peptide in all VPS13 proteins (although this is not always present in transmembrane proteins) and the functional data from their yeast homologue (see above) support the "soluble" option.

One motif that is detected in chorein by computational analysis is the Tetratrico Peptide Repeat (TPR), a structural motif of 34 aa defined by a pattern of small and large hydrophobic residues, where no positions are completely invariant, present in a wide range of proteins, and which mediates protein-protein interactions and assembly of multiprotein complexes [7]. Ten TPR motifs are detected in chorein (see Swiss-Prot entry Q96RL7) but, interestingly, none is detected in the closely related VPS13C protein (nor in VPS13B, VPS13D, or the yeast Vps13p proteins) and only six of these motifs are detected in the mouse chorein homologue (Swiss-Prot entry Q5H8C4). These data suggest that the detection of these motifs in the chorein sequence is most probably just a coincidence due to the high degree of degeneration of the TPR consensus sequence.

2.4 *What Disease Mutations Tell Us*

Analysis of mutations that cause disease might be important to obtain valuable insights into specific regions of the protein encoded by the corresponding gene, or to understand the way in which the altered protein leads to disease. In the case of ChAc, 92 mutations have been described so far (see Table 2). Examination of these mutations leads to the following conclusions, as outlined by Dobson-Stone *et al.* [12]:

1. ChAc shows strong allelic heterogeneity, with no single mutation causing the majority of cases.
2. Mutations are distributed throughout the gene, with no detectable clustering.
3. Most mutations are predicted to cause absence of chorein by changes leading to premature termination codons (PTC), such as nonsense and frameshift mutations (also including large deletions and splice-site mutations leading to frameshift) that would probably trigger the NMD response.
4. Despite the large size of chorein, only six missense mutations (6.52% of the described ChAc mutations) have been found. Another seven single amino-acid substitutions are known, which are non-pathogenic (29.16% of the reported polymorphisms detected during mutational screening) [11]. These data suggest that this protein is very tolerant to substitution.

Table 2 *VPS13A* disease mutations described in ChAc patients

#	Location	DNA change[a]	Protein change[a]	Genotype[b]	Refs.
		Missense			
1	Exon 4	c.269T>A	p.I90K	ht	[26]
2	Exon 31	c.3283G>C	p.A1095P	ht	[12]
3	Exon 37	c.4354T>C	p.S1452P	hm	[26]
4	Exon 53	c.7378T>C	p.W2460R	hm	[12]
5	Exon 57	c.8016G>C	p.K2672N	ht	[43]
6	Exon 59	c.8162A>G	p.Y2721C	ht	[26]
		Nonsense			
7	Exon 9	c.622C>T	p.R208X	ht, hm	[12, 26]
8	Exon 13	c.1078C>T	p.Q360X	ht	[11]
9	Exon 17	c.1549G>T	p.E517X	ht	[12]
10	Exon 18	c.1616C>G	p.S539X	ht	[12]
11	Exon 22	c.2191C>T	p.R731X	ht	[11]
12	Exon 23	c.2347C>T	p.Q783X	hm	[11]
13	Exon 25	c.2593C>T	p.R865X	hm	[12]
14	Exon 29	c.3109A>T	p.K1037X	ht	[12]
15	Exon 30	c.3157C>T	p.Q1053X	hm	[12]
16	Exon 34	c.3889C>T	p.R1297X	ht	[12]
17	Exon 37	c.4355C>G	p.S1452X	ht	[12, 14]
18	Exon 37	c.4411C>T	p.R1471X	hm, ht	[11, 12]
19	Exon 41	c.5084T>A	p.L1695X	ht	[11]
20	Exon 45	c.5920G>T	p.E1974X	hm	[12]
21	Exon 46	c.6094C>T	p.R2032X	ht	[12]
22	Exon 48	c.6419C>G	p.S2140X	ht	[12]
23	Exon 48	c.6494G>A	p.W2165X	hm	[12]
24	Exon 48	c.6700C>T	p.R2234X	ht	[26]
25	Exon 50	c.7005G>A	p.W2335X	ht	[12]
26	Exon 56	c.7867C>T	p.R2623X	ht	[12]
27	Exon 68	c.9109C>T	p.R3037X	ht	[12, 26]
28	Exon 70	c.9219C>G	p.Y3073X	ht	[12]
		Small insertion/deletion			
29	Exon 4	c.237del	p.E80KfsX11	ht	[26]
30	Exon 5	c.365_372dupAACAAAAA	p.V125NfsX4	ht	[11]
31	Exon 13	c.994del	p.A332LfsX10	ht	[12]
32	Exon 13	c.1115del	p.K372SfsX2	hm	[26]
33	Exon 13	c.1125_1128del	p.S375RfsX23	ht	[12]
34	Exon 14	c.1187_1188del	p.F396X	hm	[12]
35	Exon 14	c.1208_1211del	p.Q403RfsX6	ht	[12]
36	Exon 17	c.1592del	p.I531KfsX7	hm	[26]
37	Exon 20	c.2029_2031delins27	p.H677delinsIYX	ht	[12]
38	Exon 27	c.2833_2834del	p.K945EfsX11	ht	[26]
39	Exon 33	c.3556_3557dupAC	p.V1187LfsX12	hm	[11, 26]
40	Exon 34	c.3847del	p.L1283WfsX7	hm	[12]
41	Exon 35	c.3995_3996delinsA	p.F1332X	ht	[11]
42	Exon 36	c.4216del	p.V1406CfsX20	ht	[12]
43	Exon 37	c.4346del	p.S1449FfsX5	ht	[12]

(continued)

Table 2 (continued)

#	Location	DNA change[a]	Protein change[a]	Genotype[b]	Refs.
		Small insertion/deletion			
44	Exon 38	c.4419dupA	p.G1474RfsX7	ht	[26]
45	Exon 38	c.4428_4431del	p.G1478LfsX6	ht	[11]
46	Exon 39	c.4724del	p.P1575LfsX3	hm	[11]
47	Exon 39	c.4835del	p.P1612QfsX30	ht	[12]
48	Exon 40	c.4903_4906del	p.K1635VfsX6	ht	[11]
49	Exon 41	c.5253_5266del	p.F1751LfsX14	ht	[12]
50	Exon 45	c.5909_5910del	p.E1970VfsX4	ht	[14]
51	Exon 46	c.6059del	p.P2020LfsX9	hm	[11, 12]
52	Exon 47	c.6283del	p.S2095QfsX10	ht	[12]
53	Exon 48	c.6404dupT	p.S2136KfsX2	ht	[26]
54	Exon 49	c.6804dupG	p.S2269VfsX7	ht	[12]
55	Exon 49	c.6828del	p.V2277LfsX12	ht	[11, 12]
56	Exon 53	c.7339dupT	p.Y2447LfsX5	ht	[12]
57	Exon 57	c.7985_7989del	p.P2662RfsX6	ht	[12]
58	Exon 57	c.8007del	p.K2669NfsX22	ht	[12]
59	Exon 61	c.8390del	p.G2797DfsX2	ht	[12]
60	Exon 67	c.9065_9066del	p.Q3022RfsX10	ht	[11]
61	Exon 70	c.9190del	p.V3064SfsX17	ht	[12]
62	Exon 71	c.9286_9289dupTTTG	p.T3098CfsX12	ht	[26]
63	Exon 71	c.9367del	p.V3123FfsX14	ht	[43]
64	Exon 72	c.9429_9432del	p.R3143SfsX5	ht	[12, 26]
65	Exon 72	c.9431_9432del	p.E3144VfsX6	hm, ht	[11, 12]
		Gross deletion[c]			
66	Exons 2–3	c.101-?_187+?del	p.A35_G63del	hm	[11]
67	Exons 8–9	c.556-?_696+?del	p.T186_L232del	hm	[11]
68	Exon 23	c.2289-?_2427+?del	p.I766HfsX14	hm	[12]
69	Exons 46–50	c.5992-?_7026+?del	p.I1998_Q2342del	hm	[14]
70	Exon 54	c.7420-?_7652+?del	p.D2474FfsX2	hm	[11]
71	Exons 60–61	c.8211+1232_8472–245delinsTC	p.V2738AfsX5	hm	[40]
72	Exons 70–73[d]	c.9189+8647_o*GNA14*: c.723+897del	(p.V2064_L3174del)?	hm, ht	[12, 15]
		Splice site[e]			
73	Intron 3	c.188–5T > G	[SA:77 (no)][f]	ht	[14]
74	Intron 6	c.495+1G > A	[SD:no (83)]	ht	[14]
75	Intron 6	c.495+5G > A	[SD:no (83)]	ht	[11]
76	Intron 11	c.883–1_892del	[SA:no (82)]	ht	[12]
77	Intron 17	c.1595+1G > A	[SD:no (86)]	ht	[12]
78	Intron 17	c.1596–2A > C	[SA:no (89)]	ht	[12, 14]
79	Intron 17	c.1596–1G > C	[SA:no (89)]	ht	[12, 14]
80	Intron 21	c.2170+1G > A	[SD:no (85)]	ht	[12]
81	Intron 22	c.2288+2T > C	[SD:no (83)]	ht	[12]
82	Intron 36	c.4242+1G > T	A1373FfsX7; Exon 36 skipped; [SD:no (93)]	hm, ht	[12, 15]

(continued)

Table 2 (continued)

#	Location	DNA change[a]	Protein change[a]	Genotype[b]	Refs.
83	Intron 40	c.4956+1G>T	[SD:no (83)]	ht	[12]
84	Intron 48	c.6775–2A>C	[SA:no (82)]	ht	[26]
85	Exon 55	c.7806G>A	[SD:no (76)]	ht	[12]
86	Exon 57	c.8035G>A	I2652HfsX12; Exon 57 skipped; [SD:no (0.34)][g]	ht	[30]
87	Intron 61	c.8472–1G>C	[SA:no (88)]	ht	[12]
88	Intron 65	c.8907+2T>A	[SD:no (75)]	ht	[12, 14]
89	Intron 70	c.9275+1G>A	[SD:no (80)]	ht	[12]
90	Intron 70	c.9276–2A>T	[SA:no (92)]	ht	[12]
91	Intron 71	c.9399+2_+8del	[SD:no (92)]	ht	[12]
92	Exon 72	c.9474G>A	[SD:no (81)]	ht	[12]

#mutation number

[a] Nucleotides and amino acids are numbered according to the cDNA sequence of *VPS13A* variant 1A (GenBank accession no. NM_033305), with the adenosine of the initiation codon assigned position 1. Mutations are described according to the nomenclature recommended by [10] and http://www.hgvs.org/mutnomen/

[b] Mutation found homozygously (hm)/heterozygously (ht) in ChAc patients

[c] The protein change is deduced assuming that the effect of the mutation is the deletion of the affected exons from the final mRNA

[d] Mutation also affects gene *GNA14* (guanine nucleotide binding protein (G protein), alpha 14) (GenBank accession no. NM_004297) deleting exons 6 and 7. The protein change is unknown, but it only affects variant A of chorein, not variants B and D as exons 69 and 68b are not deleted

[e] Changes predicted to affect the normal splicing of *VPS13A* pre-mRNA. The protein change is only provided when evidence about the effect of the mutation on the mRNA has been reported. An estimation of the effect on splicing, originally calculated with the (no longer available) SpliceView program (http://l25.itba.mi.cnr.it/~webgene/wwwspliceview.html), is given in the same field between square brackets, indicating the affected site (SA: splice acceptor, SD: splice donor) and the obtained value (no: no score); the score of the wild-type sequence is given between brackets

[f] New SA site is predicted 4bp upstream of the normal SA of intron 3

[g] Estimation of the effect on splicing was calculated using the program NetGene2 (http://www.cbs.dtu.dk/services/NetGene2/)

Key: p.E80KfsX11 denotes a frame shifting change with Glutamic acid-80 as the first affected amino acid, changing to a Lysine and creating a new reading frame ending in a stop at position 11 (counting starts with the Lysine as amino acid 1)

5. Exons 70–73 are included in transcript A of *VPS13A* gene, but not in transcript B [26]; two mutations affecting only these exons (mutations 63 and 70 in Table 2) have been found as homozygous in ChAc patients suggesting that variant A is essential for the normal function of chorein, and other variants cannot compensate for it.

From the functional point of view, it is interesting to mention that there seems to be no significant genotype-phenotype correlation [12]. This is not surprising for most of the cases because most of the mutations lead to PTC that would probably trigger NMD, so the final result would always be the lack of chorein. However, presence of mutated chorein would be expected in some cases: (a) missense mutations, (b) frameshift mutations leading to PTC but not to NMD (when an exon-exon junction does not appear after the PTC or it does at <50–55bp: this could happen with mutations 62 and 63 in Table 2), and (c) deletion of exons (by gross deletion or by splice site mutation leading to exon skipping) that do not change the reading frame. The analysis of these kinds of mutants could provide valuable information about residues or regions of chorein that are key for its function. Unfortunately, for most of these mutations, there are no experimental data about their effect on *VPS13A* transcription, so we can only guess the most probable result. For the gross deletions, the most probable effect would be just the removal of the deleted exons in the final mRNA (providing that the deletion does not alter the normal processing of the rest of the exons), as it was reported for patients with a deletion of exons 60 + 61 [40] (the expected effect in the protein under these conditions appears in Table 2). The effect of splice-site mutations is more difficult to predict as the inactivation of the natural site can lead to exon skipping, activation of cryptic sites, or intron retention [3, 17]. The only experimental data available for these ChAc mutations are for c.4242+1G>T and c.8035G>A (mutations 82 and 86 in Table 2), where the splice donor sites of introns 36 and 57 are affected: these data support the skipping of the adjacent exons 36 and 57, respectively, in the final mRNA [15, 30].

The effect of ChAc mutations can now be assayed at the protein level from a quantitative point of view using blood samples or cellular cultures [14]. However, only a few ChAc patients have been assayed this way [14, 15]. To date, chorein has been absent or its level very reduced in all reported cases. No chorein is detected in patients with mutations 17, 26, 33, 35, 36, 69, 73, 79, 81, 82 and 87 (as listed in Table 2), suggesting that the result of all these mutations is the lack of chorein. Some protein is detected in eight mutation combinations: (3 + 3), (22 + 61), (29 + 64), (50 +?), (72 + 72), (72 + 82), (74 +?) and (78 + 88), suggesting that at least one of the mutations in each combination allows the synthesis of chorein, albeit at low levels. With this assay, however, it is not possible to discriminate between wild-type or mutant chorein harbouring small-to-medium deletions due to the large size of this protein. The detection of low levels of protein can have several explanations: (a) instability of the mutant protein, (b) infrequent alternative splicing events that remove the mutated exon(s) and restore the reading frame, thus avoiding NMD, (c) a naturally occurring isoform, not affected by the mutation, that is expressed at low levels in the analysed samples. This last option would explain the results in the combinations (22 + 61), (29 + 64), (72 + 72) and (72 + 82), where mutations affecting exons 70–73 are present and the detected protein could correspond to the unaffected variants B and/or D.

The result for (3 + 3) was unexpected as it corresponds to a homozygous missense mutation, one of the situations in which a normal level of (mutant) protein was anticipated. The low protein levels detected could be explained by instability of the resulting protein but it is also possible that this mutation could actually affect pre-mRNA splicing. If the later is true, any of the other possible options mentioned above could explain this result (there is increasing evidence that many exonic mutations initially described as nonsense, missense or silent mutations actually affect pre-mRNA splicing [1, 4, 5]). Another unexpected result is the absence of detectable chorein in a patient homozygous for mutation 69, where the prediction was just the deletion of residues 1998–2342. The fact that this deletion protein is not detected means that such a deletion makes the protein unstable or that there is some alteration in the normal splicing of the remaining exons. These examples illustrate the need of functional assays in order to establish the effect that a specific mutation has. Taken together, the available data suggest that very few mutations would result in the presence of normal levels of mutated chorein. In any case, with or without mutant chorein present, the final phenotypic result always seems to adjust to a loss-of-function situation.

A family with a dominant inheritance pattern of ChAc has been reported [30]. This family carries a single mutation in the *VPS13A* gene that causes skipping of exon 57, leading to frameshift (mutation 86 in Table 2; reported as 8295G>A in [30]). The predicted result of this mutation is either absence of chorein due to NMD or presence of mutated chorein. As haploinsufficiency of chorein does not normally lead to a ChAc phenotype, mutated chorein must be present in these patients in order for this mutation to exert a dominant effect in this family. As no data are available about chorein levels in these patients, it is not possible to say if this mutation is dominant (in which case its analysis from the functional point of view would be very interesting) or just a "normal" recessive ChAc mutation. This latter option, in fact, is not that unlikely; heterozygous whole-exon deletions or rearrangements cannot be detected using the usual mutation screening methods [12], so the affected individuals in this family could be harbouring a second undetected *VPS13A* mutation. The inheritance pattern in this family, apparently dominant, could also be compatible with a recessive pattern due to the fact that ChAc appears to be more common in Japan [27] and no information about consanguinity or ChAc mutations is available for some of the family members.

Very similar results have been obtained in mutation screening of the *VPS13B/COH1* gene in patients with Cohen syndrome [16, 19, 21, 22, 24, 33], with most of the mutations leading to PTC. Six putative missense mutations (p.L2193R, p.Y2341C, p.G2645D, p.S2773L, p.I2820T, and p.N2993S) have been reported, although the possibility of some substitutions affecting pre-mRNA splicing or being rare non-pathogenic variants was also suggested. Interestingly, all of them are clustered in the central region of the protein, including and upstream of the C2 region (see Fig. 1); mutation p.N2993S alters a well conserved residue in the C2 region (see Fig. 3E in [41]), adding supporting evidence for the importance of this region in the VPS13 proteins. These studies also provide good examples of the different possible effects that splice site mutations can lead to, as discussed above. Exon skipping, intron retention or activation of cryptic splice sites have been found [19, 33]. Furthermore, effects in the processing of the pre-mRNA (skipping of one or two whole exons leading to frameshift or to non-truncating in-frame deletions) were described in some patients where the genomic mutation could not be found [21, 22]. Unfortunately, in contrast with the situation in ChAc, no procedure is yet available to assay the result of these mutations at the protein level. It is therefore not possible to know if their deleterious effect is due to absent/low levels of protein or to alteration/deletion of a specific region of the protein. The high tolerance to substitution mentioned for chorein seems to occur also in VPS13B as the high rate of non-pathogenic non-synonymous polymorphisms suggests [33]. Two mutations have been described to affect the last exon, predicted to cause frameshift (and theoretically not targeted by NMD) and, therefore, to change the very C-terminus of the protein [22]. This result agrees with the importance of variant A of chorein mentioned before, and suggests that the C-terminal region of the VPS13 proteins is essential for their function.

3 Over-Expression of VPS13 Proteins in Human Cell Lines

All the data reported here, unless specified otherwise, correspond to a Ms in preparation by Velayos-Baeza et al.

The functional characterisation of chorein represents a challenge from the experimental point of view for several reasons, mainly due to its large size. Additionally, the initial efforts to obtain an antibody to detect endogenous chorein were not very successful. To overcome these difficulties, we chose to perform these analyses by over-expressing chorein in mammalian cell lines, a very commonly used approach in functional and cell biology. We cloned the full-length cDNA of the human *VPS13* genes, with or without tag (an extra sequence coding for a short peptide that can be recognised by commercially available antibodies). The resulting expression plasmids have a considerable size, which results in a low transformation/transfection efficiency. A specific antiserum was obtained against the N-terminus of chorein [14] which works in western blotting (WB), immunofluorescence (IF) and immunoprecipitation (IP), providing additional tools for the characterisation of chorein.

Another advantage of the over-expression approach is the ability to introduce specific mutations to the expression constructs to check their functional effects. This is especially useful for the analysis of missense and deletion mutations.

3.1 Basic Protein Characterisation

The above-mentioned expression plasmids were used to transfect different mammalian cell lines (HEK293T, MRC5, HeLa, COS-7, etc), obtaining similar results in all cases. The addition of a tag (myc + His, EGFP, etc) at the C-terminus of chorein seems not to affect the protein. The apparent size of the bands detected by WB agrees with the expected size for the full-length proteins in all cases.

When assayed by IF, chorein presents several patterns, depending on individual cells (Fig. 2). The most characteristic is a vesicular-like pattern that is very easy to detect, although not all transfected cells present it. The second pattern suggests a cytoplasmic localisation. No general co-localisation with different sub-cellular markers was detected. The "vesicles" are not aggresomes [18], structures originat-

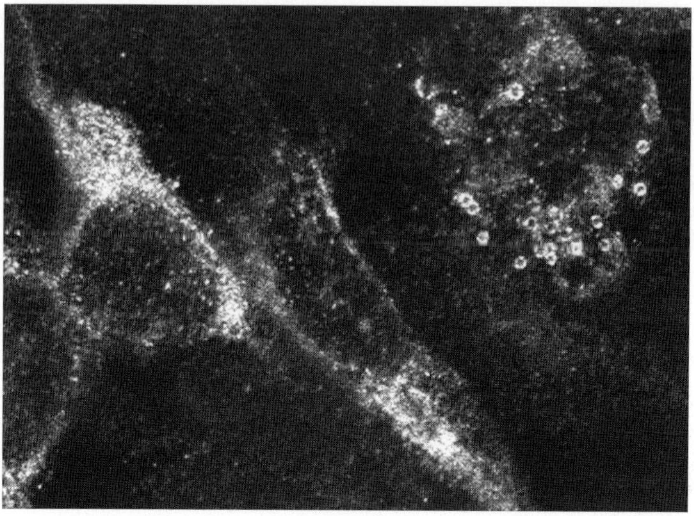

Fig. 2 Subcellular localisation of chorein. HEK293T cells overexpressing chorein with a myc + His tag at the C-terminus and detected with a monoclonal antibody against the myc tag. The cell on the right shows a typical vesicular-like pattern (see text)

ing after the aggregation of misfolded proteins (data not shown). In fact, similar structures can sometimes be detected in human cell lines with the chorein-specific antibody, suggesting that this result is not an artefact due to over-expression. For the rest of the human VPS13 proteins, a similar cytoplasmic localisation is detected with the over-expressed tagged proteins, and only with VPS13B are some vesicle-like structures detected, but at a much lower rate than for chorein.

The yeast homologue Vps13p has been reported to form high molecular weight complexes [2]. Co-IP experiments gave negative results for the detection of multimerisation of chorein and the other human VPS13 proteins, although this does not necessarily mean that such complexes are not formed.

One of the many unresolved issues about these proteins, the presence or absence of transmembrane domains (see Sect. 2.3), can now be addressed. Chorein can be detected in the soluble fraction of a protein lysate after high-speed centrifugation (data not shown), which indicates that it is a soluble and not an integral protein. It can also be detected in the precipitate, and it can be partially solubilised from this precipitate in several conditions, suggesting that it interacts with membranes. The same result was obtained with both over-expressed and endogenous chorein and with the other human VPS13 proteins, and it is consistent with the peripheral membrane-associated location reported for their yeast homologue Vps13p. This implies that the proposed model of COH1/VPS13B containing ten TM domains [21] would not be probable; this is an example of the care that must be taken with predictions obtained *in silico*.

Five missense ChAc mutations (1, 2, 3, 4, and 6 in Table 2) were introduced in a chorein expression plasmid by site-directed mutagenesis and the mutant proteins assayed by IF and WB. Protein of the normal expected size was detected in all cases by WB. The "vesicular" sub-cellular localisation pattern, however, was only detected with the p.Y2721C mutant but not with any of the other four mutants (data not shown). These results suggest that mutations p.I90K, p.A1095P, p.S1452P, and p.W2460R alter the ability of chorein to attach to membranes and, therefore, that this process is probably very important for the function of chorein. However, we have to be careful with this interpretation. As previously discussed (Sect. 2.4), chorein is detected at low levels in samples from patients homozygous for the c.4354T>C mutation (missense mutation leading to p.S1452P) and the effect of this mutation could affect the stability of chorein or maybe alter the normal splicing of the *VPS13A* gene. The result obtained with over-expressed p.S1452P may not therefore be the actual cause of disease. The fact that this mutant protein is easily detected suggests that stability is unaffected, implying that an effect on splicing may be more likely *in vivo*. In this case, the result will be absence of chorein or a deletion protein but not the S1452P substitution. The sub-cellular localisation defect found in p.S1452P could be due to a structural change introduced by the substitution, a common effect of proline residues. A similar explanation may apply for the p.A1095P mutation, i.e., an alteration of the secondary structure of chorein, while mutations p.I90K and p.W2460R might represent modification of key residues in the interaction of chorein with membranes. However, more experimental analyses would be needed to test these hypotheses and without patient cell samples we cannot know if these substitutions also lead simply to a reduction in chorein levels *in vivo*, as for p.S1452P.

3.2 Looking for Interacting Partners

The functional data available from Vps13p show that this protein is involved in the trafficking of membrane proteins such as Kex2p (see Sect. 2.1). A logical assumption would be that the human VPS13 proteins are also involved in the trafficking of different proteins. Similarly, the proteins potentially trafficked by VPS13 family members could include the human Kex2p homologues, the subtilisin/kexin-like proprotein convertases (PC). The PC protein family comprises seven members (furin, PC1, PC2, PC4, PC5, PC7 and PACE4) involved in the generation of biologically active polypeptides (including hormones, growth factors, transcription factors or cell adhesion molecules) by the cleavage of their precursors at specific sites. They present a complex expression pattern with temporal and spatial specificity [32, 36, 38]. These proteins are also good candidates for interaction with chorein because they may be involved in neurodegenerative diseases [31, 37]. Among the possible VPS13-PC interactions, we have checked so far those involving chorein, VPS13D, and some for VPS13B. However, no positive results were obtained. This may mean only that the co-IP approach used might not be sensitive enough to detect such interactions or that there is no direct interaction between VPS13 and PC proteins. Both explanations are possible; the putative interaction between both kinds of proteins may be weak or may only last for a very short time, making the detection of this direct interaction quite difficult. It is also possible that the regulatory effect of VPS13 proteins (see Sect. 2.1) does not involve a direct interaction with the trafficked proteins. More experiments are needed to check these options and, if applicable, to detect the specific combinations of VPS13-PC proteins that are biologically significant.

Even if chorein or VPS13B are involved in the sorting of any of the candidate proteins, it is possible that the actual protein(s) responsible for the ChAc or Cohen syndrome conditions is something else. Thus, a different approach must also be used to detect unknown interacting proteins. One way is the use of the yeast-two-hybrid (Y2H) technique (see [6, 25] for recent reviews on this system). However, due to the large size of the VPS13 proteins, the full-length protein cannot be used in these assays and has to be split in several fragments and, therefore, it is possible that functional domains are destroyed or structural issues introduced in the process. This approach has been tried for chorein [11], and only a few potential binding partners were found, showing all of them weak or very weak interactions (data not shown). Other Y2H screenings might be needed to find new interacting partners, and further experimental analyses performed to confirm or reject the candidates detected so far.

4 Chorein Function: A Hypothetical Model

Functional data available for homologous proteins in a different species are usually a good guide to find the actual role of a given protein. Thus, an involvement in intracellular trafficking of transmembrane proteins (as reported for Vps13p in yeast) has been hypothesised for chorein [13, 26, 27] as well as for VPS13B/COH1 [21]. The experimental results described above for the human VPS13 proteins, suggesting a soluble cytoplasmic localisation with ability to bind to membranes, support this hypothesis.

The symptomatic differences between ChAc and Cohen Syndrome indicate the existence of specific, non-overlapping functions for chorein and VPS13B, and the same is probably true for the other two members of the VPS13 family. All the *VPS13* genes are widely expressed [41] but the effects of single gene alteration, as the example of these two diseases suggests, seem to be reduced to specific cells or tissues. An integrating hypothesis is that there might be some redundancy in the function of VPS13 proteins and, therefore, the defect of any of them in a particular cell type and/or for a particular function would be compensated by other member(s) of the family, although this compensatory effect will not occur for some specific functions that each protein may have. Another option, not mutually exclusive with the previous one, is derived from the regulatory effect of Vps13p in protein sorting; absence of a specific VPS13 protein might alter the trafficking of proteins in a different way depending on the trafficked protein or the cell type. This could lead to the development of the defining symptoms of each disease.

The range of proteins whose trafficking is regulated by VPS13 proteins is potentially quite wide and examples such as that of the PC family (probably included among those proteins) give an idea of the complexity that possibly exists in the interacting network of chorein and similar proteins. Time and space components in the expression of all the proteins involved in this network add to such complexity. Under this scenario, it is not difficult to expect that the different symptoms that present in ChAc (or in Cohen syndrome) might be the result of alteration in the trafficking of different, unrelated proteins.

5 Conclusions

Although the basic characterisation of chorein and other similar proteins is technically difficult due to the size of the genes and proteins, a number of advances have been made to overcome these difficulties and some experimental results have been obtained that are compatible with previous data on related proteins. Different pieces of information from several studies can be put together to draw a general picture, but we still have a long way ahead before we can understand how the loss of chorein unleashes the effects detected in ChAc. A more detailed basic characterisation of VPS13 proteins, as well as the identification of specific processes in which they are involved, would be needed to accomplish this goal. The use of animal models, such as that for ChAc [39], is of invaluable importance for this task. The similarity with other diseases (McLeod syndrome or Huntington's disease in the case of ChAc) indicates that their respective pathogenic pathways probably overlap and some functional insights from those diseases could be brought to the functional study of chorein.

References

1. Baralle D, Baralle M (2005) Splicing in action: assessing disease causing sequence changes. J Med Genet 42:737–748
2. Brickner JH, Fuller RS (1997) *SOI1* encodes a novel, conserved protein that promotes TGN-endosomal cycling of Kex2p and other membrane proteins by modulating the function of two TGN localization signals. J Cell Biol 139:23–36
3. Buratti E, Baralle M, Baralle FE (2006) Defective splicing, disease and therapy: searching for master checkpoints in exon definition. Nucleic Acids Res 34:3494–3510
4. Caceres JF, Kornblihtt AR (2002) Alternative splicing: multiple control mechanisms and involvement in human disease. Trends Genet 18:186–193
5. Cartegni L, Chew SL, Krainer AR (2002) Listening to silence and understanding nonsense: exonic mutations that affect splicing. Nat Rev Genet 3:285–298
6. Coates PJ, Hall PA (2003) The yeast two-hybrid system for identifying protein-protein interactions. J Pathol 199:4–7
7. D'Andrea LD, Regan L (2003) TPR proteins: the versatile helix. Trends Biochem Sci 28:655–662
8. Danek A, Walker RH (2005) Neuroacanthocytosis. Curr Opin Neurol 18:386–392

9. Danek A, Jung HH, Melone MA, Rampoldi L, Broccoli V, Walker RH (2005) Neuroacanthocytosis: new developments in a neglected group of dementing disorders. J Neurol Sci 229–230:171–186
10. den Dunnen JT, Antonarakis SE (2000) Mutation nomenclature extensions and suggestions to describe complex mutations: a discussion. Hum Mutat 15:7–12
11. Dobson-Stone C (2004) Molecular genetics of chorea-acanthocytosis. Thesis, University of Oxford, Oxford
12. Dobson-Stone C, Danek A, Rampoldi L, Hardie RJ, Chalmers RM, Wood NW, Bohlega S, Dotti MT, Federico A, Shizuka M, Tanaka M, Watanabe M, Ikeda Y, Brin M, Goldfarb LG, Karp BI, Mohiddin S, Fananapazir L, Storch A, Fryer AE, Maddison P, Sibon I, Trevisol-Bittencourt PC, Singer C, Caballero IR, Aasly JO, Schmierer K, Dengler R, Hiersemenzel LP, Zeviani M, Meiner V, Lossos A, Johnson S, Mercado FC, Sorrentino G, Dupre N, Rouleau GA, Volkmann J, Arpa J, Lees A, Geraud G, Chouinard S, Nemeth A, Monaco AP (2002) Mutational spectrum of the *CHAC* gene in patients with chorea-acanthocytosis. Eur J Hum Genet 10:773–781
13. Dobson-Stone C, Rampoldi L, Monaco AP (2004) The spectrum of mutations and possible function of the *CHAC* gene. In: Danek A (ed) Neuroacanthocytosis syndromes. Springer, Dordrecht, pp 169–175
14. Dobson-Stone C, Velayos-Baeza A, Filippone LA, Westbury S, Storch A, Erdmann T, Wroe SJ, Leenders KL, Lang AE, Dotti MT, Federico A, Mohiddin SA, Fananapazir L, Daniels G, Danek A, Monaco AP (2004) Chorein detection for the diagnosis of chorea-acanthocytosis. Ann Neurol 56:299–302
15. Dobson-Stone C, Velayos-Baeza A, Jansen A, Andermann F, Dubeau F, Robert F, Summers A, Lang AE, Chouinard S, Danek A, Andermann E, Monaco AP (2005) Identification of a *VPS13A* founder mutation in French Canadian families with chorea-acanthocytosis. Neurogenetics 6:151–158
16. Falk MJ, Feiler HS, Neilson DE, Maxwell K, Lee JV, Segall SK, Robin NH, Wilhelmsen KC, Traskelin AL, Kolehmainen J, Lehesjoki AE, Wiznitzer M, Warman ML (2004) Cohen syndrome in the Ohio Amish. Am J Med Genet 128A:23–28
17. Faustino NA, Cooper TA (2003) Pre-mRNA splicing and human disease. Genes Dev 17:419–437
18. Garcia-Mata R, Gao YS, Sztul E (2002) Hassles with taking out the garbage: aggravating aggresomes. Traffic 3:388–396
19. Hennies HC, Rauch A, Seifert W, Schumi C, Moser E, Al-Taji E, Tariverdian G, Chrzanowska KH, Krajewska-Walasek M, Rajab A, Giugliani R, Neumann TE, Eckl KM, Karbasiyan M, Reis A, Horn D (2004) Allelic heterogeneity in the *COH1* gene explains clinical variability in Cohen syndrome. Am J Hum Genet 75:138–145
20. Ho M, Chelly J, Carter N, Danek A, Crocker P, Monaco AP (1994) Isolation of the gene for McLeod syndrome that encodes a novel membrane transport protein. Cell 77:869–880
21. Kolehmainen J, Black GC, Saarinen A, Chandler K, Clayton-Smith J, Traskelin AL, Perveen R, Kivitie-Kallio S, Norio R, Warburg M, Fryns JP, de la Chapelle A, Lehesjoki AE (2003) Cohen syndrome is caused by mutations in a novel gene, *COH1*, encoding a transmembrane protein with a presumed role in vesicle-mediated sorting and intracellular protein transport. Am J Hum Genet 72:1359–1369
22. Kolehmainen J, Wilkinson R, Lehesjoki AE, Chandler K, Kivitie-Kallio S, Clayton-Smith J, Traskelin AL, Waris L, Saarinen A, Khan J, Gross-Tsur V, Traboulsi EI, Warburg M, Fryns JP, Norio R, Black GC, Manson FD (2004) Delineation of Cohen syndrome following a large-scale genotype-phenotype screen. Am J Hum Genet 75:122–127
23. Lejeune F, Maquat LE (2005) Mechanistic links between nonsense-mediated mRNA decay and pre-mRNA splicing in mammalian cells. Curr Opin Cell Biol 17:309–315
24. Mochida GH, Rajab A, Eyaid W, Lu A, Al-Nouri D, Kosaki K, Noruzinia M, Sarda P, Ishihara J, Bodell A, Apse K, Walsh CA (2004) Broader geographical spectrum of Cohen syndrome due to *COH1* mutations. J Med Genet 41:e87

25. Parrish JR, Gulyas KD, Finley RL Jr (2006) Yeast two-hybrid contributions to interactome mapping. Curr Opin Biotechnol 17:387–393
26. Rampoldi L, Dobson-Stone C, Rubio JP, Danek A, Chalmers RM, Wood NW, Verellen C, Ferrer X, Malandrini A, Fabrizi GM, Brown R, Vance J, Pericak-Vance M, Rudolf G, Carre S, Alonso E, Manfredi M, Nemeth AH, Monaco AP (2001) A conserved sorting-associated protein is mutant in chorea-acanthocytosis. Nat Genet 28:119–120
27. Rampoldi L, Danek A, Monaco AP (2002) Clinical features and molecular bases of neuroacanthocytosis. J Mol Med 80:475–491
28. Redman CM, Russo D, Lee S (1999) Kell, Kx and the McLeod syndrome. Baillieres Best Pract Res Clin Haematol 12:621–635
29. Redman CM, Russo DCW, Pu JJ, Lee S (2004) The Kell blood group protein, its relation to XK and its function as an endothelin-3-converting enzyme. In: Danek A (ed) Neuroacanthocytosis syndromes. Springer, Dordrecht, pp 197–203
30. Saiki S, Sakai K, Kitagawa Y, Saiki M, Kataoka S, Hirose G (2003) Mutation in the *CHAC* gene in a family of autosomal dominant chorea-acanthocytosis. Neurology 61:1614–1616
31. Scamuffa N, Calvo F, Chretien M, Seidah NG, Khatib AM (2006) Proprotein convertases: lessons from knockouts. FASEB J 20:1954–1963
32. Seidah NG, Chretien M (1999) Proprotein and prohormone convertases: a family of subtilases generating diverse bioactive polypeptides. Brain Res 848:45–62
33. Seifert W, Holder-Espinasse M, Spranger S, Hoeltzenbein M, Rossier E, Dollfus H, Lacombe D, Verloes A, Chrzanowska KH, Maegawa GH, Chitayat D, Kotzot D, Huhle D, Meinecke P, Albrecht B, Mathijssen I, Leheup B, Raile K, Hennies HC, Horn D (2006) Mutational spectrum of *COH1* and clinical heterogeneity in Cohen syndrome. J Med Genet 43:e22
34. Stamm S, Ben-Ari S, Rafalska I, Tang Y, Zhang Z, Toiber D, Thanaraj TA, Soreq H (2005) Function of alternative splicing. Gene 344:1–20
35. Stege JT, Laub MT, Loomis WF (1999) *tip* genes act in parallel pathways of early *Dictyostelium* development. Dev Genet 25:64–77
36. Steiner DF (1998) The proprotein convertases. Curr Opin Chem Biol 2:31–39
37. Taylor NA, Van De Ven WJ, Creemers JW (2003) Curbing activation: proprotein convertases in homeostasis and pathology. FASEB J 17:1215–1227
38. Thomas G (2002) Furin at the cutting edge: from protein traffic to embryogenesis and disease. Nat Rev Mol Cell Biol 3:753–766
39. Tomemori Y, Ichiba M, Kusumoto A, Mizuno E, Sato D, Muroya S, Nakamura M, Kawaguchi H, Yoshida H, Ueno S, Nakao K, Nakamura K, Aiba A, Katsuki M, Sano A (2005) A gene-targeted mouse model for chorea-acanthocytosis. J Neurochem 92:759–766
40. Ueno S, Maruki Y, Nakamura M, Tomemori Y, Kamae K, Tanabe H, Yamashita Y, Matsuda S, Kaneko S, Sano A (2001) The gene encoding a newly discovered protein, chorein, is mutated in chorea-acanthocytosis. Nat Genet 28:121–122
41. Velayos-Baeza A, Vettori A, Copley RR, Dobson-Stone C, Monaco AP (2004) Analysis of the human *VPS13* gene family. Genomics 84:536–549
42. Walker RH, Danek A, Dobson-Stone C, Guerrini R, Jung HH, Lafontaine AL, Rampoldi L, Tison F, Andermann E (2006) Developments in neuroacanthocytosis: expanding the spectrum of choreatic syndromes. Mov Disord 21:1794–1805
43. Walker RH, Liu Q, Ichiba M, Muroya S, Nakamura M, Sano A, Kennedy CA, Sclar G (2006) Self-mutilation in chorea-acanthocytosis: manifestation of movement disorder or psychopathology? Mov Disord 21:2268–2269
44. Walker RH, Jung HH, Dobson-Stone C, Rampoldi L, Sano A, Tison F, Danek A (2007) Neurologic phenotypes associated with acanthocytosis. Neurology 68:92–98

Recent Studies of Kell and XK: Expression Profiles of Mouse Kell and XK mRNA

S. Lee(✉), X. Zhu, and Q. Sha

Abstract Absence of XK protein, the McLeod phenotype, is responsible for red cell acanthocytosis and neuromuscular abnormalities known as the McLeod syndrome (MLS). XK is predicted to be a membrane transport protein but its substrate is unknown. Kell is an endothelin-3-converting enzyme, and in erythroid tissues, where Kell and XK are expressed in near equal amounts, Kell and XK are linked by a disulfide bond. Absence of XK is accompanied by a reduced amount of Kell on red cells, however, in non-erythroid tissues their expressions differ. Northern blot analyses indicated that human XK is largely expressed in erythroid tissues, skeletal muscle, heart, testis and brain with less amounts in many other tissues. By contrast, human Kell is predominantly expressed in erythroid tissues with small amounts in non-erythroid tissues. The different tissue expressions of Kell and XK suggest that XK may function by itself in non-erythroid tissues, and in conjunction with Kell in erythroid tissues. To obtain further information regarding the expressions of Kell and XK, as a prerequisite to understand their cellular functions, we performed in situ hybridization in newborn mouse whole body sagittal section and in adult mouse brain, spleen and testis. Mouse XK (mXK) mRNA expression was detected in brain, spleen, bone marrow, testis, spinal cord, stomach, small intestine, pancreas kidney and bladder. mXK mRNA was expressed in most regions of the brain, with large amounts in Purkinje cells of cerebellum, magnocellular neurons in the pontine region, Cornu Ammonis fields of hippocampus and mitral cell layer of olfactory lobe, and noticeable amounts

S. Lee
The New York Blood Center, 310 East 67th Street, New York, NY 10065, USA
solee@nybloodcenter.org

R.H. Walker et al. (eds.), *Neuroacanthocytosis Syndromes II*.

in caudate/putamen, nucleus lateralis thalami and cerebral cortex. Northern blots of samples from different regions of the human brain support the in situ hybridization results of mouse brain. Large amounts of mouse Kell (mKell) mRNA expression were noted only in spleen, bone marrow and newborn mouse liver, but RT-PCR analysis detected small amounts of mRNA in testis and placenta.

1 Introduction

Absence of XK leads to the McLeod syndrome, a late onset multi-system disorder which involves hematological, neuromuscular and central nervous systems [4, 30]. XK, a 50 kDa protein, is a putative membrane transport protein that is predicted to have ten transmembrane regions [6] and belongs to the XK family, composed of three homologous proteins, XK, XPLAC and XTES. On red cells, XK forms a covalent complex, through a single disulfide bond, to another protein, Kell, which is a type II membrane glycoprotein [8, 26]. Kell is an endothelin-3-converting enzyme and belongs to M13 family of zinc dependent endopeptidases [11].

Kell expresses over 30 different Kell alloantigens that are important in transfusion medicine since some are highly immunogenic [9, 10, 13]. XK carries a single antigen, termed Kx, which is clinically important in identification of the McLeod phenotype [12, 24]. Absence of Kx antigen and reduced levels of all Kell antigens, as detected by serology, are indications of the McLeod phenotype. The McLeod genotypes vary, ranging from single nucleotide mutations to deletions of different sizes in the *XK* gene and the deletions can extend to neighboring genes [29]. In almost all cases the gene defects results in absence of the normal XK protein.

Expression of Kell and XK in human tissues, generated mainly by Northern blots and dot blot analyses, indicated that XK is ubiquitously present in all tissues examined, with highest levels in erythroid tissues, and skeletal muscle [2, 3, 6, 27]. Expression of human Kell was originally thought to be restricted to erythroid tissues [16], but later reports indicated that Kell was also present in testis and in small amounts in many other tissues [3, 27]. Because Kell and XK exhibit marked differences in distribution of expression, it has been speculated that Kell and XK are not always covalently linked, and that the functions of XK may be different depending on whether XK is complexed with Kell or if it is present by itself.

An immunohistochemical study of skeletal muscle from normal and McLeod patients showed that in normal, but not in McLeod, tissues, XK is expressed in the sarcoplasmic reticulum of type II muscle cells. Kell was detected in the sarcoplasmic membrane of normal muscle cells [7]. Studies with transfected COS cells constructed to co-express Kell and XK demonstrated the presence of a Kell/XK complex in the endoplasmic reticulum, and some Kell/XK complex traveled to the cell surface [25]. However, expression of either XK or Kell alone also allowed the single proteins to be transported to the plasma membrane indicating that linkage of Kell and XK is not required for cell surface expression. The transcriptional activity of the *KEL* promoter is stronger in K562 cells that are of erythroid origin, than in

HeLa cells, indicating a preference for Kell expression in erythroid cells, rather than non-erythroid, cells [3].

Although we know that Kell is an endothelin-3-converting enzyme and that XK is probably a membrane transport protein, their physiological functions are not yet well understood. Kell cleaves an inactive peptide, big endothelin-3, releasing endothelin-3, a potent bioactive peptide that is involved in many biological processes including the regulation of blood pressure, by affecting the contraction and proliferation of vascular smooth muscle, and developmental processes of the enteric nervous system, by affecting migration and differentiation of neural crest derived cells [17–20]. However, the rare human Kell null phenotype that lacks Kell protein [14], does not express any obvious clinical abnormalities, perhaps due to redundancy with other zinc metalloproteases with overlapping enzyme specificities. By contrast, although XK, like Kell, is a member of a family of proteins with similar structural characteristics, absence of XK as occurs in the McLeod phenotype, manifests a distinct set of clinical symptoms known as MLS that indicates a role for XK in neuromuscular, peripheral and central nervous system, and hematological functions [4, 30].

Further information on the expression patterns of Kell and XK in mouse should prove useful in the eventual understanding of their physiological functions. To this end, we have studied the expression profiles of XK and Kell in mouse tissues by in situ hybridization histochemistry (ISHH) and by RT-PCR analysis where needed to verify the ISHH results [15].

2 Kell and XK Expression Profiles in Mouse

2.1 Erythroid Tissues

When Kell and XK proteins are expressed in equal amounts, as in red cells, the cysteine at position 347 of XK is linked to the cysteine at Kell 72 [26]. Because Kell and XK are covalently linked it is speculated that they may complement each other's functions. From an evolutionary point of view, *XK* evolved as early as in vertebrate fish [2] while *KEL*, in contrast to *XK*, is only found in birds and higher species after duplication from its ancestor gene *ECE1/ECE2* (unpublished data). This suggests that if the Kell/XK complex has a separate function than the individual XK or Kell, then the new function of the complex started in warm-blood species.

In rare phenotypes when either Kell (KELnull) or XK (McLeod) are absent from red cells, the amount of the complex partner is reduced. However, paradoxically, in Kell null red cells, Kx antigen, as determined by serology, is increased. It has been speculated that Kell may partially cover the Kx antigen epitope and when Kell is not present Kx antigen is fully exposed. Absence of XK on red cells results in varying degrees of acanthocytosis but Kell null red cells have normal shape.

Studies of Kell surface antigen expression in cell culture systems showed that Kell is expressed early during erythropoiesis [1, 28]. XK is also expressed early during erythropoiesis and RT-PCR studies detected XK and Kell transcripts in multipotential hematopoietic progenitor cells [23]. However, whether the expression of Kell and XK on the cell surface is synchronized or not has not been determined due to the nonspecific nature of available anti-Kx alloantibodies.

Both Kell and XK are abundantly expressed in human erythroid tissues such as bone marrow and fetal liver. mXK and mKell mRNA levels are both high in mouse erythroid tissues, specifically in bone marrow, spleen and liver of new born mice, and spleen and bone marrow of adult mice (Fig. 1B, C for mXK mRNA, and E and F for mKell mRNA). In spleen, the expression was mainly in red pulp which is peripherally located [15].

As determined by RT-PCR, *XK* and especially its homologue, *XPLAC* are expressed in mouse lymph nodes and thymus implying their possible involvement in immune responses. Kell was minimally detected, but this could be due to erythroid cell contamination since the erythroid specific gene *GPA* was also detected.

2.2 Testis

ISHH of adult mouse testis gave a positive signal for mXK mRNA throughout the seminiferous tubules, but mKell mRNA was not detected. mKell was detected only by RT-PCR indicating that the amount may not be high enough to be detected by ISHH [15]. The discrepancy of the results obtained by the two techniques may be due to differences in sensitivity.

GATA-1 transcriptional factor is specific, not only to erythroid tissues, but is also expressed in testis and is important in the expression of proteins in testis [5, 21, 22]. The Kell promoter contains multiple conserved GATA-1 binding sites and is expected to regulate expression in both erythroid tissues and testis. Both Kell and XK are expressed in high levels in human testis as studied by Northern blot and dot blot analyses [3, 27]. However, there may be species difference as Kell expression in mouse testis appears to be lower than in humans. An XK homologue, XTES, is only present in primate testis and not in mouse [2]. The expression of Kell and XK may not be equal in testis and may not be necessarily in the same cells. mKell was shown by immunohistochemistry to be expressed in sertoli cells of human testis [3].

2.3 Other Non-Erythroid Tissues

In human tissues Kell expression was originally thought to be restricted to erythroid tissues, but later two reports, employing dot blot and Northern blot analyses, showed that Kell is also expressed in testis and in small amounts in other non-erythroid tissues including brain. It is important to know whether Kell is expressed in non-erythroid tissues such as brain since the cleaved product generated by Kell, endothelin-3, has many biological functions that may affect XK's function.

To determine the expression profiles of mKell and mXK, a whole body sagittal section of newborn mouse and adult brain were analyzed for mXK and mKell mRNA expression (Fig. 1) [15]. mXK mRNA was ubiquitously present in non-erythroid tissues including spinal cord, transitional epithelium and smooth muscles of the bladder, stomach, renal cortex, villi of small intestine, and brain (Fig. 1B). There was weak staining in mouse pancreas, in contrast to strong expression in human pancreas as shown by Northern blots [2, 6].

Adult brain ISHH showed the presence of mXK mRNA in many different regions such as cerebellum, pons, hippocampus, hypothalamus, cerebral cortex, caudate-putamen and olfactory lobe. The magnocellular neurons in the pontine region, cerebellar Purkinje cells, Cornu Ammonis fields of hippocampus, nucleus lateralis thalami and mitral cell layer of olfactory lobe were noticeably stained. Northern blot analyses of various regions of human brain confirmed the findings of mouse brain ISHH except that the corpus callosum, which showed low expression in human, did not yield a positive signal in mouse by ISHH.

mKell mRNA ISHH of a newborn sagittal section, adult testis and brain did not give a distinct positive signal in non-erythroid tissues [15]. mKell was not detected in brain, even in the caudate-putamen or pons, where mXK expression was moderate to high. RT-PCR of mouse cerebellum showed a very weak mKell band, but *GPA* was also detected, indicating that the mKell band may be the result of residual blood in the tissue samples. We conclude that the expression of Kell in mouse brain may be either less than in humans and not detectable by the methods employed, or that the earlier reports on the expression of Kell in human brain may be due to non-specific signals.

mKell is not detected by ISHH or RT-PCR in skeletal muscle or in C2C12 cells derived from a mouse myoblast cell line. Earlier reports showed that Kell was expressed in human muscle as measured by immunohistochemistry and by immunoprecipitation of Kell/XK complex [27]. The discrepancy between the mouse and human Kell expression in muscle may be due to species difference, or more likely, in the immunoprecipitation study, by the possible presence of residual blood in the human muscle tissue.

3 Summary

mXK and mKell are co-expressed in erythroid tissues, and mXK, but probably not mKell, is expressed in different regions of brain. The mXK expression pattern in brain indicates that it is mainly present in presumptive neuronal cells. The fact that both mXK and mKell are co-expressed in the spleen, liver (in newborn mouse) and bone marrow cells, and that they are covalently linked in red cells, suggests that they may play complementary roles in a hematopoietic function. However, in non-erythroid tissues, such as neurons and bladder epithelial cells, where mKell is not expressed, mXK may have a separate neuronal or epithelial function. To understand XK's physiological roles it is necessary to find the substrate for its presumptive transport function.

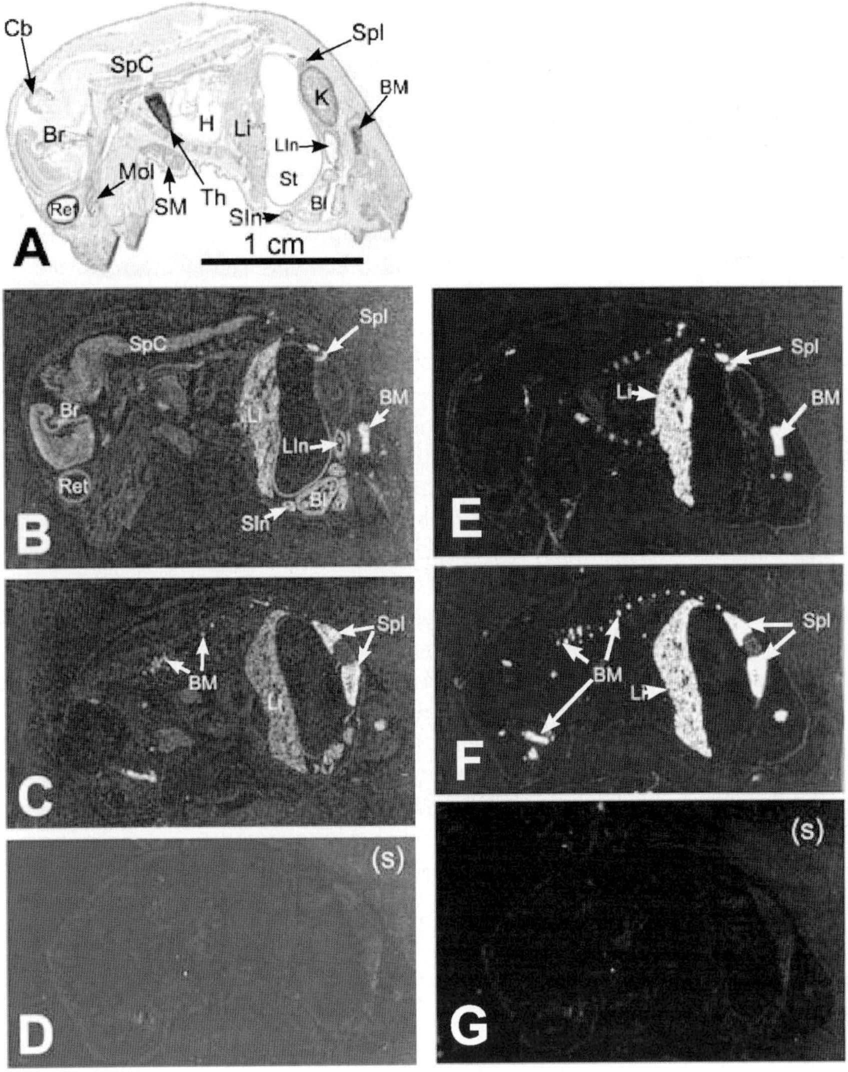

Fig. 1 Expression of mXK and mKell mRNA. (**a**) Anatomical view of newborn mouse whole-body sagittal section stained with hematoxylin as a reference for the ISHH of the adjacent similar section (**b** and **e**). (**b–d**) Expression of mXK mRNA: (**b**) x-Ray film autoradiography following hybridization with the mXK antisense riboprobe after 5 days exposure time showing heterogeneous pattern of mXK mRNA distribution throughout several structures including the brain and spinal cord, liver, small and large intestine, bladder and spleen. (**c**) Lateral-most section of newborn mouse displaying spleen tissue. (**d**) Sense control results for (**b**). (**e–g**) Expression of mKell mRNA: (**e**) Adjacent similar section (**a**) subjected to ISHH with antisense probe following x-ray film autoradiography showing a presence of mKell mRNA in the liver and spleen. (**f**) Lateral-most section of newborn mouse displaying mKell labeled spleen tissue. (**g**) Sense control result for (**e**). [*Bl*, bladder; *BM*, bone marrow; *Br*, brain; *Cb*, cerebellum; *H*, heart; *K*, kidney; *Li*, liver; *Lin*, large intestine; *Mol*, molar tooth; *Ret*, retina; *Sin*, small intestine; *SM*, submaxillary gland; *Spc*, spinal core; *Spl*, spleen; *St*, stomach; *Th*, thymus]. (Reproduced, with permission, from [15].)

Acknowledgments We thank Dr. Colvin M. Redman, Emeritus member of the New York Blood Center, for valuable discussions in preparation of the manuscript and Dr. Ruth H. Walker, James J. Peters Veterans Affairs Medical Center, Bronx, NY and Mount Sinai School of Medicine, New York, NY, whose expertise in the McLeod syndrome guided our selection of brain sections for the ISHH study. This study was supported in part by an NIH Specialized Center of Research (SCOR) grant in Transfusion Biology and Medicine, (HL54459), and by an NIH grant (RO1 HL075716).

References

1. Bony V, Gane P, Bailly P, Cartron JP (1999) Time-course expression of polypeptides carrying blood group antigens during human erythroid differentiation. Br J Haematol 107:263–274
2. Calenda G, Peng J, Redman CM, Sha Q, Wu X, Lee S (2006) Identification of two new members, XPLAC and XTES, of the XK family. Gene 370:6–16
3. Camara-Clayette V, Rahuel C, Lopez C, Hattab C, Verkarre V, Bertrand O, Cartron JP (2001) Transcriptional regulation of the *KEL* gene and Kell protein expression in erythroid and non-erythroid cells. Biochem J 356:171–180
4. Danek A, Rubio JP, Rampoldi L, Ho M, Dobson-Stone C, Tison F, Symmans WA, Oechsner M, Kalckreuth W, Watt JM, Corbett AJ, Hamdalla HH, Marshall AG, Sutton I, Dotti MT, Malandrini A, Walker RH, Daniels G, Monaco AP (2001) McLeod neuroacanthocytosis: genotype and phenotype. Ann Neurol 50:755–764
5. Feng ZM, Wu AZ, Zhang Z, Chen CL (2000) GATA-1 and GATA-4 transactivate inhibin/activin beta-B-subunit gene transcription in testicular cells. Mol Endocrinol 14:1820–1835
6. Ho M, Chelly J, Carter N, Danek A, Crocker P, Monaco AP (1994) Isolation of the gene for McLeod syndrome that encodes a novel membrane transport protein. Cell 77:869–880
7. Jung HH, Russo D, Redman C, Brandner S (2001) Kell and XK immunohistochemistry in McLeod myopathy. Muscle Nerve 24:1346–1351
8. Khamlichi S, Bailly P, Blanchard D, Goossens D, Cartron JP, Bertrand O (1995) Purification and partial characterization of the erythrocyte Kx protein deficient in McLeod patients. Eur J Biochem 228:931–934
9. Lee S (1997) Molecular basis of Kell blood group phenotypes. Vox Sang 73:1–11
10. Lee S, Debnath AK, Wu X, Scofield T, George T, Kakaiya R, Yogore MG, Sausais L, Yacob M, Lomas-Francis C, Reid ME (2006) Molecular basis of two novel high-prevalence antigens in the Kell blood group system, KALT and KTIM. Transfusion 46:1323–1327
11. Lee S, Lin M, Mele A, Cao Y, Farmar J, Russo D, Redman C (1999) Proteolytic processing of big endothelin-3 by the Kell blood group protein. Blood 94:1440–1450
12. Lee S, Russo D, Redman C (2000) Functional and structural aspects of the Kell blood group system. Transfus Med Rev 14:93–103
13. Lee S, Russo D, Redman CM (2000) The Kell blood group system: Kell and XK membrane proteins. Semin Hematol 37:113–121
14. Lee S, Russo DC, Reiner AP, Lee JH, Sy MY, Telen MJ, Judd WJ, Simon P, Rodrigues MJ, Chabert T, Poole J, Jovanovic-Srzentic S, Levene C, Yahalom V, Redman CM (2001) Molecular defects underlying the Kell null phenotype. J Biol Chem 276:27281–27289
15. Lee S, Sha Q, Wu X, Calenda G, Peng J (2007) Expression profiles of mouse Kell, XK, and XPLAC mRNA. J Histochem Cytochem 55:365–374.
16. Lee S, Zambas ED, Marsh WL, Redman CM (1991) Molecular cloning and primary structure of Kell blood group protein. Proc Natl Acad Sci U S A 88:6353–6357
17. Masaki T (1995) Possible role of endothelin in endothelial regulation of vascular tone. Annu Rev Pharmacol Toxicol 35:235–255
18. McCallion AS, Chakravarti A (2001) EDNRB/EDN3 and Hirschsprung disease type II. Pigment Cell Res 14:161–169

19. Motohashi T, Aoki H, Yoshimura N, Kunisada T (2006) Induction of melanocytes from embryonic stem cells and their therapeutic potential. Pigment Cell Res 19:284–289
20. Nagy N, Goldstein AM (2006) Endothelin-3 regulates neural crest cell proliferation and differentiation in the hindgut enteric nervous system. Dev Biol 293:203–217
21. Onodera K, Takahashi S, Nishimura S, Ohta J, Motohashi H, Yomogida K, Hayashi N, Engel JD, Yamamoto M (1997) GATA-1 transcription is controlled by distinct regulatory mechanisms during primitive and definitive erythropoiesis. Proc Natl Acad Sci U S A 94:4487–4492
22. Onodera K, Yomogida K, Suwabe N, Takahashi S, Muraosa Y, Hayashi N, Ito E, Gu L, Rassoulzadegan M, Engel JD, Yamamoto M (1997) Conserved structure, regulatory elements, and transcriptional regulation from the GATA-1 gene testis promoter. J Biochem (Tokyo) 121:251–263
23. Pu JJ, Redman CM, Visser JW, Lee S (2005) Onset of expression of the components of the Kell blood group complex. Transfusion 45:969–974
24. Redman CM, Russo D, Lee S (1999) Kell, Kx and the McLeod syndrome. Baillieres Best Pract Res Clin Haematol 12:621–635
25. Russo D, Lee S, Redman C (1999) Intracellular assembly of Kell and XK blood group proteins. Biochim Biophys Acta 1461:10–18
26. Russo D, Redman C, Lee S (1998) Association of XK and Kell blood group proteins. J Biol Chem 273:13950–13956
27. Russo D, Wu X, Redman CM, Lee S (2000) Expression of Kell blood group protein in nonerythroid tissues. Blood 96:340–346
28. Southcott MJ, Tanner MJ, Anstee DJ (1999) The expression of human blood group antigens during erythropoiesis in a cell culture system. Blood 93:4425–4435
29. Walker RH, Danek A, Uttner I, Offner R, Reid M, Lee S (2007) McLeod phenotype without the McLeod syndrome. Transfusion 47:299–305
30. Walker RH, Jung HH, Tison F, Lee S, Danek A (2007) Phenotypinc variation among brothers with the McLeod neuroacanthocytosis syndrome. Mov Disord 22:244–248.

Questions of Cell Shape

G.W. Stewart(✉), S.M.S. Wilmore, S. Ohno, and N. Terada

Abstract This chapter considers existing knowledge regarding the molecular basis of the mechanisms of bending of cell membranes, and then attempts to relate this knowledge to the possible pathophysiology underlying the neuroacanthocytosis syndromes. Curvature of animal cell membranes can be induced by a series of different mechanisms: by insertion or deletion of phospholipid from inner or outer leaflet; by contraction or expansion of the underlying cytoskeleton; by binding of soluble, typically cytoplasmic, proteins to either the lipid bilayer directly or to adaptor integral proteins; or (theoretically) by change in conformation of integral membrane proteins. Any of these, alone or in combination, could be active in chorea-

G.W. Stewart
Department of Medicine, University College London, Rayne Building, University Street, London WC1E 6JJ, UK
g.stewart@ucl.ac.uk

R.H. Walker et al. (eds.), *Neuroacanthocytosis Syndromes II.*

acanthocytosis/McLeod syndrome. Striking electron micrographs suggest that there is a degree of cytoskeletal disorganisation in both of these conditions.

1 Introduction

This book mainly concerns two conditions, both of which show the unusual combination of a progressive neurological disorder and a haemolytic anaemia in which the red cells have a spiky appearance, known as acanthocytosis. The first condition is known as 'chorea-acanthocytosis' (ChAc) [30]. This is an autosomal recessive disorder, while the condition known as 'McLeod' is X-linked. ChAc is caused by changes in the *VPS13A* (formerly *CHAC*) gene [48], one of the so-called '*VPS13*' family, while McLeod syndrome [15] is caused by changes in the gene *XK* [31]. Although the mutant genes have both been cloned, the diseases remain difficult to treat, at least in part because the underlying pathological mechanisms are poorly understood. The fact that two completely different proteins can cause virtually identical diseases suggests that the protein products of these two genes are components of the same pathway.

The combination of red cell and neurological pathology is an unusual one, although not unknown [24]. Presumably there is some common molecular mechanism that is shared between red cell and basal ganglia neurons. The point of this chapter is to look at the disease from the point of view of the red cell (and in particular its abnormal shape), to see what knowledge about the neuropathology might be gleaned from considerations of the simpler haematological cell.

We will look at the red cells from the point of view of what can be called 'membrane bending'. As is well known, biological membranes consist of a lipid bilayer, composed of phospholipids in two sheets, their hydrophobic acyl chains facing each other. To the phospholipids is added a considerable proportion of cholesterol, which is thought to lie alongside the acyl chains. The whole is studded with proteins that penetrate the bilayer, and in turn is underpinned and supported by an interconnecting protein scaffold or cytoskeleton that confers order and structure on the lipid bilayer, which forms a flexible seal. As will be seen below, the cytoskeleton forms a geodesic filamentous meshwork on the cytoplasmic surface of the membrane.

The lipids are present in fixed proportions (Fig. 1). The mechanism that controls these proportions in the face of free (if slow) exchange with the plasma is unclear [67]. The lipids have a sidedness, in that essentially all of the phosphatidylserine (PS) and phosphatidylinositol (PI) are internally facing (Fig. 1). Sphingomyelin (SM) and phosphatidylcholine (PC) are largely in the outer leaflet. This asymmetry is maintained at least partly by a kind of pump known as a flipase (also spelt 'flippase'), whose molecular identity remains elusive [18]. In the red cell and probably all other human cells, the lipids are laterally organised into domains relatively enriched or depleted in cholesterol and sphingomyelin.

The plasma membrane of a neurone is basically similar, but is of course much more complex. Like the red cell it must be flexible and durable, but it has a huge signalling job, receiving, transmitting and relaying action potential signals. It must

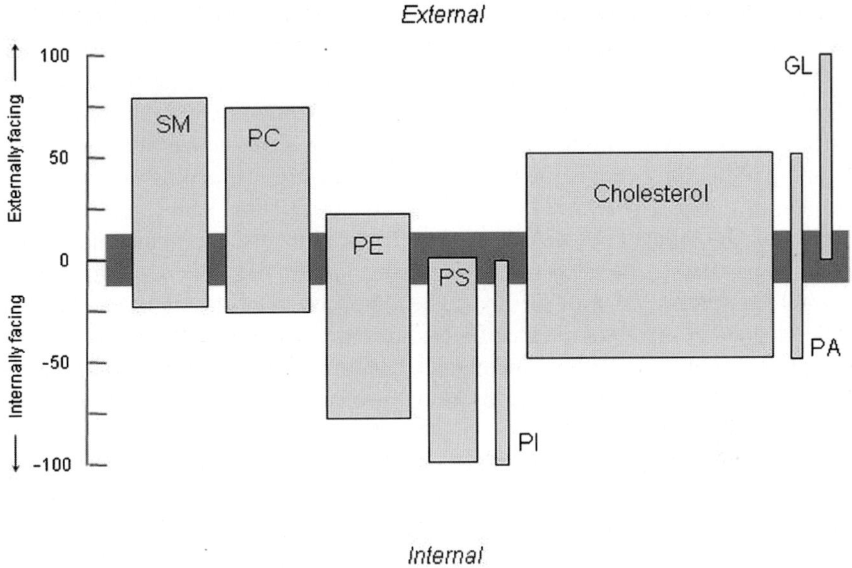

Fig. 1 Relative proportions of lipids in the inner and outer leaflets of the human red cell membrane. The different bars represent the different lipids (*SM*, sphingomyelin; *PC*, phosphatidylcholine; *PE*, phosphatidylethanolamine; *PS*, phosphatidylserine; *PI*, phosphatidylinositol and phosphorylated derivatives, *PA*, phosphatidic acid; *GL*, glycolipids). The vertical position denotes the relative distribution of the lipid between inner and outer leaflets. The width of the bar denotes the relative molar proportion of each lipid. There is a very substantial amount of cholesterol (nearly half of all lipid). The phospholipids with negatively charged headgroups (*PS, PI*) are entirely confined to the inner leaflet, while the glycolipids are all external. *SM, PC* and *PE* can be found in either leaflet, although *PC* and *SM* are predominantly in the outer and *PE* in the inner

make correct connections with many other neurones. The neuronal membrane has electrical activity related to its many ion channels; it has receptors to receive information; and a synaptic system to release packets of chemicals to tell other neurones what to do. One feature that is common to both red cells and neurones is their comparatively long life in relation to their housekeeping machinery. Neurones must last the life of the organism and cannot, in general, be replaced, although they have a nucleus with apparatus for protein and lipid synthesis to allow internal renewal. The red cell has no nucleus but nevertheless must survive a punishing 3-month existence without access to new protein synthesis.

The idea of 'membrane bending' was originated by Sheetz and Singer [61], working on the red cell. Devoid of intracellular organelles and equipped with a simple two dimensional cytoskeleton which simply underpins the lipid bilayer, the easily accessible red cell can be deformed by simple forces acting within the membrane. Sheetz and Singer doctored the composition of the lipid sheet by choosing compounds that intercalated selectively into either the outer leaflet or the inner. If the outer layer was expanded the cell developed externally facing protrusions to become either an acanthocyte or an echinocyte; if the inner leaflet was enlarged, the cell took on an invaginated shape known as a stomatocyte (Fig. 2). The theoretical basis of this idea was later extended [64].

Sheetz and Singer interpreted these data in terms of the 'bilayer couple'. In mechanics, a 'couple' is a pair of forces acting in opposite, anti-parallel directions which typically serve to rotate an object about an axis. In the membrane context, the anti-parallel forces act in the planes of the membrane, expanding or contracting one leaflet compared to the other. Instead of the rotation that would occur in mechanics, the 'bilayer couple' deforms, or folds, or bends, the membrane.

For some years these ideas remained confined to the red cell community. More recently the processes behind membrane bending have been closely investigated by cell biologists in more complex cell types, recently reviewed [34, 40, 70]. In cell biology, the main interest in membrane bending centres on the processes of budding, vesiculation and subsequent fusion involved in the trafficking between the different membrane components of the intracellular compartment, mainly endoplasmic reticulum and Golgi. Studies of these organelles have revealed many insights into different aspects of membrane bending. The principles are generic; they are applicable to all biological membranes.

Modern ideas on membrane bending have recently been reviewed [40]. In broad terms, bending can occur by three main mechanisms; (a) manipulation of lipid, (b) bending by the concerted action of the cytoskeleton, and (c) by individual integral or directly membrane-associated proteins, whose conformations endow curvature on a lipid membrane. The ideas and systems in these modern appreciations of membrane bending will be compared with the latest and most comprehensive study of the proteomics of the red cell, which describes a total of nearly 600 proteins in the cytosol

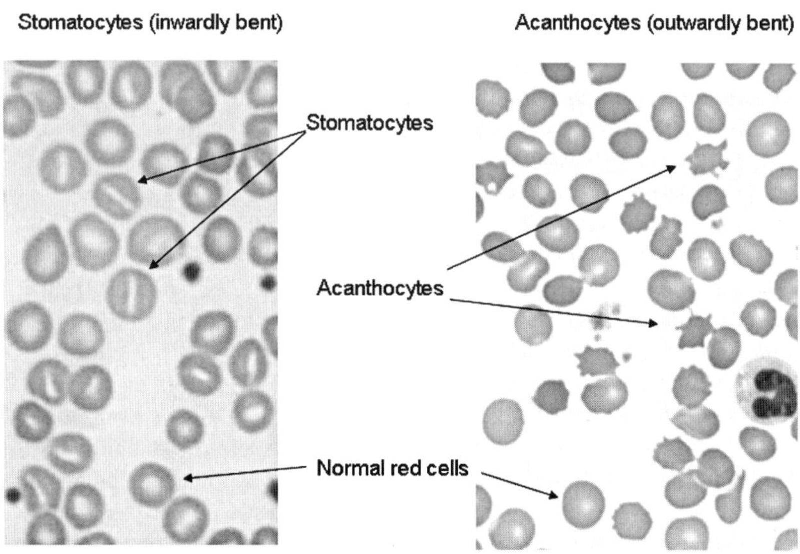

Fig. 2 Pathological erythrocytes showing the two extremes of membrane bending in the red cell membrane. Stomatocytes (*left*) are inwardly folded, while the spiculated acanthocytes (*right*) are outwardly folded

and membrane [46]. In spite of this very major effort, this proteomic collection is apparently incomplete, for neither chorein nor the mutant protein in McLeod, XK, was listed. Presumably these are present at too low an expression level for detection.

2 Bending by Variation in Lipid Bilayer Composition and its Modifications by Proteins

2.1 *Bending by Variation in Lipid Composition*

The early work of Sheetz and Singer work emphasised the role of lipid composition in bending of membranes (Fig. 3a). Natural membranes do it in different ways. The simplest method to conceive of is enzymic modification of existing lipids, reducing the headgroup size or charge, or by removing an acyl chain, for instance. One example of this is the induction of stomatocytic change in red cells by treatment of red cells with external sphingomyelinase, that presumably reduces the lipid content of the external leaflet by destruction of one of the main externally-facing lipids [3]. Other workers have shown a similar effect in giant liposomes [32]. Interestingly, the intracellular pathogen *Neisseria gonorrhoeae* employs a similar mechanism to gain entry to cells by prompting inward membrane bending, and subsequent endocytosis, by digesting external sphingomyelin [27].

Another way of changing the relative composition of the leaflets of the bilayer is to pump lipids from one leaflet to the other. All human cells contain at least one 'aminophospholipid flipase', an ATP-dependent membrane pump that is capable of transferring phospholipids from one leaflet to the other. The easiest phospholipid to study is PS. Although the existence of this system has been known for decades, its molecular identity remains enigmatic [18]. An 'ATP-binding cassette' (ABC) type protein is the obvious choice, but although such proteins do exist in the red cell membrane, proof of its function has been hard to find.

The composition of the bilayer could be influenced by either an abnormality in the lipids present in the surrounding plasma, which slowly exchange with the red cell, or conceivably in the completely uncharacterised system which preserves the relative proportions of PC, SM, PS, PI, phosphatidylethanolamine (PE) and cholesterol in the membrane. Presumably this regulatory system involves some kind of lipid exchange protein, but this has never been identified. The human anaemia once known as 'hereditary hyperphosphatidylcholine haemolytic anaemia' [33], now known to be identical to 'dehydrated hereditary stomatocytosis' or 'hereditary xerocytosis' [14] shows an excess of PC compared to PE and PS. The exact molecular cause remains unknown, but the condition is known to map to a locus on chromosome 16 [11].

The metabolic condition 'phytosterolaemia' [7] causes very marked stomatocytosis [49]. In this condition, there is an enteric abnormality in which the absorption of sterols, including both cholesterol and the plant-derived phytosterols, is unselective and unrestricted. Phytosterols resemble cholesterol but have added alkyl or alkenyl

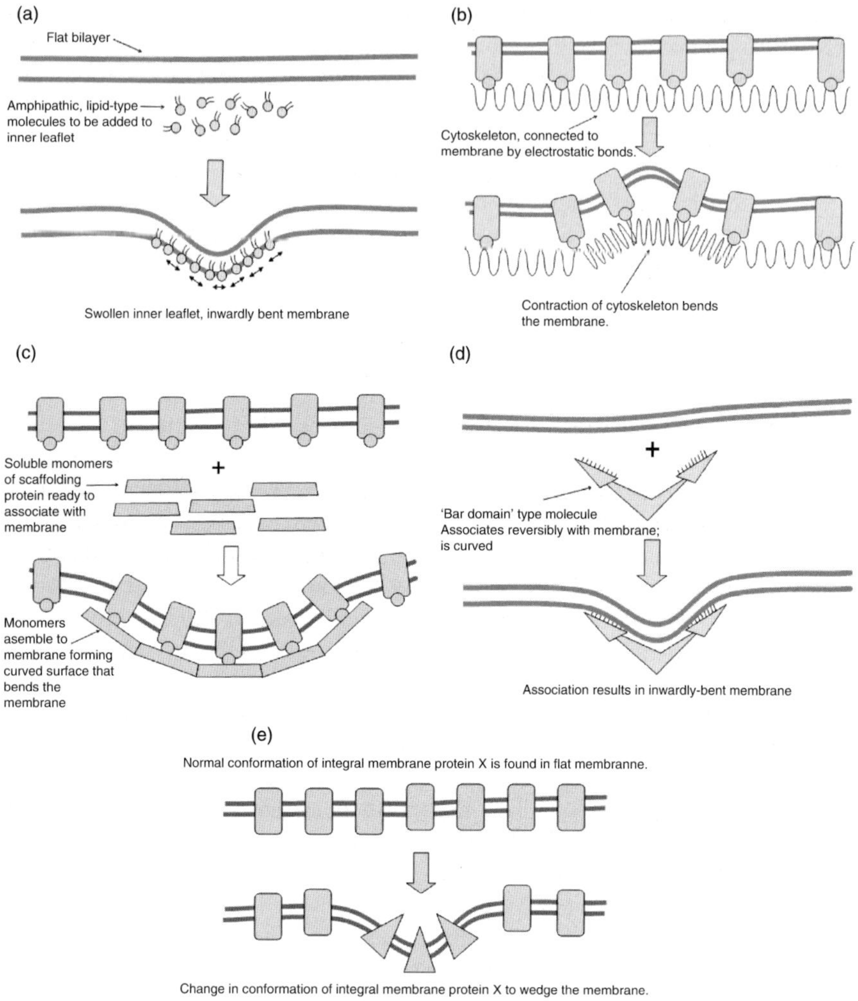

Fig. 3 Different mechanisms of membrane bending in red cells and other cell types. (**a**) Alteration of lipid composition of one leaflet of the bilayer. The double grey line denotes the bilayer. In this cartoon, a new lipid is added to the inner leaflet, causing inward membrane bending. (**b**) Bending by contraction of the underlying cytoskeleton. The cytoskeleton, denoted by a spring-shaped meshwork underlying the membrane, contracts, in this case causing outward bending, which would be manifest in the red cell as acanthocytosis. (**c**) Bending by assembly into membrane of new cytoskeletal proteins that associate via electrostatic interactions with integral proteins. Clathrin is the paradigm for this mechanism (**d**) Bending by reversible association of 'BAR domain' protein with one leaflet of the membrane. This protein has angled ends that directly associate with the lipid bilayer of the membrane, joined by an angled crank. As it binds to the membrane, it causes inward bending. (**e**) Membrane bending due to change in conformation of an integral membrane protein. The membrane-spanning protein denoted by a rectangle changes conformation to become wedge-shaped, changing the shape of the cell. This has never been shown directly but is theoretically feasible

groups at position C24 on the side chain. The nuclear part of the sterol is identical to that in cholesterol. These phytosterols, which are found in the 'oily' plants such as olive and avocado, circulate in the blood and can partition into the red cell membrane. The shape change caused by phytosterols begs the question, what role do sterols play in membrane bending? This is very hard to answer. There is a major technical difficulty; exchange of cholesterol between the inner and outer leaflets is very fast, quite beyond measurement by current methods [63]. Is it pumped? This is impossible to know. Is there more cholesterol associated with the inner leaflet than the outer? Again, this is very difficult to know, but this was the conclusion reached by Devaux and others some years ago, using spin-labelled cholesterol [59].

2.2 The Role of Rafts in Membrane Bending

It is impossible to mention membrane cholesterol without considering the idea of rafts, laterally demarcated domains in the membrane. Some early studies on red cells did suggest that the membrane was not simply a fluid mixture, but that it had a patchwork quality [52, 53]. The study of the tiny invaginations in the plasma membrane of the endothelial cell known as caveolae started the trail [20, 54–56]. Rothberg and others showed that caveolae were rich in the membrane protein caveolin, a 21 kDa hairpin protein that has both N-and C-termini in the cytosol and a loop of hydrophobic sequence that enters the membrane and does a U-turn within it. Simons and others then showed that animal cell membranes could be fractionated using a simple but novel technique in which membranes were exposed to cold non-ionic detergent such as Triton X-100, then centrifuged on a sucrose gradient [22]. A fraction of the membrane is insoluble under these conditions, and floats on the sucrose gradient. This fraction is rich in the lipids cholesterol and sphingomyelin, and selectively enriched in some proteins, notably caveolin. Thus the idea emerged that the anatomical caveolus structure was associated with lipids that were insoluble in cold detergent and represented a laterally demarcated domain in the membrane. The sucrose gradient technique, which probably distinguishes between 'ordered' and 'disordered' lipids [38], can be applied to other cell types (e.g. red cells, which do not have caveolae). This work has led to a renewed appreciation of the structure of biological membranes, summed up in the title of Engelman's review, 'more mosaic than fluid' [21], in which the author emphasises that the membrane is a very busy place packed with organised proteins, more like the busy traffic in London than the empty roads of northern Scotland.

From the point of view of membrane bending, it is clear that rafts, and the caveolus in particular, can function as localised foci for membrane bending [41]. The fact that the sterol condition phytosterolaemia leads to such marked deformation of the red cell emphasises the role of sterols (in which the caveolus is rich) in membrane bending. Exactly how this all works remains to be elucidated. Given the very rapid rate at which cholesterol can cross membranes, it seems unlikely that there is a pump pushing it, for it should just flip back.

2.3 *The Important Role of Phosphatidylinositol Lipids*

Although they comprise only 1% of the membrane lipids, PI and its phosphorylated derivatives play a crucial role in membrane bending, quite apart from their important roles in cell signalling. It is of both historical and practical interest that the first real inroads into the fertile ground that became Pl biology were made in studies of the human red cell, by Michell and colleagues in Birmingham, UK. This group recognised that when the intracellular [Ca^{2+}] in red cells was elevated, the cells became echinocytic, and that this shape change was associated with a rise in 1,2-diacylglycerol [1]. Later they determined that this change was due to the activation by internal calcium of phospholipases, which were acting on the minority phospholipid, PI [2, 19]. Backman later revisited these points and calculated that the echinocytic effect of internal calcium could be directly attributed to the loss of intracellular PI consequent on the activation by internal calcium of a phospholipase C [5].

More recently it has become recognised that PI lipids are key binding sites for a series of proteins within the cell. The lipid can be seen as a controllable adaptor link at the membrane. The key point about this class of lipid seems to lie in the ability of the body to modify the inositol headgroup by phosphorylation. This may reach its most sophisticated form in the idea that the intracellular lipids are identified by a different variant of Pl, a very interesting principle [6].

Aside from the early work of Allan and Michell, later red cell workers have emphasised the point that the well-characterised red cell cytoskeletal protein known as 'protein 4.1' has a PI binding site [26]. More recent work has focussed upon the role of the twice-phosphorylated PI known as PIP2 in controlling the interactions of protein 4.1 with other proteins [4].

These red cell studies are now only a tiny part of PI biology. The crucial role of PI in signalling is well known [39, 42]. PI is a focus for the membrane attachment of many cytoskeletal proteins, quite apart from 4.1 [12, 36]. The role of the cytoskeleton will be discussed next.

3 Bending by the Contractile, Remodelling Cytoskeleton

3.1 *Bending by Contractions and Expansions in the Cytoskeleton*

The paragraphs above have emphasised the role of lipids. But a lipid bilayer has little structure; lipids are sheep which must be shepherded. Proteins are required to do this work. In the section above we have referred to enzymes that may alter lipid composition on one side of the membrane or the other, and to phospholipid-flipping pumps that move the phospholipids between one leaflet and the other. Other proteins can bend membranes in different ways. All cells have a 'cytoskeleton', a multimolecular assembly of interacting proteins that has tensile strength, is flexible and has the ability to remodel. Unlike other cell types, the red cell cytoskeleton is confined to two dimensions rather than three [43]. Its actin filament length is comparatively short. In the red cell it can be pulled apart and will reform, the links being largely electrostatic in origin, partner proteins matching each other in shape and charge at interacting sites. In red cells this assembly is attached to the lipid bilayer by 'studs', adaptor molecules which penetrate the bilayer, including band 3 and glycophorin C. While the role of cytoskeleton in human red cells is mainly passive – it simply resists the battering that it receives in the circulation – in other cell types, it is a much more proactive structure, with major remodelling powers that actively change the shape of the cell (Fig. 3b). One obvious example is the phagocyte, where strands of actin monomers are actively extended, bundled for strength and subsequently disassembled as the cell moves forward to engulf its prey. 'Treadmilling', which has a similar basis, is seen by migrating cells, for instance in the genesis of axonal growth cones [29].

Recent proteomic surveys show the presence in the red cell of a series of proteins identified as components of the cytoskeleton in other cell types, but as yet uncharacterised in human red cells. Although tubulin is present in red cells [46] it is not as prominent as actin. Nevertheless it is equipped with a diversity of actin-associated proteins and membrane-attaching proteins which could be involved in the active bending of the membrane. Ankyrin, adducin, protein 4.1, tropomyosin, tropomodulin, and dematin (protein 4.9, actin bundling) are well known, as are the main membrane links, which include band 3 and glycophorin C.

Others have been associated with actin and the cytoskeleton; actin-related protein 3 (ARP3, involved in the regulation of actin filaments); anillin, an actin binding protein [23]; cofilin, that bundles actin; ezrin and radixin, members of the ERM (ezrin radixin moesin) family of actin binding proteins [37]; the actin bundling protein fascin [45].

A series of proteins have been identified in the red cell, that have not been extensively characterised in this cell type, but which are well known elsewhere. Several have been associated with vesicle transporting processes in other cell types: the 'nipsnap' protein, associated with vesicle fusion [60]; the vesicle-associated membrane protein (VAMP); a 'sec1 family domain containing protein', possibly involved in the docking of vesicles [68]; an isoform of dynamin, possibly involved in the nipping off of endocytic vesicles [58]; pantophysin [9]; the small GTP-binding, vesicle-associated proteins Rabs 10, 14, 21, 33B, 35, 5B, 5C, 8A, 8B, Ral A, Rap 1A, 1B, 2B, key signalling molecules in these processes [13, 50]. Knock-out mice for Rac1 and Rac2 GTPases showed major abnormalities in the cytoskeleton with anisocytosis and poikilocytosis, variability in size and shape of the red cell [35], and Rap2 has been associated with exovesicles liberated from human red cells when internal [Ca^{2+}] is raised [28].

3.2 Bending by Scaffolding Proteins at Localised Points Forming an 'Exoskeleton'

Overlapping with the ideas above about the role of the cytoskeleton, there are other membrane bending processes which involve localised blebbing of the membrane by the assembly of specialised proteins recruited to patches of membrane from the cytosol (Fig. 3c). The main example here is clathrin, a cytosolic protein which can polymerise around lipid membrane to form a spherical geodesic cage [57] and which is present in human red cells [16]. It forms a protein exoskeleton, a buckyball lined by a sheet of lipid on the inside. It does this in company with other proteins, some of which (AP50, AP180) have been identified in the red cell proteome [46].

4 Bending by Presence of Potentially Wedge-Shaped Integral Proteins or by the Insertion of Amphiphilic Helices

4.1 BAR Domains, Amphipathic Helices

There is another major way in which proteins can bend membranes. It is possible for proteins that penetrate the membrane to occupy the leaflets asymmetrically, expanding one leaflet at the expense of the other. A well-studied example is the reversible association between membranes and proteins containing a so-called 'amphipathic helix'. The idea here is that the protein adopts a helical secondary structure (a common conformation) but the amino acid sequence of this helix is such that one side of the cylindrical structure is hydrophobic, and therefore naturally associates with the membrane, while the other side is hydrophilic, and does not. This protein can therefore embed in the surface of the membrane to which it is presented (typically cytosolic), and expand that surface, typically bending the membrane inwards at that point (Fig. 3d).

This motif is often associated with a curved banana-shaped strut-type structure known as a BAR (bin, amphiphysin Rvs) domain [17], whose crystal structure was elucidated by the McMahon lab [25, 47]. This domain is found in many different proteins (amphiphysin, endophilins, BRAP1/bin2, nadrins, tuba, oligophrenin, centaurin, nexins, and arfaptins) but none of these has ever been associated with the red cell [46]. However, other proteins that have an amphiphilic helix could be hard to find. This is a structural feature rather than a sequence feature. It requires two characteristics, first that the protein adopts a helical structure, and second that one side of the helix is hydrophobic and the other hydrophilic. It can be imagined that many proteins could fulfil this condition; they need have no sequence similarity to those that are already characterised, simply that roughly every third amino acid should be hydrophobic in character.

4.2 *Wedge Shaped Integral Proteins*

It can be imagined that some integral proteins could cause membrane bending by virtue of their shape. If the protein had a wedge-type configuration, e.g. larger at the cytosolic surface than the external one, then that protein would tend to promote curvature (Fig. 3e). The crystal structure of only a very few integral membrane proteins has been determined, but among membrane proteins those of the caveolin structure are favourites. These have a 'hairpin' configuration; both N- and C-termini face the cytosol, and the protein appears to form a 'hairpin loop' within the membrane. From this format (which has not yet been proved by crystallography) it would be predicted that the protein would occupy a greater area in the cytoplasmic leaflet than the external, and would tend to cause inward bending. It is striking that in endothelial cells where caveolin is present, this is exactly the case.

Caveolins are not found in red cells. However, the stomatin–prohibitin–flotillin proteins are all found in red cells. Although there is no sequence similarity between these proteins and the caveolins, they share other similarities, such as the hairpin loop structure, palmitoylation, and raft association. Interestingly, stomatin was first identified because it is missing from invaginated, stomatocytic membranes in the condition 'overhydrated hereditary stomatocytosis'. Since the protein is cytoplasmically oriented, one might expect the protein to expand the inner leaflet of the membrane if it was present, and to allow this inner membrane to shrink if it was absent, giving acanthocytosis in the pathological cells, the exact opposite of what is seen. It is becoming clear that the deficiency of stomatin is only one aspect of the pathophysiology in these cells. The cause of the membrane curvature could lie in other proteins or in the cation leak that is so prominent. The true function of these proteins has never been elucidated. They may be involved in intracellular lipid movements.

Other integral membrane proteins could also change conformation within the membrane to exert a bending moment. The obvious choice here is band 3 protein (as suggested by Wong [69]), whose membrane-associated domain has not yet been crystallised. One potential theory is that mutations in band 3 cause some of the cation-leaky 'hereditary stomatocytosis' disorders, and the multiple amino-acid substitutions that are found are consistent with a conformational change rather than a change in an charge selectivity filter [10].

5 Implications for the Diseases Chorea-Acanthocytosis and McLeod Syndrome, and the Proteins Chorein and XK

It is evident from the above that there are many proteins involved in membrane bending. It might be expected that in an acanthocytic condition there would be some expansion of the outer leaflet at the expense of the inner leaflet. As has been said, there could be an abnormality in lipid composition, either a phospholipid or a sterol, or there could be an abnormality in a lipid-pumping protein that has the task of

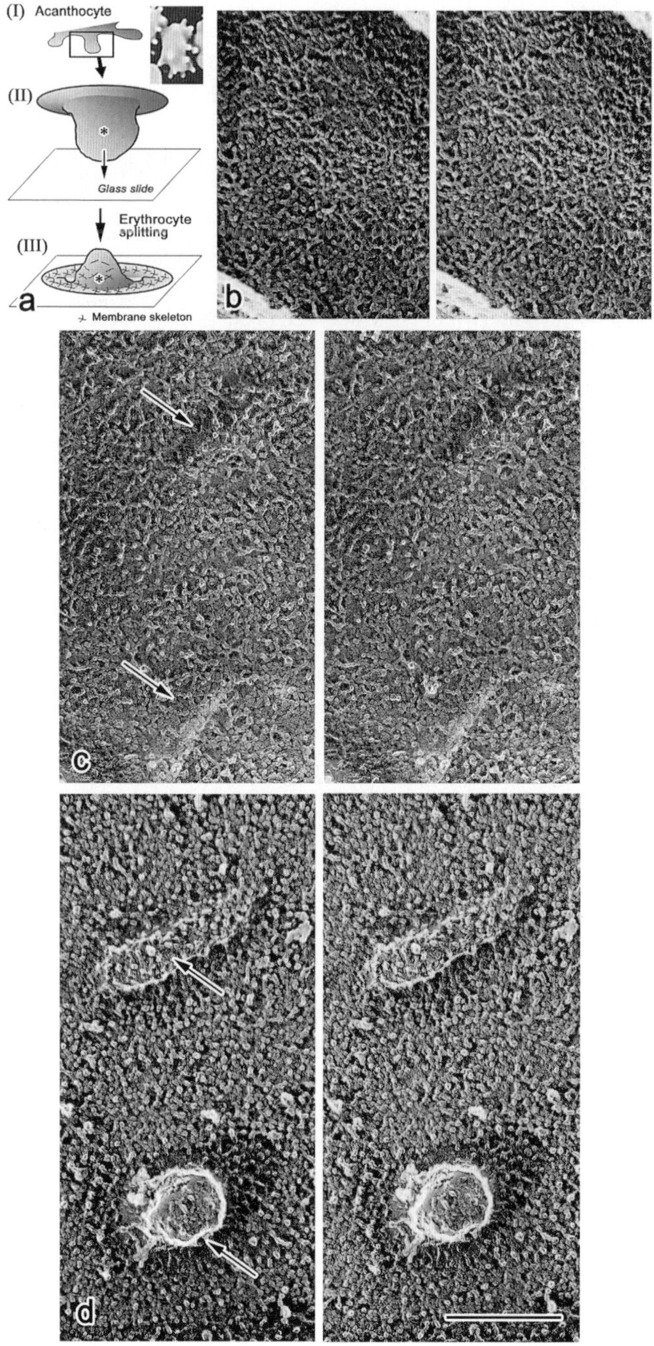
(I) Acanthocyte
(II)
Glass slide
Erythrocyte splitting
(III)
a
Membrane skeleton
b
c
d

pushing lipids into the inner leaflet. There could be a defect in a membrane bending protein that enhanced inward membrane bending. It is easy to see how such a red cell abnormality, when present in the much more complex neurone, could lead to abnormalities in not only the configuration of the plasma membrane but also in the trafficking of intracellular materials or in the development of growth cones.

It is unlikely to be a coincidence that chorein is homologous in sequence to the yeast protein Vps13p (vacuolar protein sorting 13p), which, although poorly understood, is clearly associated in some way the kind of intracellular trafficking events in which membrane bending is such a prominent feature. The gene was originally identified in the baker's yeast *Saccharomyces cerevisiae*. Emr and others made a search for mutant strains that caused mistrafficking of a marker protein, carboxypeptidase-Y (which was in fusion with the enzyme invertase for identification purposes). Their studies revealed dozens of strains, of which Vps13p was but one [51]. The same gene, this time called *SOI1*, cropped up is another such genetic screen, in which Brickner and Fuller searched for yeast strains that could correct, or 'suppress' in genetic parlance, a problem in the gene *Kex2p*, which codes for a protein that should be trafficked to the vacuole in the cell [8]. The work on SOI1 suggested that the protein might have role in the recognition of a sequence within Kex2p that acted as a targeting label for its residence in the so-called trans-Golgi network, another subcellular membrane system within the yeast cell. The human genome contains four *VPS13*-like genes [66] (discussed in detail in Chapter X). Recessive changes in the related *VPS13B* cause the human condition known as the Cohen syndrome (MIM, 216550), in which there is psychomotor retardation with a degree of clumsiness and problems in facial appearance, the joints, the eyes and the white cells of the blood. This is a very different condition to ChAc/McLeod syndrome.

XK, changed in McLeod syndrome, is similar to a gene from the nematode *Caenorhabditis elegans*. Mutations in this gene, *ced-8*, cause a problem is so-called 'programmed cell death', or apoptosis, the mechanism by which cells that are no longer required (for one reason or another) are cleanly tidied away [62]. These genes (*ced-8. XK*) appear to code for a plasma membrane protein that could act as a channel, but what the channel transports is an unanswered question.

Fig. 4 Three-dimensional ultrastructure of erythrocyte membrane skeletons in McLeod syndrome revealed by the quick-freezing and deep-etching replica method. (**a**) Schematic representation of protrusions of acanthocytic erythrocytes [inset in a-(I)]. By sandwich-splitting between two silane-glutaraldehyde-coated glass slides, a protrusion of the acanthocyte [*asterisk* in a-(II)] is easily turned inside-out [*asterisk* in a-(III)], enabling us to directly observe membrane skeletal structures with the quick-freezing and deep-etching replica method [65] Stereo-pictures were routinely obtained by a pair of pictures of the replica membranes with ± 5°. (**b**) Stereo-picture of normal erythrocyte membrane skeletons. Compact membrane skeletons can be seen on the lipid cell membrane. (**c, d**) Three-dimensional membrane skeletons of acanthocytes from McLeod syndrome patients are viewed as stereo-pictures. In some acanthocytes, loosely arranged membrane skeletons [*arrows* in (**c**)] are focally observed. Other completely inside-out protrusions [*arrows* in (**d**)] have less filamentous structures, whereas mixed filamentous and granular structures still remain around the protrusions. Bar; 500 nm

6 Electron Microscopic Studies of McLeod Membranes

Studies of red cell membranes from an individuals with both ChAc and McLeod syndrome have been performed using electron microscopy, in an attempt to better understand the structural membrane abnormalities in neuroacanthocytosis [65]. In these studies, a novel 'sandwich-splitting' technique (that lays the cytoplasmic surface open to the scanning electron microscope, allowing it to be seen in relief) was employed to visualise the cytoskeleton underlying the bilayer. In the images from the normal subject, the cytoskeleton can be seen to form a filamentous network, as expected (Fig. 4b). In the subject with McLeod syndrome, the cytoskeleton forms a looser meshwork (Fig. 4c, d). Further, the cytoskeleton has a rather inhomogeneous, unevenly spread, granular appearance, consisting in part of knob-like structures.

These pictures suggest quite a drastic alteration in cytoskeletal structure in these conditions, and under the considerations above this could easily be related to the acanthocytic shape change. But whether these changes are due to a malfunctioning protein that is part of the cytoskeleton itself, or is part of the membrane attachment apparatus, or is secondary to yet another more fundamental abnormal process (such as a lipid attachment), is another matter.

7 Malaria

In the wider world, membrane bending in erythrocytes is not simply about diagnostic changes in rare inherited conditions. The invasion of the normal red cell by the malarial merozoite involves some complicated bending events. As the merozoite invades the red cell, it forms around itself an invaginated membrane known as the 'parasitophorous vacuolar membrane' (PVM). This is known to involve the cholesterol-rich raft elements mentioned above, which in other contexts have roles in endocytosis. Kasturi Haldar's group has shown that among the proteins known to be associated with rafts in the red cell membrane, invasion is associated with an interesting process of selection. While flotillin clearly moves from the plasma membrane to become part of the PVM, the sequence-related protein stomatin (first identified because of its deficiency in 'hereditary stomatocytosis' red cells, does not [44]. Stomatin remains in the plasma membrane of the infected cell. How this happens, and why, and what is signifies, are all quite unknown.

8 Conclusions

How can we tie up the discussions above? We have two very similar diseases (ChAc/McLeod) in which the membrane of the red cell shows abnormal bending; we have a series of known mechanisms by which membrane are bent; we have the two mutated genes in ChAc and McLeod syndrome. How are these related? The *VPS13A* protein seems to be intimately involved in trafficking, which as we have said often involves bending of membranes, budding of vesicles and subsequent fusion with other membranes. Domains within this large protein could fulfil any of the membrane-bending roles described in the earlier sections, but it is also possible that the abnormal membrane bending results from the mistrafficking of a cargo protein trafficked by the mechanism in which VPS13A is a component. The role of XK is even harder to delineate. It may be a membrane transporter, but that fact is not yet proved, and its substrate is unknown. As has been pointed out in this chapter, abnormal calcium transport could explain acanthocytosis.

The bottom line is that membrane bending is a complex business; many proteins and lipids are involved. Matching up the functions of chorein and XK with the pathophysiology of ChAc and McLeod syndrome will require extensive further study.

Acknowledgements We thank Advocacy for Neuroacanthocytosis for support, and Prof Hugh Pelham and Dr Ruth Walker for helpful discussions and advice.

References

1. Allan D, Michell RH (1975) Accumulation of 1,2-diacylglycerol in the plasma membrane may lead to echinocyte transformation of erythrocytes. Nature 258:348–349
2. Allan D, Thomas P, Michell RH (1978) Rapid transbilayer diffusion of 1,2-diacylglycerol and its relevance to control of membrane curvature. Nature 276:289–290
3. Allan D, Walkin C (1988) Endovesiculation of human erythrocytes exposed to sphingomyelinase C: a possible explanation for the enzyme-resistant pool of sphingomyelin. Biochim Biophys Acta 938:403–410
4. An X, Zhang X, Debnath G, Baines AJ, Mohandas N (2006) Phosphatidylinositol-4,5-biphosphate (PIP2) differentially regulates the interaction of human erythrocyte protein 4.1 (4.1R) with membrane proteins. Biochemistry 45:5725–5732
5. Backman L (1986) Shape control in the human red cell. J Cell Sci 80:281–298
6. Behnia R, Munro S (2005) Organelle identity and the signposts for membrane traffic. Nature 438:597–604
7. Bhattacharyya AK, Connor WE (1974) Beta-sitosterolemia and xanthomatosis. A newly described lipid storage disease in two sisters. J Clin Invest 53:1033–1043
8. Brickner JH, Fuller RS (1997) SOI1 encodes a novel, conserved protein that promotes TGN-endosomal cycling of Kex2p and other membrane proteins by modulating the function of two TGN localization signals. J Cell Biol 139:23–36
9. Brooks CC, Scherer PE, Cleveland K, Whittemore JL, Lodish HF, Cheatham B (2000) Pantophysin is a phosphoprotein component of adipocyte transport vesicles and associates with GLUT4-containing vesicles. J Biol Chem 275:2029–2036
10. Bruce LJ, Robinson H, Guizouarn H, Harrison P, King M-J, Goede JS, Coles SE, Gore DM, Lutz H, Ficarella R, Layton M, Iolascon A, Ellory JC, Stewart GW (2005) Monovalent cation leaks in human red cells caused by single amino-acid substitutions in the transport domain of the band 3 chloride-bicarbonate exchanger, AE1. Nat Genet 37:1258–1263
11. Carella M, Stewart GW, Ajetunmobi JF, Perrotta S, Grootenboer S, Tchernia G, Delaunay J, Totaro A, Zelante L, Gasparini P, Iolascon, A (1998) Genomewide search for dehydrated hereditary stomatocytosis (hereditary xerocytosis): mapping of locus to chromosome 16 (q23-qter). Am J Hum Genet 63:810–816
12. Caroni P (2001) Actin cytoskeleton regulation through modulation of PI(4,5)P(2) rafts. EMBO J 20:4332–4336

13. Charest PG, Firtel RA (2007) Big roles for small GTPases in the control of directed cell movement. Biochem J 401:377–390
14. Clark MR, Shohet SB, Gottfried EL (1993) Hereditary hemolytic disease with increased red blood cell phosphatidylcholine and dehydration: one, two, or many disorders? Am J Hematol 42:25–30
15. Danek A, Rubio JP, Rampoldi L, Ho M, Dobson-Stone C, Tison F, Symmans WA, Oechsner M, Kalckreuth W, Watt JM, Corbett AJ, Hamdalla HH, Marshall AG, Sutton I, Dotti MT, Malandrini A, Walker RH, Daniels G, Monaco AP (2001) McLeod neuroacanthocytosis: genotype and phenotype. Ann Neurol 50:755–764
16. Davis J, Bennett V (1985) Human erythrocyte clathrin and clathrin-uncoating protein. J Biol Chem 260:14850–14856
17. Dawson JC, Legg JA, Machesky LM (2006) Bar domain proteins: a role in tubulation, scission and actin assembly in clathrin-mediated endocytosis. Trends Cell Biol 16:493–498
18. Devaux PF, Lopez-Montero I, Bryde S (2006) Proteins involved in lipid translocation in eukaryotic cells. Chem Phys Lipid 141:119–132
19. Downes CP, Michell RH (1982) The control by Ca^{2+} of the polyphosphoinositide phosphodiesterase and the Ca^{2+}-pump ATPase in human erythrocytes. Biochem J 202:53–58
20. Dupree P, Parton RG, Raposo G, Kurzchalia TV, Simons K (1993) Caveolae and sorting in the trans-Golgi network of epithelial cells. EMBO J 12:1597–1605
21. Engelman DM (2005) Membranes are more mosaic than fluid. Nature 438:578–580
22. Fiedler K, Kobayashi T, Kurzchalia TV, Simons K (1993) Glycosphingolipid-enriched, detergent-insoluble complexes in protein sorting in epithelial cells. Biochemistry 32:6365–6373
23. Field CM, Alberts BM (1995) Anillin, a contractile ring protein that cycles from the nucleus to the cell cortex. J Cell Biol 131:165–178
24. Fricke B, Jarvis HG, Reid CD, Aguilar-Martinez P, Robert A, Quittet P, Chetty M, Pizzey A, Cynober T, Lande WF, Mentzer WC, During M, Winter S, Delaunay J, Stewart GW (2004) Four new cases of stomatin-deficient hereditary stomatocytosis syndrome: association of the stomatin-deficient cryohydrocytosis variant with neurological dysfunction. Br J Haematol 125:796–803
25. Gallop JL, Jao CC, Kent HM, Butler PJ, Evans PR, Langen R, McMahon HT (2006) Mechanism of endophilin N-BAR domain-mediated membrane curvature. EMBO J 25:2898–2910
26. Gascard P, Pawelczyk T, Lowenstein JM, Cohen CM (1993) The role of inositol phospholipids in the association of band 4.1 with the human erythrocyte membrane. Eur J Biochem 211:671–681
27. Grassme H, Gulbins E, Brenner B, Ferlinz K, Sandhoff K, Harzer K, Lang F, Meyer TF (1997) Acidic sphingomyelinase mediates entry of *N. gonorrhoeae* into nonphagocytic cells. Cell 91:605–615
28. Greco F, Ciana A, Pietra D, Balduini C, Minetti G, Torti M (2006) Rap2, but not Rap1 GTPase is expressed in human red blood cells and is involved in vesiculation. Biochim Biophys Acta 28:28
29. Haglund K, Ivankovic-Dikic I, Shimokawa N, Kruh GD, Dikic I (2004) Recruitment of Pyk2 and Cbl to lipid rafts mediates signals important for actin reorganization in growing neuritis. J Cell Sci 117:2557–2568
30. Hardie RJ, Pullon HW, Harding AE, Owen JS, Pires M, Daniels GL, Imai Y, Misra VP, King RH, Jacobs JM et al. (1991) Neuroacanthocytosis. A clinical, haematological and pathological study of 19 cases. Brain 114:13–49
31. Ho M, Chelly J, Carter N, Danek A, Crocker P, Monaco AP (1994) Isolation of the gene for McLeod syndrome that encodes a novel membrane transport protein. Cell 77:869–880
32. Holopainen JM, Angelova MI, Kinnunen PK (2000) Vectorial budding of vesicles by asymmetrical enzymatic formation of ceramide in giant liposomes. Biophys J 78:830–838
33. Jaffe ER, Gottfried EL (1968) Hereditary nonspherocytic hemolytic disease associated with an altered phospholipid composition of the erythrocytes. J Clin Invest 47:1375–1388
34. Janmey PA, Kinnunen PK (2006) Biophysical properties of lipids and dynamic membranes. Trends Cell Biol 16:538–546

35. Kalfa TA, Pushkaran S, Mohandas N, Hartwig JH, Fowler VM, Johnson JF, Joiner CH, Williams DA, Zheng Y (2006) Rac GTPases regulate the morphology and deformability of the erythrocyte cytoskeleton. Blood 1:1
36. Kwik J, Boyle S, Fooksman D, Margolis L, Sheetz MP, Edidin M (2003) Membrane cholesterol, lateral mobility, and the phosphatidylinositol 4,5-bisphosphate-dependent organization of cell actin. Proc Natl Acad Sci U S A 100:13964–13969
37. Mangeat P, Roy C, Martin M (1999) ERM protein in cell adhesion and membrane dynamics. Trends Cell Biol 9:187–192
38. Maxfield FR, Tabas I (2005) Role of cholesterol and lipid organization in disease. Nature 438:612–621
39. McLaughlin S, Wang J, Gambhir A, Murray D (2002) PIP(2) and proteins: interactions, organization, and information flow. Annu Rev Biophys Biomol Struct 31:151–175
40. McMahon HT, Gallop JL (2005) Membrane curvature and mechanisms of dynamic cell membrane remodeling. Nature 438:590–596
41. McNiven MA (2006) Big gulps: specialized membrane domains for rapid receptor-mediated endocytosis. Trends Cell Biol 16:487–492
42. Michell RH, Heath VL, Lemmon MA, Dove SK (2006) Phosphatidylinositol 3,5-bisphosphate: metabolism and cellular functions. Trends Biochem Sci 31:52–63
43. Mohandas N, An X (2006) New insights into function of red cell membrane proteins and their interaction with spectrin-based membrane skeleton. Transfus Clin Biol 13:29–30
44. Murphy SC, Samuel BU, Harrison T, Speicher KD, Speicher DW, Reid ME, Prohaska R, Low PS, Tanner MJ, Mohandas N, Haldar K (2004) Erythrocyte detergent-resistant membrane proteins: their characterization and selective uptake during malarial infection. Blood 103:1920–1928
45. Otto JJ, Kane RE, Bryan J (1979) Formation of filopodia in coelomocytes: localization of fascin, a 58,000 dalton actin cross-linking protein. Cell 17:285–293
46. Pasini EM, Kirkegaard M, Mortensen P, Lutz HU, Thomas AW, Mann M (2006) In-depth analysis of the membrane and cytosolic proteome of red blood cells. Blood 108:791–801
47. Peter BJ, Kent HM, Mills IG, Vallis Y, Butler PJ, Evans PR, McMahon HT (2004) BAR domains as sensors of membrane curvature: the amphiphysin BAR structure. Science 303:495–499
48. Rampoldi L, Dobson-Stone C, Rubio JP, Danek A, Chalmers RM, Wood NW, Verellen C, Ferrer X, Malandrini A, Fabrizi GM, Brown R, Vance J, Pericak-Vance M, Rudolf G, Carre S, Alonso E, Manfredi M, Nemeth AH, Monaco AP (2001) A conserved sorting-associated protein is mutant in chorea-acanthocytosis. Nat Genet 28:119–120
49. Rees D, Iolascon A, Carella M, O'Marcaigh A, Kendra J, Jowitt S, Wales J, Vora A, Nicolaou A, Fisher J, Mann A, Chetty MC, Clayton PT, Gasparini P, Stewart GW (2005) Stomatocytic haemolysis and macrothrombocytopenia (Mediterranean stomatocytosis/macrothrombocytopenia) is the haematological presentation of phytosterolaemia. Br J Haematol 130:297–309
50. Ridley AJ (2006) Rho GTPases and actin dynamics in membrane protrusions and vesicle trafficking. Trends Cell Biol 16:522–529
51. Robinson JS, Klionsky DJ, Banta LM, Emr SD (1988) Protein sorting in *Saccharomyces cerevisiae*: isolation of mutants defective in the delivery and processing of multiple vacuolar hydrolases. Mol Cell Biol 8:4936–4948
52. Rodgers W, Glaser M (1991) Characterization of lipid domains in erythrocyte membranes. Proc Natl Acad Sci U S A 88:1364–1368
53. Rodgers W, Glaser M (1993) Distribution of proteins and lipids in the erythrocyte membrane. Biochemistry 32:12592–12598
54. Rothberg K, Ying Y, Kolhouse J, Kamen B, Anderson R (1990) The glycosphospholipid linked folate receptor internalizes folate without entering the clathrin-coated pit endocytic pathway. J Cell Biol 110:637–649
55. Rothberg K, Ying Y-S, Kamen B, Anderson R (1990) Cholesterol controls the clustering of the glycolipid-anchored membrane receptor for 5-methyltetrahydrofolate. J Cell Biol 111:2931–2938

56. Rothberg KG, Heuser JE, Donzell WC, Ying YS, Glenney JR, Anderson RG (1992) Caveolin, a protein component of caveolae membrane coats. Cell 68:673–682
57. Schmid EM, Ford MG, Burtey A, Praefcke GJ, Peak-Chew SY, Mills IG, Benmerah A, McMahon HT (2006) Role of the AP2 beta-appendage hub in recruiting partners for clathrin-coated vesicle assembly. PLoS Biol 4:e262
58. Schmid S (1999) Dynamin. In: Kreis T, Vale R (eds) Guidebook to cytoskeletal and motor proteins. Sambrook and Tooze, Oxford, pp 524–527
59. Schroeder F, Nemecz G, Wood WG, Joiner C, Morrot G, Ayraut-Jarrier M, Devaux PF (1991) Transmembrane distribution of sterol in the human erythrocyte. Biochim Biophys Acta 1066:183–192
60. Seroussi E, Pan HQ, Kedra D, Roe BA, Dumanski JP (1998) Characterization of the human NIPSNAP1 gene from 22q12: a member of a novel gene family. Gene 212:13–20
61. Sheetz M, Singer SJ (1974) Biological membranes as bilayer couples: a molecular mechanism of drug-erythrocyte interactions. Proc Natl Acad Sci U S A 71:4457
62. Stanfield GM, Horvitz HR (2000) The ced-8 gene controls the timing of programmed cell deaths in *C. elegans*. Mol Cell 5:423–433
63. Steck TL, Ye J, Lange Y (2002) Probing red cell membrane cholesterol movement with cyclodextrin. Biophys J 83:2118–2125
64. Svetina S, Zeks B (1983) Bilayer couple hypothesis of red cell shape transformations and osmotic hemolysis. Biomed Biochim Acta 42:S86–S90
65. Terada N, Fujii Y, Ueda H, Kato Y, Baba T, Hayashi R, Ohno S (1999) Ultrastructural changes of erythrocyte membrane skeletons in chorea-acanthocytosis and McLeod syndrome revealed by the quick-freezing and deep-etching method. Acta Haematol 101:25–31
66. Velayos-Baeza A, Vettori A, Copley RR, Dobson-Stone C, Monaco AP (2004) Analysis of the human VPS13 gene family. Genomics 84:536–549
67. Walensky LD, Narla M, Lux SE (2003) Disorders of the red blood cell membrane. In: Handin R, Lux S, Stossel T (eds) Blood: principles and practice of Hematology. Lippincott, Philadelphia, pp 1709–1858
68. Weimer RM, Richmond JE (2005) Synaptic vesicle docking: a putative role for the Munc18/Sec1 protein family. Curr Top Dev Biol 65:83–113
69. Wong P (2004) A basis of the acanthocytosis in inherited and acquired disorders. Med Hypotheses 62:966–969
70. Zimmerberg J, Kozlov MM (2006) How proteins produce cellular membrane curvature. Nat Rev Mol Cell Biol 7:9–19

Neuroacanthocytosis-Related Changes in Erythrocyte Membrane Organization and Function

G.J.C.G.M. Bosman(✉) and L. de Franceschi

Abstract The presence of thorny erythrocytes in patients with mutations in chorein, XK protein, and junctophilin-3 suggests that abnormalities in the organization of the erythrocyte membrane are related to the neurodegeneration process. Band 3, the major erythrocyte membrane protein, is at the center of a multiprotein complex that regulates erythrocyte metabolism, erythrocyte shape, and erythrocyte survival. Immunoblot analyses indicate that alterations in band 3 structure are involved in the formation of acanthocytes, probably through phosphorylation-regulated interaction with the cytoskeleton components beta spectrin, actin and p55. In addition, preliminary data from a proteomic analysis suggest the occurrence of neuroacanthocytosis-related changes in the association of various classes of cytosolic proteins with the membrane. The relationship between these observations and the altered vesicle formation in vitro is not clear, but together these findings suggest an inborn flaw in the membrane organization of erythrocytes and possibly neurons. These data provide a foundation for further research on the identity of the mechanism of acanthocyte formation, which may be a first step towards intervention in the neurodegenerative process.

G.J.C.G.M. Bosman
Department of Biochemistry, Nijmegen Center for Molecular Life Sciences and University Medical Center-Nijmegen, the Netherlands
g.bosman@ncmls.ru.nl

R.H. Walker et al. (eds.), *Neuroacanthocytosis Syndromes II.*

1 Introduction

Acanthocytes are one of the clinical hallmarks of chorea-acanthocytosis (ChAc), McLeod syndrome (MLS) and Huntington disease-like 2 (HDL2), suggesting that abnormalities in erythrocyte membrane might be related to the neurodegeneration-associated mutations in chorein, XK protein, and junctophilin-3, respectively [26]. These are all membrane or membrane-associated proteins, but their role in physiology is a matter of speculation for neuronal cells, and a mystery in erythrocytes. A comparison of membrane structure and function of erythrocytes from neuroacanthocytosis (NA) patients, from patients with erythrocyte abnormalities, and from control donors may provide clues to the identity of the mechanism that causes the neurodegeneration-associated occurrence of acanthocytes, and thereby help in unravelling the mechanisms that cause the neurodegeneration.

2 The Band 3 Network

In NA, acanthocytes are associated with various erythrocyte membrane abnormalities, in particular with the band 3 protein. Band 3 is the major protein of the erythrocyte membrane. The membrane domain of band 3 catalyzes the transport of anions, mainly chloride and bicarbonate, and carries the antigens of the Diego blood group system. The cytoplasmic domain of band 3 anchors the spectrin-actin cytoskeleton to the lipid bilayer. In addition, band 3 seems to be the center of a multiprotein complex containing membrane proteins such as the glucose transporter, Rhesus proteins, glycophorin A, and CD47, as well as cytosolic proteins such as carbonic anhydrase II, and the enzymes of glycolysis glyceraldehyde 3-phosphate dehydrogenase (GAPDH), phosphofructokinase, and aldolase [25]. Recent data indicate that abnormalities in the cytoplasmic domain of band 3 can modify the ability of band 3 to bind enzymes such as aldolase [21]. In addition, the red cell signal transduction pathways that are active in response to various cell stresses such as oxidation or osmotic stimuli involve changes in the phosphorylation state of band 3. In fact, band 3 is not only the substrate for tyrosine kinases of the Src-family [3, 16], but also the substrate for protein tyrosine phosphatases [17; de Franceschi, unpublished observations].

The Pro 868→Leu mutation in the membrane domain of erythrocyte band 3 is associated with acanthocytosis, as well as with restriction of rotational diffusion of band 3 in the membrane, increased anion transport, and a decrease in the number of high-affinity ankyrin-binding sites. These changes were not observed in acanthocytes from patients with lipid disorders, suggesting a functional connection between NA and red cell membrane protein abnormalities [13].

Specific changes in band 3 also provide the signal for removal of old and damaged erythrocytes from the circulation by the immune system, and are involved in age-related vesiculation [6, 12]. Thus, band 3 occupies a central position in the regulation of function, shape and removal of the erythrocyte. Members of the band 3 gene family, the Na^+-independent Cl^--HCO_3^- SLC4 family, are also present in neurons [10].

The function of neuronal band 3-related proteins has not been elucidated completely, but they are involved in the regulation of intracellular pH [4]. Polymorphisms in SLC4A3 have been associated with seizures [1], and aging-related changes in band 3 are associated with neuronal degeneration in the brain [12].

We conclude that changes in band 3 protein sequence, structure, conformational state and phosphorylation state may affect the band 3 network and participate in the generation of acanthocytes.

3 The Band 3 Network in Acanthocytosis

Using an antibody against the cytoplasmic domain of band 3, we found specific immunoblot patterns in the membrane fractions of erythrocytes from patients with ChAc, MLS and HDL2 (Fig. 1). Aberrant immunoblot patterns were also obtained with anti-Diego and anti-phosphotyrosine antibodies [6]. These data suggest that the presence of acanthocytes in patients with various forms of NA is associated with, as well as characterized by, specific changes in the conformation of band 3 [5]. Based on immunoblot data obtained with antibodies of various specificities, these changes are likely to affect not only the interaction between cytoskeleton and lipid bilayer, but also the interaction of band 3 with other integral membrane proteins, and with cytosolic proteins. These changes may affect erythrocyte metabolism, either through the anion transport activity, or through binding and inactivation of enzymes such as aldolase and GAPDH.

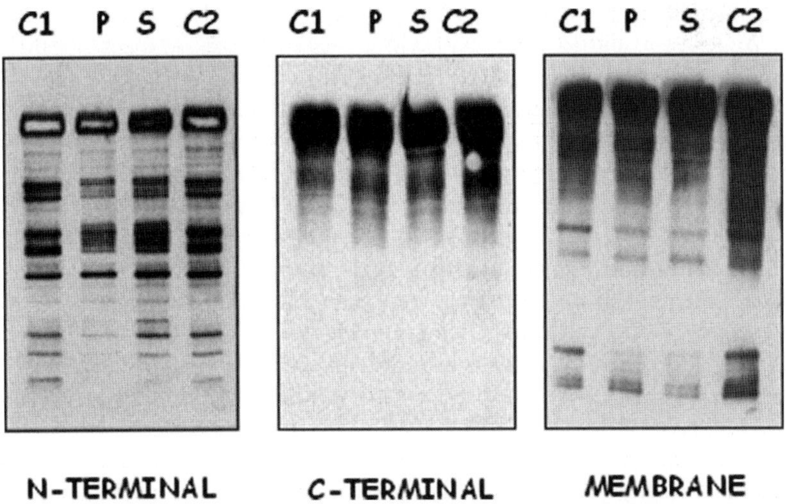

Fig. 1 Immunoblots of erythrocyte membranes from HDL2 patients (P and S) show abnormal band 3 patterns with anti-band 3 antisera directed against the cytoplasmic, N-terminal and the membrane domain of band 3, but not with antisera against the C-terminal domain of band 3

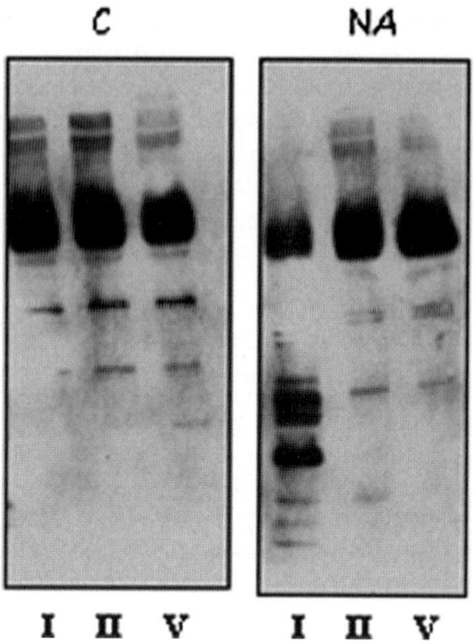

Fig. 2 Immunoblots of erythrocyte membranes from age-separated erythrocytes show an aging-related breakdown of band 3 in a control donor (C), and an abnormal band 3 pattern especially in the youngest fraction (fraction I) of a HDL2 patient (NA). I, II and V, erythrocytes of increasing age, consisting of 5%, 15% and 95% acanthocytes, respectively, in the HDL2 patient

In order to determine whether the NA-specific changes in band 3 structure are specific for the acanthocytes or for all erythrocytes of a NA patient, we separated the erythrocytes from an HDL2 patient into different populations using a combination of volume (counterflow elutriation) and subsequent density (Percoll gradient) centrifugation, that has been developed for the isolation of erythrocytes of various ages [7]. The acanthocytes were concentrated in the fraction which, in control donors, consists of the oldest, most dense and smallest erythrocytes (fraction V in [7]). In the samples from the HDL2 patient, this fraction consisted of 95% acanthocytes, whereas the other fractions consisted for maximally 15% erythrocytes with an echinocytic/acanthocytic morphology. This is in agreement with previous analyses showing that acanthocytes from a patient with chorea-acanthocytosis were concentrated in the high-density layers of density gradients [8]. Immunoblot analysis showed an aging-related increase in band 3 degradation in control erythrocytes (Fig. 2), as reported before [7]. The NA-related changes observed in the total erythrocyte populations were observed in all fractions, but especially in the fraction that contains hardly any acanthocytes and that, in control donors, contains the youngest erythrocytes (fraction I in Fig. 2). These data suggest that NA-related alterations in band 3 structure are not associated with the presence of acanthocytic morphology per se. We hypothesize that these alterations may reflect a NA-related imbalance in the erythrocyte membrane stability that, during the lifespan of the erythrocyte, leads to the appearance of acanthocytes.

The red cell membrane stability depends on the maintenance of protein bridges connecting the erythrocyte membrane with the spectrin-based cytoskeleton, which are mainly established by band 3 and ankyrin in the band 3-ankyrin-band 4.2 complex. Additional anchoring sites are organized in the Rh-RhAG-ankyrin complex, and the glycophorin C-band 4.1-p55 complex [2, 18]. The red cell membrane tyrosine phosphorylation pattern is significantly different in ChAc red cells when compared to normal controls [19] (Fig. 3). Using a proteomic approach, we excised the bands in twin gels stained with colloidal Coomassie Blue. We identified some of the proteins that were differently phosphorylated (bands 1 to 4 in Fig. 3) by mass spectrometric analyses (MALDI-TOF). We confirmed that band 3 tyrosine phosphorylation is increased in ChAc compared to normal controls. We also showed that the tyrosine phosphorylation of protein 55 kDa, β actin and GAPDH is higher in ChAc than in control erythrocytes (Fig. 3). We also observed differences in the red cell mem-

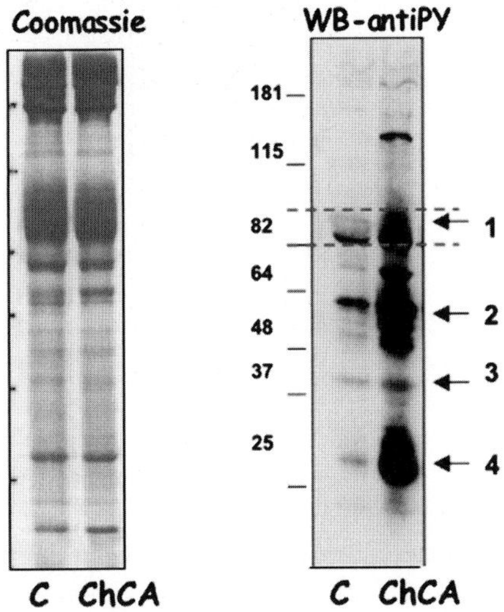

Bands	Accession No.	Protein description	Matching peptides	Coverage (%)
1	gi14277739	Cytoplasmatic domain band 3 (B3)	11	33
2	gi105150	55kDa erythrocyte membrane protein	6	18
3	gi14250401	Actin beta	10	34
4	P04406	Glyceraldehyde-3-P-dehydrogenase	9	36

Fig. 3 The tyrosine-phosphorylation (P-Tyr) membrane protein pattern was evaluated in normal control and in ChAc erythrocytes. The membrane proteins were solubilized and separated by gel electrophoresis. The gels were then transferred to nitrocellulose membrane and subsequently blotted with anti-phosphotyrosine antibodies (WB-anti-PY). In order to identify the proteins showing a different tyrosine phosphorylation state, the corresponding bands in twin gels were excised and trypsinized before undergoing MALDI-TOF analysis. The bands identified to date are reported in the table (indicated in the figure by an *arrow* with a corresponding number). To identify the proteins we carried out a database search using the peptide maa volumes against the Swiss-Prot database (*taxa human*) using the Mascot search engine (Matrix Science Ltd, London, UK). A mass accuracy of 0.3 Da and a single missed cleavage allowed for each matching peptide, Here, we show one Western blot analysis and the corresponding colloidal Coomassie-stained gel that are representative for three experiments with similar results

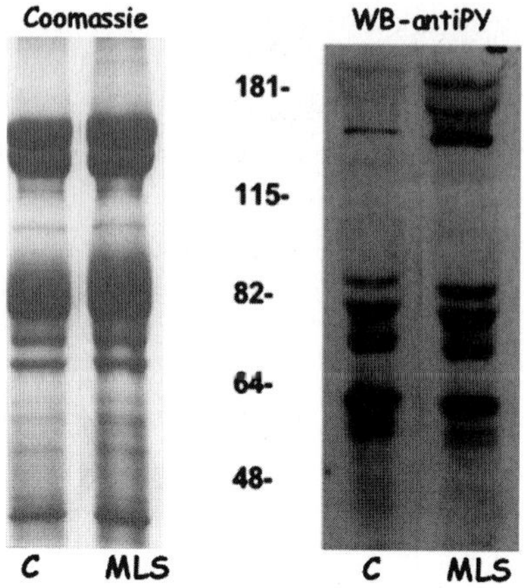

Fig. 4 The tyrosine-phosphorylation (P-Tyr) membrane protein pattern was evaluated in normal control and in MLS erythrocytes as described in the legend to Fig. 3. The bands are now being identified by proteomic analysis. Here, we show one Western blot analysis and the corresponding colloidal Coomassie-stained gel that are representative for four other experiments with similar results

brane tyrosine phosphorylation pattern in MLS erythrocytes compared to normal controls (Fig. 4). In particular, we found increased tyrosine phosphorylation of proteins with molecular weights between 115 and 181 kDa and a band around 82 kDa, without major differences of bands in the lower molecular weight regions. The bands that are differently phosphorylated are now being identified by MALDI-TOF analysis. We have already identified one of the bands as β-spectrin. It is interesting to note that abnormal tyrosine phosphorylation of spectrin is also present in ChAc erythrocytes (Fig. 3), suggesting a perturbation of the functional connections between the red cell membrane and the cytoskeleton in both disorders. Changes in the tyrosine phosphorylation state might alter spectrin stability and thereby the spectrin network and cytoskeleton organization [24]. Recently, a three-dimensional computational study of the equilibrium between shape and deformation in red cells, using spectrin-level energetics, has shown that the spectrin network is constantly remodelled in any red cell shape [14, 15]. These data implicate a crucial role for the membrane-anchoring sites, including the band 3-ankyrin bridges between the erythrocyte membrane and the spectrin-based cytoskeleton, in maintaining optimal cell morphology.

These data further support the hypothesis that the generation of acanthocytes is related to a perturbation of the erythrocyte membrane network, associated with abnormalities in the tyrosine phosphorylation state of various proteins, which may result in an abnormal modulation of either membrane protein-protein or membrane protein–lipid bilayer cross-talk.

4 Membrane-Associated Proteins

Apart from the indications for the association of changes in band 3 conformation with NA, there are no structural data that might help explain the formation of acanthocytes in these patients. The known association of band 3 with other proteins [1, 12, 25], however, leads to the suggestion that the association of cytosolic proteins with the membrane might be specifically altered in the erythrocytes of patients with NA.

Preliminary findings support this suggestion. Membrane proteins were separated by gel electrophoresis, digested with trypsin, and peptide sequences were established using a nano-HPLC system connected to a LTQ-Fourier mass spectrometer. Proteins were identified by searching peak lists containing fragmentation spectra against the NCBI *Homo sapiens* protein sequence data-base, and annotated by function, biological process and/or cellular localization. A first analysis of a proteomic inventory of the membrane fraction of erythrocytes from HDL2 patients shows no obvious changes in the number of membrane and cytoskeleton proteins. However, these data do indicate a NA-associated increase in the membrane association of metabolic enzymes, components of the proteasome, and small G proteins (Fig. 5). These results are in general agreement with the growing awareness, based on comparable proteomic analyses, of the complexity of the erythrocyte proteome, and more particularly of the presence of an active signal transduction network in the erythrocyte membrane regulating erythrocyte homeostasis [20]. The increase in the number of various small G proteins in the patients' erythrocytes is especially intriguing since, in nucleated cells, small G proteins coordinate formation of vesicles, their motility and tethering to their target compartment [30]. In addition, recent data on erythrocytes of mice genetically lacking Rac GTPase show that these proteins are also involved in the dynamic regulation of red cell membrane network, subsequently affecting red cell morphology [11]. Thus, in erythrocytes, these proteins may mediate aging-related membrane restructuring and vesiculation (Bosman et al., unpublished observations). The latter process may be of relevance to acanthocyte formation as well as to neurodegeneration, in view of the postulated role of chorein and junctophilins in vesicle-mediated protein trafficking and/or internal membrane-plasma membrane interactions [26].

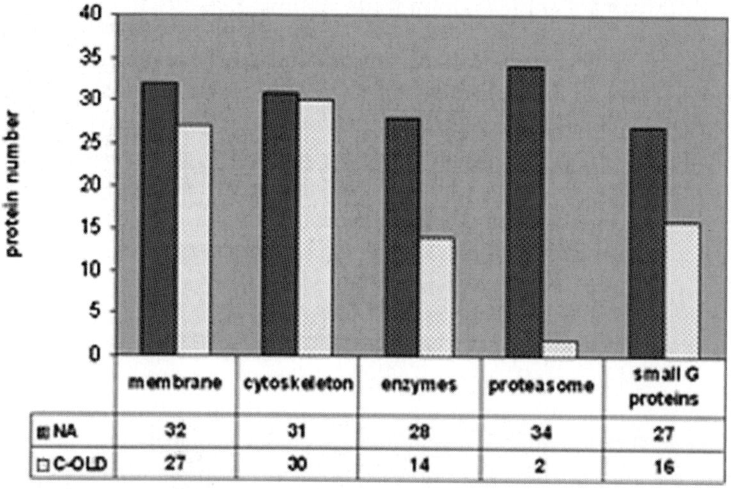

Fig. 5 Proteomic analysis shows an increased membrane recruitment of metabolic enzymes, components of the ubiquitin-proteasome system, and small G proteins in erythrocytes from an HDL2 patient. NA, neuroacanthocytosis; C-OLD, old erythrocytes (fraction V; see also the legend to Fig. 2) from a control donor

5 Vesiculation

We have demonstrated that spleen-mediated vesiculation is an integral part of the physiological aging process, and we postulate that vesicle-mediated removal of damaged hemoglobin occurs mostly in the liver during the second half of the erythrocyte lifespan [27, 28]. We have recently obtained data suggesting that vesicles may also serve to remove damaged membrane components, and that alterations in band 3 structure are involved in the vesiculation process (Bosman et al., unpublished observations). The mechanism by which vesicles are formed is largely unknown, but it is known that an artificial increase in the intracellular calcium concentration results in the formation of various types of vesicles [22]. In a series of pilot experiments we observed that upon incubation with a calcium ionophore in the presence of low concentrations of Ca^{2+}, erythrocytes from various NA patients produced both microvesicles and nanovesicles that were qualitatively and quantitatively different from the vesicles that were produced by erythrocytes from control donors (Bosman and Salzer, unpublished observations). This conclusion, although mostly based on preliminary findings, constitutes another indication for a disease-related alteration in the membrane organization of erythrocytes of NA patients.

6 Conclusions

Based on the recent developments from the study of erythrocytes of various NA patients, mainly communicated in the symposia on NA in Seeon [9], Montreal [6, 26] and Kyoto (this volume), we draw the following conclusions: (1) NA is associated with changes in the structure of erythrocyte band 3, probably by NA-related changes in membrane organization; (2) these changes affect membrane bridges and the anchoring sites between the erythrocyte membrane and the spectrin-based-cytoskeleton; (3) these changes affect the recruitment of cytosolic proteins to the plasma membrane, are associated with altered signal transduction pathways, and probably determine the propensity for vesiculation in vivo. Especially the latter conclusion may not be restricted to erythrocytes, but could also apply to neurons. In neurons, altered vesiculation is likely to affect primarily intracellular trafficking, and will have pronounced effects on neuronal physiology and function.

Formation of acanthocytes is likely to precede the formation and release of vesicles. Expansion of the outer monolayer by putative NA-related changes in (phospho)lipid metabolism could, according to the bilayer-couple hypothesis, result in acanthocyte formation. However, recent modeling of membrane shedding suggests that a decrease of anchorage of the lipid bilayer, resulting in contraction and stiffening of the cytoskeleton, may by itself induce buckling and lead to vesiculation [23]. In view of the current knowledge summarized here, we propose that abnormalities in the interaction between band 3 and ankyrin, and between band 3 and protein complexes that anchor the membrane lipid bilayer to the cytoskeleton, are the main cause for the appearance of acanthocytes in patients with NA. It has already been proposed that alterations in band 3 that change the ratio between the inward-facing and outward-facing conformations of band 3 – determined by the Donnan equilibrium of anions and protons – contract and relax the cytoskeleton, which would make band 3 the sole determinant of erythrocyte shape [29].

In addition, we speculate that the disease-related mutations in *VPS13A, XK*, and *JPH3* genes, that all code for membrane-related proteins, may play a role especially in the erythroid maturation process. This would result in the appearance in the circulation of erythrocytes with an unstable membrane structure. Periods of stress, such as inflammation, may induce acanthocyte formation either by a direct effect on the erythrocyte membrane, or indirectly by disturbing the balance between erythrocyte formation and removal. This hypothesis is supported by the observation that (1) the number of acanthocytes may vary over time in individual patients, and varies between patients with the same genotype; (2) that splenomegaly, which is a sign of disturbed erythrocyte homeostasis and vesiculation, is sometimes – but not invariably – observed in NA patients.

In conclusion, these data provide a concrete foundation for further research on the mechanism of acanthocyte formation. Elucidation of this mechanism may pave the way for intervention in the neurodegeneration process.

References

1. Alper SL (2006) Molecular physiology of SLC4 anion exchangers. Exp Physiol 91:153–161
2. Anong WA, Weis TL, Low PS (2006) Rate of rupture and reattachment of the band 3-ankyrin bridge on the human erythrocyte membrane. J Biol Chem 281:22360–22366
3. Bordin L, Brunati AM, Donella-Deana A, Baggio B, Toninello A, Clari G (2002) Band 3 is an anchor protein and target for SHP-2 tyrosine phosphatase in human erythrocytes. Blood 100:276–282
4. Bosman GJCGM, Albers-Reijnen A, Renkawek K, De Grip WJ (1997) Anion exchange proteins and regulation of intracellular pH in cultured rat astrocytes and neurons. Neuroreport 8:427–430
5. Bosman GJCGM, Bartholomeus IGP, De Grip WJ, Horstink MWIM (1994) Erythrocyte anion transporter and antibrain immunoreactivity in chorea-acanthocytosis. A contribution to etiology, genetics and diagnosis. Brain Res Bull 33:523–528
6. Bosman GJCGM, Walker RH (2005) Acanthocytosis-related changes in erythrocyte band 3: clues for a mechanism and inspiration for future research. Mov Disord 20:1675–1676
7. Bosman GJCGM, Willekens FLA, Werre JM (2005) Erythrocyte aging: a more than superficial resemblance to apoptosis? Cell Physiol Biochem 16:1–8
8. Clark MR, Aminoff MJ, Chiu DT, Kuypers FA, Friend DS (1989) Red cell deformability and lipid composition in two forms of acanthocytosis: enrichment of acanthocytic populations by density gradient centrifugation. J Lab Clin Med 113:469–481
9. Danek A (2006) Neuroacanthocytosis syndromes. Springer, The Netherlands
10. Havenga MJE, Bosman GJCGM, Appelhans H, De Grip WJ (1995) Expression of the anion exchanger (AE) gene family in human brain. Identification of a new AE protein: AE0. Mol Brain Res 25:97–104
11. Kalfa TA, Pushkaran S, Mohandas N, Harting JH, Fowler VM, Johnson JF, Joiner CH, Williams DA, Zheng Y (2006) Rac GTPases regulate the morphology and deformability of the erythrocyte cytoskeleton. Blood 108:3637–3645
12. Kay MMB (2005) Immunoregulation of cellular life span. Ann N Y Acad Sci 1057:85–111

13. Kay MMB, Bosman GJCGM, Lawrence C (1988) Functional topography of band 3: specific structural alteration linked to functional aberration in human erythrocytes. Proc Natl Acad Sci U S A 85:492–496
14. Li J, Dao M, Lim CT, Suresh S (2005) Spectrin level modelling of the cytoskeleton and optical tweezers stretching of the erythrocyte. Biophys J 88:3707–3719
15. Lin LC, Brown FL (2005) Dynamic simulations of membranes with cytoskeleton interactions. Phys Rev E Stat Nonlin Soft Matter Phys 72:011910 Epub
16. Malozzi C, De Franceschi L, Brugnara C, Di Stasi AMM (2005) Protein phosphatase 1A is tyrosine phosphorylated and inactivated by peroxynitrite in erythrocyte through the *src* family kinase *fgr*. Free Radic Biol Med 38:1625–1636
17. Mallozzi C, Di Stasi AMM, Minetti M (1997) Peroxynitrite modulates tyrosine-dependent signal transduction pathway of human erythrocyte band 3. FASEB J 11:1281–1290
18. Nicolas V, Mouro-Chateloup I, Lopez C, Gane P, Gimm A, Mohandas N, Cartron JP, Le Van Kim C, Colin Y (2006) Functional interaction between Rh proteins and spectrin-based skeleton in erythroid and epithelial cells. Transfus Clin Biol 13:23–28
19. Olivieri O, De Franceschi L, Bordin L, Manfredi M, Miraglia de Giudice E, Perrotta S, De Vito M, Guarini P, Corrocher R (1997) Increased membrane protein phosphorylation and anion transport activity in chorea-acanthocytosis. Haematologica 82:648–653
20. Pasini EM, Kirkegaard M, Mortensen P, Lutz HU, Thomas AW, Mann M (2006) In-depth analysis of the membrane and cytosolic proteome of red blood cells. Blood 108:791–801
21. Perrotta S, Borriello A, Scaloni A, De Franceschi L, Brunati AM, Turrini F, Nigro V, Miraglia del Giudice E, Nobili B, Conte ML, Rossi F, Iolascon A, Donella-Deana A, Zappia V, Poggi V, Anong W, Low P, Narla M, Della Ragione F (2005) The N-terminal 11 amino acids of human erythrocyte band 3 are critical for aldolase binding and protein phosphorylation: implications for band 3 function. Blood 106:4359–4366
22. Salzer U, Hinterdorfer P, Hunger U, Borken C, Prohaska R (2002) Ca(++)-dependent vesicle release from erythrocytes involves stomatin-specific lipid rafts, synexin (annexin VII), and sorcin. Blood 99:2569–2577
23. Sens P, Gov N (2007) Force balance and membrane shedding at the red-blood-cell surface. Phys Rev Lett 98:1–4
24. Tang Hy, Speicher DW (2004) In vivo phosphorylation of human erythrocyte spectrin occurs in a sequential manner. Biochemistry 43:4251–4262
25. Tanner MJ (2002) Band 3 anion exchanger and its involvement in erythrocyte and kidney disorders. Curr Opin Hematol 9:133–139
26. Walker RH, Danek A, Dobson-Stone C et al. (2006) Developments in neuroacanthocytosis: expanding the spectrum of choreatic syndromes. Mov Disord 21:1794–1805
27. Willekens FLA, Roerdinkholder-Stoelwinder B, Groenen-Döpp YAM, Bos HJ, Bosman GJCGM, Van den Bos AG, Verkleij AJ, Werre JM (2003) Hemoglobin loss from erythrocytes in vivo results from spleen-facilitated vesiculation. Blood 101:747–751
28. Willekens FLA, Werre J, Kruijt JK, Roerdinkholder-Stoelwinder B, Groenen-Döpp YAM, Van den Bos AG, Bosman GJCGM, Van Berkel TJ (2005) Liver Kupffer cells rapidly remove red blood cell-derived vesicles from the circulation by scavenger receptors. Blood 105:2141–2145
29. Wong P (2004) A basis of the acanthocytosis in inherited and acquired disorders. Med Hypotheses 62:966–969
30. Zerial M, McBride H (2001) Rab proteins as membrane organizers. Nat Rev Mol Cell Biol 2:107–117

McLeod Syndrome: A Perspective from Japanese Blood Centers

Y. Tani(✉), J. Takahashi, M. Tanaka, and H. Shibata

Abstract McLeod phenotype is one of the rare blood cell types, defined as those that occur at a frequency of 1:1,000 or less. We have screened donors for rare cells since 1987 using mouse monoclonal antibodies (MAbs). In the western regions of Japan a total of 16,160,714 donor red cells were screened using mouse anti-Kell MAbs (anti-k, anti-Ku and anti-K14) during 1987–2005. We found 182 blood cells with McLeod phenotype or Kmod and 286 with Ko. To identify McLeod phenotype, we examine the expression level of Kell antigens by flow cytometry, cell morphology (acanthocytes) by scanning electron microscopy, red cells of family members and the *XK* gene. As a result, we identified three donors with McLeod phenotype (two subsequently developed McLeod syndrome) and two with McLeod-like phenotype. Nine donors are now registered as McLeod phenotype in all Japan.

Additionally, we have collaborated with several medical institutes and have identified five cases of McLeod syndrome, four of chronic granulomatous disease (CGD) with McLeod phenotype, ten of chorea-acanthocytosis (ChAc) and seven of chorea (without acanthocytosis). Two patients with CGD and McLeod phenotype underwent hematopoietic stem cell transplantation, and one survives with improvement of his immunity.

1 Blood Donors with McLeod Phenotype

More than 300 antigens are recognized on the red cell surface and can be classified into 4 separate categories; these are 29 genetically determined systems, 6 collections of related antigens, and the series of lo w incidence (700 series) and high incidence

Y. Tani
Japanese Red Cross Osaka Blood Center, 2-4-43, Morinomiya Joto-ku, Osaka, 536-8505, Japan, tani@osaka.bc.jrc.or.jp

(901 series) antigens [2]. Among the 29 systems (Table 1), red cells which lack XK protein on the cell surface are called McLeod phenotype. McLeod phenotype is one of the rare blood cell types, defined as those that occur at a frequency of 1:1,000 or less, and was first reported by Allen et al. in 1961 [1].

In Japan, we have screened donors for rare cell types since 1987 using monoclonal antibodies (MAbs) which we have developed. Our screening strategy for McLeod phenotype is shown in Fig. 1. We do not have good MAbs against XK protein, but Kell glycoprotein is covalently linked at Cys 72 to Cys 347 of the XK protein in the red cell membrane (Fig. 2) [8] and it is well known that Kell antigen expression is markedly reduced in the absence of normal XK protein. We screen red cells by agglutination methods in an automated system such as Olympus PK7200

Table 1 The blood group systems, the genes that encode them, and their chromosomal location (ISBT, 2004) [2]

No.	System name	System symbol	Gene name(s)	Chromosomal location	CD numbers
001	ABO	ABO	*ABO*	9q34.2	
002	MNS	MNS	*GYPA, GYPB, GYPE*	4q31.21	CD235
003	P	P1		22q11.2-qter	
004	Rh	RH	*RHD, RHCE*	1p36.11	CD240
005	Lutheran	LU	*LU*	19q13.32	CD239
006	Kell	KEL	*KEL*	7q34	CD238
007	Lewis	LE	*FUT3*	19p13.3	
008	Duffy	FY	*FY*	1q23.2	CD234
009	Kidd	JK	*SLC14A1*	18q12.3	
010	Diego	DI	*SLC4A1*	17q21.31	CD233
011	Yt	YT	*ACHE*	7q22.1	
012	Xg	XG	*XG, MIC2*	Xp22.32, Yp11.3	CD99b
013	Scianna	SC	*ERMAP*	1p34.2	
014	Dombrock	DO	*DO*	12p12.3	
015	Colton	CO	*AQP1*	7p14.3	
016	Landsteiner–Wiener	LW	*ICAM4*	19p13.2	CD242
017	Chido/Rodgers	CH/RG	*C4A, C4B*	6p21.3	
018	Hh	H	*FUT1*	19q13.33	CD173
019	Kx	XK	*XK*	Xp21.1	
020	Gerbich	GE	*GYPC*	2q14.3	CD236
021	Cromer	CROM	*DAF*	1q32.2	CD55
022	Knops	KN	*CR1*	1q32.2	CD35
023	Indian	IN	*CD44*	11p13	CD44
024	Ok	OK	*BSG*	19p13.3	CD147
025	Raph	RAPH	*CD151*	11p15.5	CD151
026	John Milton Hagen	JMH	*SEMA7A*	15q24.1	CD108
027	I	I	*GCNT2*	6p24.2	
028	Globoside	GLOB	*B3GALT3*	3q26.1	
029	GIL	GIL	*AQP3*	9p13.3	

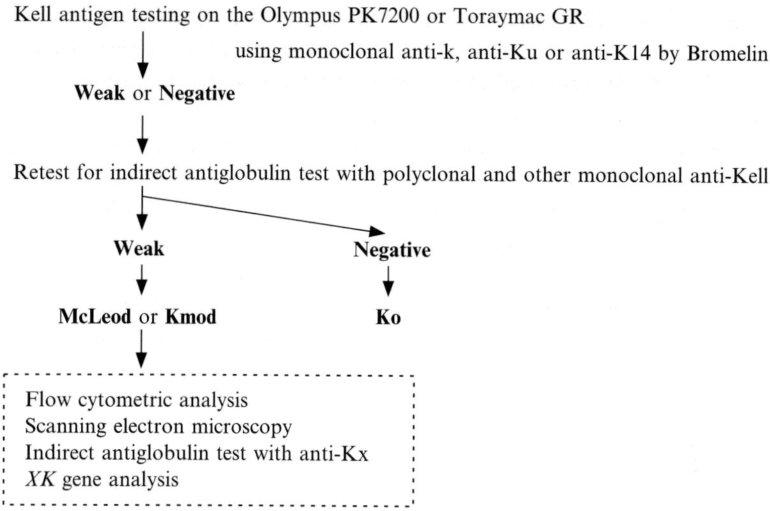

Fig. 1 Flow chart of screening of red cells for McLeod phenotype

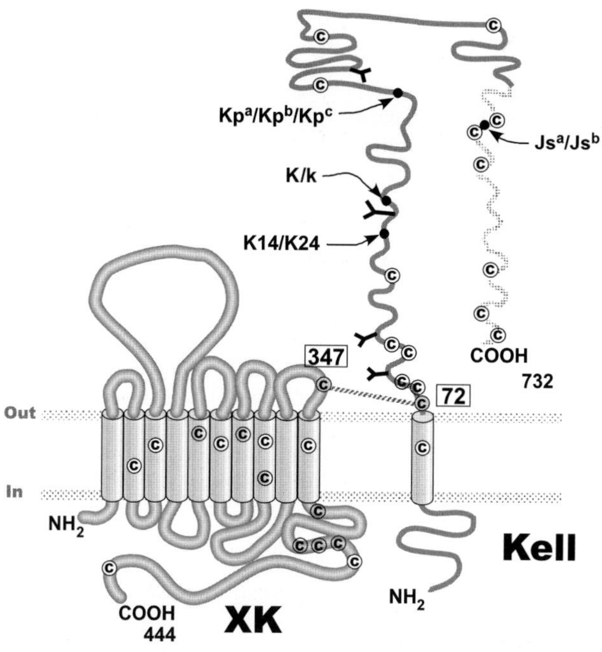

Fig. 2 Kell/XK complex (modified from Lee et al. [8]). Kell antigen is 93 kDa type II glycoprotein consisted of 732 amino acid polypeptides and XK is 37 kDa, 444 amino acid polypeptides. They are covalently linked by a disulphide bond. (C: cysteine)

or Toraymac GR using anti-Kell MAbs (anti-k, anti-Ku and anti-K14 etc.). In the western regions of Japan from 1987 to 2005, a total of 16,160,714 donations of red blood cells (including repeat donors) were screened using mouse monoclonal anti-k (OSK5), anti-Ku (OSK32) and anti-K14 (OSK25).

We found 182 cell donations with McLeod phenotype or Kmod and 286 with Ko. Kmod red cells express Kell antigens weakly, and Ko express no Kell antigens. XK protein expression is intact or even increased in these two phenotypes as opposed to the McLeod phenotype. To identify McLeod phenotype after screening, we examine the expression level of Kell antigens by flow cytometry (Fig. 3); cell morphology (acanthocytes) by scanning electron microscopy (Fig. 4); serum creatine phosphokinase (CPK) and haptoglobin (Hp); red cells of family members (especially mothers), and the *XK* gene.

Although we have not examined all 182 donors, we have identified 3 donors with McLeod phenotype (2 subsequently developed McLeod syndrome [13, 16], 2 with McLeod-like phenotype (McLeod-like red cells show the same serological reactions and acanthocytes as McLeod except they weakly react with anti-Kx) [10], and more than 50 with Kmod. Nine donors are currently registered as McLeod phenotype in the whole of Japan [3, 11, 14]. Their CPK is usually high and Hp usually low (Table 2). Of note, one donor (M-1) developed McLeod syndrome 15 years after we identified him, although his CPK was within the normal range at initial screening.

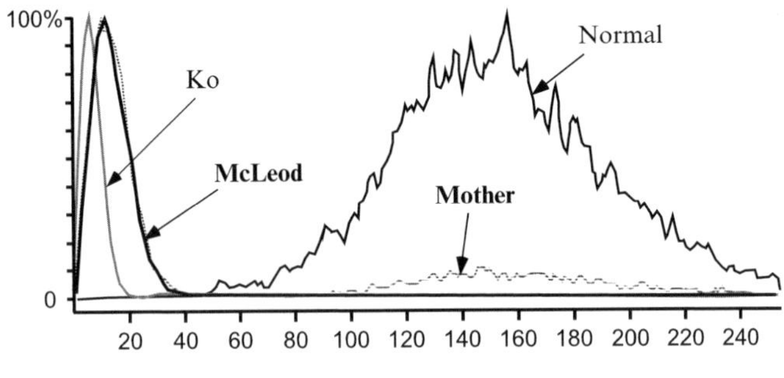

Red cells	Fluorescence/cell	
Ko	4.5	
Normal	156.0	
McLeod	12.9	
Mother	55.6	two peaks

Fig. 3 Kell antigen expression in McLeod male, female carrier (his mother) and Ko donors. Flow cytometry using anti-Kp^b (OSK36) shows that Kell antigens are weak on red blood cells from the McLeod male and that the female carrier has a mixed population (two peaks) because *XK* is subject to X-chromosome inactivation

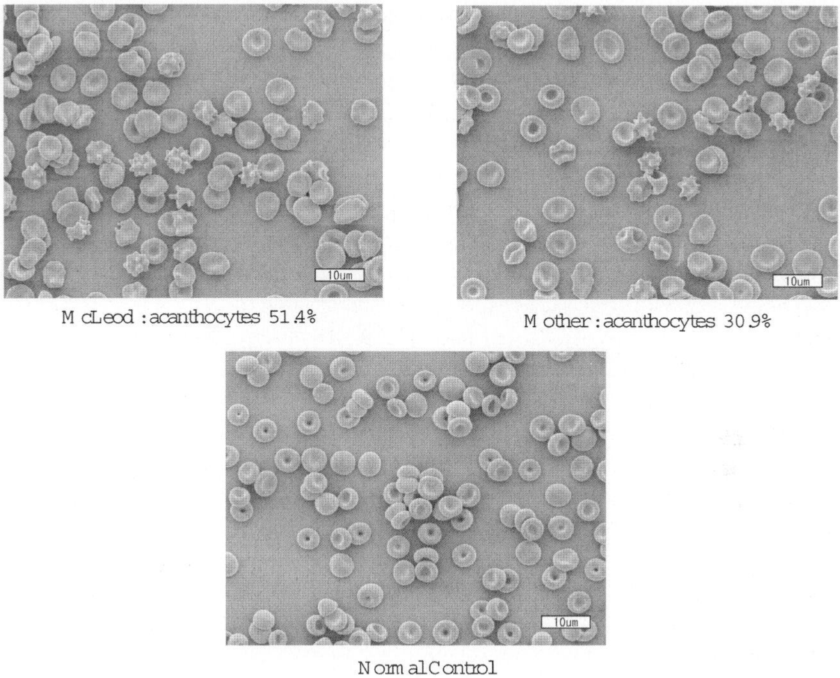

Fig. 4 Scanning electron micrographs of acanthocytes in McLeod male and female carrier (mother). A lower percentage of acanthocytes is observed in the female carrier (JSM-5410, ×1,500)

Table 2 CPK and haptoglobin in McLeod phenotype donors

	Donor No.	Age detected (report)	CPK (U/L)	Hp (mg/dL)	References
McLeod	M-1[a]	21 (1989)	56	n.t.	[13, 15]
	M-2	45 (1994)	921	<10	[3]
	M-3	31 (1995)	683	<10	[11]
	M-4	29 (1995)	692	51	[14]
	M-5	34 (2000)	1,937	<10	[14]
	M-6	20 (2000)	282	152	[14]
	M-7	27 (2000)	439	<10	[14]
	M-8	? (2000)	862	<10	[14]
McLeod-like[b]	M-9	21 (1989)	382	<11.5	[10]
	Normal		57–197	41–341	

CPK creatine phosphokinase, *Hp* haptoglobin

[a] Developed McLeod syndrome at 36 years old

[b] McLeod-like: the same laboratory findings as McLeod apart from weak reaction with anti-Kx

Table 3 Differences among McLeod syndrome, chorea-acanthocytosis and non-acanthocytic chorea

	Antigen expression		RBC	Possible antibodies	
	Kell	Kx[a]	Morphology	Produced	CPK
Normal Kell	N	N	N	N	N
Ko (Kell null)	(–)	Increased	N or elliptocyte	Anti-Ku	N
Kmod	Reduced	Increased	N	Anti-Ku-like	N
McLeod syndrome	Reduced	(–)	Acanthocyte	Anti-Kx, Km	N or High
McLeod carriers	Reduced to N[b]	(–) to N	Acanthocyte	N	?
Chorea-acanthocytosis	N	N	Acanthocyte	N	High
Chorea (non-acanthocytic)	N	N	N	N	High

CPK creatine phosphokinase, *N* normal
[a] Anti-Kx + Km kindly gifted by Marion Reid, New York Blood Center
[b] Flow cytometry: two peaks

2 Differential Diagnosis of Chorea

We receive requests from neurologists for expression levels of Kell antigens and for cell morphology (acanthocytes) as part of the work-up of patients with chorea. Table 3 summarizes how we diagnose McLeod syndrome or ChAc from a blood sample. Kell antigens are normally expressed in patients with ChAc and or other forms of chorea, but are reduced in those with McLeod syndrome. Acanthocytes are observed in patients with McLeod syndrome and ChAc. CPK is usually high in all. To date, blood samples of 22 patients from 13 medical institutes have been examined. We have identified five cases of McLeod syndrome [4, 5, 7] and ten of ChAc [5, 6] (by exclusion of McLeod syndrome). Seven had non-acanthocytic chorea.

3 Chronic Granulomatous Disease with McLeod Phenotype

The *CYBB* (cytochrome *b* 245 oxidase) gene is responsible for CGD and this gene is located close to *XK* gene on chromosome Xp21 locus (Fig. 5). It is very important to know the range of the genetic defect in CGD patients, in case they also have McLeod phenotype and require blood transfusion. We have identified four cases of CGD with McLeod phenotype [15, 17], and two of them received cord blood transplantation to improve their immunity. One unfortunately died of subarachnoid bleeding but the other was successfully transplanted and survives [9, 12]. As hematopoietic stem cells have the potential to differentiate into nerve and muscle cells, our interest is whether this patient will develop McLeod syndrome in the future or not. If not, hematopoietic stem cell transplantation might be a useful therapy for McLeod males to prevent McLeod syndrome.

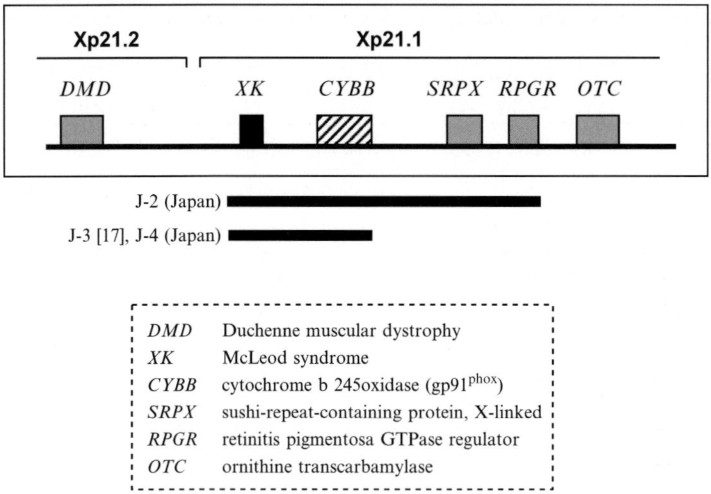

Fig. 5 Gene map of *XK* and *CYBB* genes on Xp21 locus. Four cases of deletions of both *XK* and *CYBB* genes were reported in Japanese CGD patients with McLeod phenotype. *XK*, *CYBB*, *SRPX* and *RPGR* genes were deleted in one case (J-2) and *XK* and *CYBB* genes were deleted in two cases (J-3 [17], J-4). Genetic analysis was not done in one case (J-1)

Acknowledgements We thank Taiko Seno, MT, for comments and suggestions; Yoko Tomonari, MT of Japanese Red Cross Fukuoka Blood Center for providing blood samples; and Drs. Eitou Boku, Naoko Hanaoka, Fuminari Ishibashi, Shouzou Ishikawa, Minami Yoda, Yukio Sakiyama, Norinao Hanyu, Nobuhiro Suzuki, Yoshihisa Takiyawa, Tsutomo Oishi, Shuhei Takagi, Taro Yamashita, Yasufumi Kageyama, Hiroki Takeuchi, and Masami Sada for collaboration.

References

1. Allen FH, Krabbe SM, Corcon PA (1961) A new phenotype (McLeod) in the Kell blood-group system. Vox Sang 6:555–560
2. Daniels GL, Fletcher A, Garratty G, Henry S, Jorgensen J, Judd WJ, Levene C, Lomas-Francis C, Moulds JJ, Moulds JM, Moulds M, Overbeeke M, Reid ME, Rouger P, Scott M, Sistonen P, Smart E, Tani Y, Wendel S, Zelinski T, International Society of Blood Transfusion (2004) Blood group terminology 2004: from the International Society of Blood Transfusion committee on terminology for red cell surface antigens. Vox Sang 87:304–316
3. Enomoto T, Omura K, Hando K, Hattori T, Obara T, Ishijima A, Kozaki N, Hatakeyama H, Tsunetama H, Uchikawa M (1994) A case of McLeod phenotype detected in Japanese blood donors [in Japanese]. Jpn J Transfus Med 40:361
4. Hanaoka N, Hanaoka N, Yoshida K, Nakamura A, Furihata K, Seo T, Tani Y, Takahashi J, Ikeda S, Hanyu N (1999) A novel frameshift mutation in the McLeod syndrome gene in a Japanese family. J Neurol Sci 165:6–9
5. Ishikawa S, Tachibana N, Tabata KI, Fujimori N, Hayashi RI, Takahashi J, Ikeda SI, Hanyu N (2000) Muscle CT scan findings in McLeod syndrome and chorea-acanthocytosis. Muscle Nerve 23:1113–1116

6. Kageyama Y, Kodama Y, Tadano M, Yamamoto S, Ichikawa K (2000) A case of chorea-acanthocytosis with dilated cardiomyopathy and myopathy. Clin Neurol 40:816–820
7. Kawakami T, Takiyama Y, Sakoe K, Ogawa T, Yoshioka T, Nishizawa M, Reid ME, Kobayashi O, Nonaka I, Nakano I (1999) A case of McLeod syndrome with unusually severe myopathy. J Neurol Sci 166:36–39
8. Lee S, Russo D, Redman C (2000) Functional and structural aspects of the Kell Blood group system. Transfus Med Rev 14:93–103
9. Suzuki N, Hatakeyama N, Yamamoto M, Mizue N, Kuroiwa Y, Yod M, Takahashi J, Tani Y, Tsutsumia H (2007) Treatment of McLeod phenotype chronic granulomatous disease with reduced-intensity conditioning and unrelated-donor umbilical cord blood transplantation. Int J Hematol 85:70–72
10. Takahashi J, Fuchihara T, Horie H, Ishii H, Yoshimura K, Toyota Y, Seno T, Yamano T, Nagao N, Okubo Y, Yamaguchi H (1989) A family of McLeod phenotype detected by monoclonal anti-K2 [in Japanese]. Jpn J Transfus Med 35:206
11. Takahashi Y, Tsutida N, Konishi Y, Karino I, Ozaki E, Kagami T, Tsutida H, Kojima I (1995) Two Japanese blood donors with McLeod phenotype identified [in Japanese]. Jpn J Transfus Med 41:81
12. Tani Y, Takahashi J, Tanaka M, Hirayama F, Hatakeyama N, Yamamoto M, Kuroiwa Y, Mizue N, Yoda M, Sakiyama Y, Suzuki N, Shibata H (2006) A case of a patient with McLeod phenotype and CGD who did not produce anti-Kx regardless of packed red cells transfusions from random donors during nonmyeloablative cord blood transplantation. Vox Sang 91:165(P-341)
13. Tomonari Y, Toyota T, Watanabe S, Watanabe Y, Eto K, Yoshitake N, Yoshimura T, Miyanari Y, Taguchi T, Kodama T, Seno T, Nagao N, Okubo Y, Yamaguchi H (1989) Two blood donors of McLeod phenotype found in blood donors [in Japanese]. J Jpn Blood Programme 12:205
14. Tsutida H, Fujisawa N, Ozaki E, Takahashi Y, Kagami T, Mito S, Murata G, Tsuneyama H, Uchikawa M, Nakajima I (2000) A donor red cell considered as McLeod phenotype [in Japanese]. Jpn J Transfus Med 46:25
15. Uchida K, Nakajima K, Shima H, Ichiba K, Seno T, Okubo Y, Yamaguchi H (1992) The first example of the McLeod phenotype in a Japanese baby with chronic granulomatous disease. Transfusion 32:691
16. Ueyama H, Kumamoto T, Nagao S, Masuda T, Sugihara R, Fujimoto S, Tsuda T (2000) A novel mutation of the McLeod syndrome gene in a Japanese family. J Neurol Sci 176:151–154
17. Yoda M, Yoshida K, Ohyanagi N, Ohsaki M, Yoshimura H, Hatano N, Ariga T, Sakiyama Y (2006) A case of X-linked chronic granulomatous disease associated with McLeod syndrome. J Jpn Pediatr Soc 110:38–41

Part III
Animal Models

A Mouse Model of Chorea-Acanthocytosis

M. Nakamura(✉), Y. Katoh, K. Yutaka, Y. Kurano, M. Ichiba, M. Matsuda, M. Katoh, S. Ueno, and A. Sano

Abstract We have recently generated a model mouse of chorea-acanthocytosis (ChAc), which carries a targeted deletion mutation in the mouse *VPS13A* gene corresponding to a human disease mutation. Although these model mice exhibited a normal life span, they begin to display a disease phenotype with red cell acanthocytosis and behavioral abnormalities in old age. Neuropathological examination reveals prominent apoptotic neuronal cells and astrogliosis in the striatum. These are thought to be very similar to the features of the disease course and neuropathology seen in ChAc patients. Thus, this model mouse may be useful for pathogenic studies related to ChAc. Here, we review the ChAc-model mouse and recent findings on the upregulation of gephyrin and its related proteins in ChAc-model mouse striatum.

M. Nakamura
Department of Psychiatry, Kagoshima University Graduate School of Medical and Dental Sciences, 8-35-1 Sakuragaoka, Kagoshima 890-8520, Japan
nakamu36@m.kufm.kagoshima-u.ac.jp

R.H. Walker et al. (eds.), *Neuroacanthocytosis Syndromes II.*

1 Introduction

Chorea-acanthocytosis (ChAc) is a rare autosomal recessive neurodegenerative disorder characterized by the gradual onset in adulthood of hyperkinetic movements with peripheral blood red cell acanthocytosis. Neuropathologically, neuronal loss and astrogliosis were observed in the striatum, and variably in the substantia nigra [4]. In 2001, we and another group independently identified the *CHAC* gene (now renamed *VPS13A*) mutations of which are responsible for ChAc [8, 11]. *VPS13A* is located on human chromosome 9q21 spanning 250 kb region and consists of 73 exons. The exon–intron organization of the mouse *VPS13A* gene is identical to the human *VPS13A* [10]. The *VPS13A* gene encodes a protein named chorein, which has a predicted molecular mass of approximately 360 kDa. Immunoblot studies showed that chorein was highly expressed in testis, kidney, spleen and various brain regions [5]. The null mutant of the *Saccharomyces cerevisiae* homologue *VPS13* is viable but shows defects in vacuolar protein sorting and a *TipC* gene mutant of *Dictyostelium discoideum* has aberrant cell-sorting behavior [2, 9], suggesting that chorein may play a role in the cytoskeleton and cell transport. However, the function of chorein is still unclear. In order to understand the molecular pathogenesis of ChAc, we have used gene-targeting techniques to generate a ChAc-model mouse which carries the ChAc-causative Ehime deletion mutation, found in some Japanese ChAc patients [10].

2 Generation of the ChAc-Model Mouse

The constructs used for the *VPS13A* mutation in this mouse model are shown in Fig. 1a [6]. We assembled a targeting vector to replace from the middle of intron 59 to the middle of intron 61 by a neomycin-resistance cassette. Electroporation of the targeting vector into embryonic stem (ES) cells (CCE, 129/Sv background), selection of ES cells containing the properly targeted gene, generation of 129/Sv/C57BL/6J-chimeric mice, and germline transmission for the generation of heterozygous mutant F1 mice in a 129/Sv/C57BL/6J background were carried out. A fragment including exons 60 and 61, which is deleted in human ChAc-causing Ehime-deletion mutation, was replaced with the neomycin resistance gene in ChAc-model mouse. This change is predicted to introduce a premature stop codon leading to the production of a truncated protein product (approximately 310 kDa). Subsequent immunoblot analysis using anti-chorein antibody showed little expression of chorein in the *VPS13A* mutant mouse, confirming that the deletion in *VPS13A* led to the production of truncated proteins (Fig. 1b) [5]. This implies the degradation of the mutant *VPS13A* mRNA before the translation.

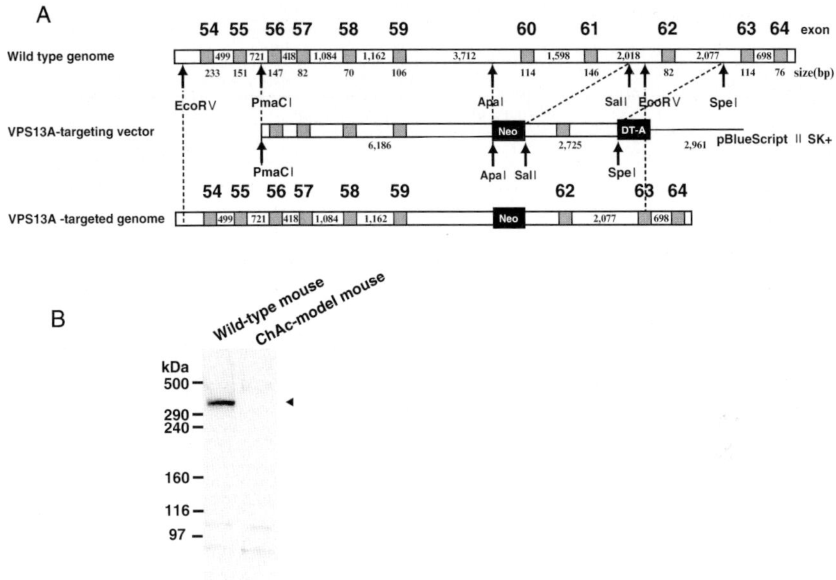

Fig. 1 (**A**) Construction of the ChAc-model mouse [10]. Homologous recombination resulted in replacement of the 3.7 kb fragment including exons 60 and 61 with the neomycin resistance gene (Neo). The diphtheria toxin gene (DT-A) was attached to the 3′ end of targeting vector for negative selection. Restriction sites are indicated by *arrows*. (**B**) Detection of endogenous chorein [5]. Anti-chorein immunoblot analysis of whole brain homogenates from a wild-type, and a ChAc-model mouse carrying homozygous deletions. The size of the band seen in Wild/Wild is consistent with the predicted molecular weight of full-length chorein (360 kDa; *arrowhead*)

3 Features of the Phenotype

Features of the phenotype of the ChAc-model mouse are shown in Table 1 [6].

3.1 Life Span, Body Weight, Age of Onset, and Reproductive Capacity

The ChAc-model mice are viable and exhibit normal life span. There was no significant difference in body weight at the age of 87 weeks. However, at older ages (>70 weeks) they displayed disturbances in motor function and erythrocyte morphological changes, similar to the features of adult onset and gradual progression seen in humans with ChAc [8]. Although as far as we know human male ChAc patients show normal reproductive capacity, our male ChAc-model mice that carried homozygous mutation were totally infertile. Immunoblot analysis demonstrated the most abundant chorein expression in mouse testis, and

Table 1 Phenotype of ChAc-model mouse [6, 10]

Characteristic	Age: <70 weeks	Age: >70 weeks
Wild-type versus ChAc-model		
Reproductive capacity		
Male	Infertile	
Female	Fertile	
Life span		n.s.
Body weight loss	n.s.	n.s.
Hematological test		
Osmotic fragility test		
50% hemolysis (C50)	n.s.	Wild-type < ChAc-model mouse*
Behavioral tests		
Foot print test		
Stride length	n.s.	Wild-type > ChAc-model mouse**
Rotarod test		
Latency to fall	n.s.	Wild-type > ChAc-model mouse*
Open field test		
Moving distance	n.s.	Wild-type > ChAc-model mouse*
Social interaction		
Contact time	n.s.	Wild-type > ChAc-model mouse*
Neurotransmitter analysis		
HVA (mid brain)	–	Wild-type > ChAc-model mouse*
NE, DA, 5-HT, 5-HIAA, and GABA	–	n.s.
Brain pathology		
Weight ratio of striatum/whole brain	–	Wild-type > ChAc-model mouse*
No. of TUNEL-positive cells	–	Wild-type < ChAc-model mouse**
GFAP staining	–	Wild-type < ChAc-model mouse
Others		
Gephyrin		
Immunoblotting		
Striatum	–	Wild-type < ChAc-model mouse**
Hippocampus	–	Wild-type < ChAc-model mouse*
Immunostaining	–	Wild-type < ChAc-model mouse
$GABA_A$ receptor α1 subunit		
Immunoblotting		
Striatum	–	n.s.
Hippocampus	–	n.s.
Immunostaining	–	Wild-type < ChAc-model mouse
$GABA_A$ receptor γ2 subunit		
Immunoblotting		
Striatum	–	Wild-type < ChAc-model mouse*
Hippocampus	–	Wild-type < ChAc-model mouse**
Immunostaining	–	Wild-type < ChAc-model mouse

5-HIAA, 5-hydroxyindoleacetic acid; 5-HT, 5-hydroxytyramine; DA, dopamine; GABA, gamma aminobutyric acid; GFAP, glial fibrillary acidic protein; HVA, homovanillic acid; NE, norepinephrine; n.s., not significant; TUNEL, Tdt-mediated dUTP nick-end labeling

*$P < 0.05$; **$P < 0.01$

immunohistochemically, variable chorein-like immunoreactivities were observed in the stages of differentiation of sperm [5]. These suggest that chorein in mouse testis may have an important role in spermatogenesis.

3.2 Acanthocytosis of ChAc-Model Mouse

Light microscopic observation of peripheral blood smears of ChAc-model mice showed heterogeneity in sizes and shapes of erythrocytes, including acanthocytes. The osmotic fragility test is the most sensitive test available to detect cells that are more sensitive to osmotic stress than normal cells [1]. The red blood cells from ChAc-model mice showed an increase in their in vitro osmotic fragility when exposed to hypotonic NaCl solutions.

3.3 Behavioral Abnormalities of ChAc Model Mouse

Gait abnormality was assessed by analyzing the footprint patterns. The hindpaws were dipped in non-toxic ink. The mouse was then placed at the one end of a dark tunnel. The bottom surface of the tunnel was lined with white paper. The resulting footprint patterns were assessed quantitatively by measuring the stride length. The mean stride length of *VPS13A* mutant mice displayed a significantly shorter length compared with that of control mice.

Motor coordination and balance were measured using a Rotarod. The mean latency to falling off was significantly shorter in ChAc-model mice compared to wild-type mice. Human ChAc patients present with chorea as the major motor symptom, but the model mice showed gait disturbance and early falling from the Rotarod without any involuntary movements. In many kinds of Huntington Disease-model transgenic or knock-in mice, similar discrepancies have often been reported [7]. Phylogenetic differences in the function of the basal ganglia may be part of the reason for disparities in motor functions.

Spontaneous locomotor activity in the open field was measured for 10 min in daytime with a behavioral tracing analyzer. Wild-type and ChAc-model mice showed significant differences between total movement distances in a novel environment.

To evaluate social interactions, the contact time between two male mice was calculated when the mice were in an unfamiliar open field environment. ChAc-model mice showed less contact time and stayed in the edge area of the field for a significantly longer time.

The results of these behavioral tests showed broad standard deviations, indicating differences between individuals. Heterogeneous behavioral phenotypes observed in ChAc-model mice may be caused by a hybrid 129/Sv and C57BL/6J background. This kind of heterogeneous phenotypes in terms of age of onset and symptoms are also observed among human patients with the same mutations [11].

3.4 Neurotransmitter Analysis in the Brain

Monoamines, their metabolites, and gamma amino butyric acid (GABA) were measured from homogenates of brain sections divided into six portions, hippocampus, striatum, cerebral cortex, cerebellum, brain stem and others (midbrain, thalamus and hypothalamus). Only the dopamine metabolite, homovanillic acid (HVA), in the portion including midbrain, showed significantly lower levels in the ChAc-model mouse.

3.5 Neuropathology

When the weight ratio of one region/whole brain was measured, we found a significant difference in the striatum. The ratio in ChAc-model mice was smaller than that in wild-type mice ($P < 0.05$), indicating selective atrophy of striatum. Astrocytic gliosis was detected by anti-glial fibrillary acidic protein (GFAP) antibody in the striatum and the substantia nigra pars reticulata to varying degrees, from mild to severe. Prominent apoptosis was detected in the striatum using the TdT-mediated dUTP nick-end labeling (TUNEL) staining as a marker for cell death. The numbers of (TUNEL)-positive cells in the striatum of mutant mice were significantly more than those of wild-type mice. The numbers of TUNEL-positive cells in the striatum of mutant mice were much greater than those of GFAP-positive astroglial cells, indicating that most of the TUNEL-positive cells are neurons. The pathological findings in the mutant mouse correlate well with those in ChAc patients [4].

3.6 Gephyrin and the $GABA_A$ Receptor $\alpha 1$ and $\gamma 2$ Subunits

A comparative microarray analysis of gene expression in the striatum revealed an increased level of gephyrin gene expression in the ChAc-model mice as compared with wild-type mice [6]. Since gephyrin is known as a $GABA_A$ receptor-anchoring protein [3], we compared the levels of protein expression and the localization of gephyrin and the $GABA_A$ receptor $\alpha 1$ and $\gamma 2$ subunits. Gephyrin and $GABA_A$ receptor $\gamma 2$ subunit immunoreactivities in the striatum and hippocampus of the ChAc-model mice were significantly higher than those in the wild-types. These findings suggest that chorein dysfunction may lead to upregulation of gephyrin and increases in the levels of expression of gephyrin-related proteins.

4 Summary

In summary, we have created a ChAc-model mouse with a deletion mutation in *VPS13A* corresponding to human disease. Similarities between human ChAc and ChAc-model mouse neuropathology, together with the phenotypes of motor disturbances and hematological abnormalities make the ChAc-model mouse an ideal animal model for understanding the molecular pathogenesis of ChAc. The mutant mice which have a hybrid C57BL/6J and 129/Sv genetic background displayed variable phenotypes, strongly suggesting the existence of modifier genes. Molecular analyses using this model mouse, including a microarray analysis and subsequent immunoblot and immunohistochemical analyses, indicate that an increase in the expression of gephyrin and its related proteins in the ChAc-model mice may be related to ChAc molecular pathogenesis.

References

1. Becker PS, Lux SE (1995) Hereditary spherocytosis and hereditary elliptocytosis. In: Scriver CR, Beaudet AL (eds) The metabolic basis of inherited disease. McGraw-Hill, New York, pp 3513–3560
2. Brickner JH Fuller RS (1997) SOI1 encodes a novel, conserved protein that promotes TGN-endosomal cycling of Kex2p and other membrane proteins by modulating the function of two TGN localization signals. J Cell Biol 139:23–36
3. Essrich C, Lorez M, Benson JA, Fritschy JM, Luscher B (1998) Postsynaptic clustering of major GABAA receptor subtypes requires the gamma 2 subunit and gephyrin. Nat Neurosci 1:563–571
4. Hardie RJ, Pullon HW, Harding AE, Owen JS, Pires M, Daniels GL, Imai Y, Misra VP, King RH, Jacobs JM et al. (1991) Neuroacanthocytosis. A clinical, haematological and pathological study of 19 cases. Brain 114(Pt 1A):13–49
5. Kurano Y, Nakamura M, Ichiba M, Matsuda M, Mizuno E, Kato M, Agemura A, Izumo S, Sano A (2007) In vivo distribution and localization of chorein. Biochem Biophys Res Commun 353:431–435
6. Kurano Y, Nakamura M, Ichiba M, Matsuda M, Mizuno E, Kato M, Izumo S, Sano A (2006) Chorein deficiency leads to upregulation of gephyrin and GABA(A) receptor. Biochem Biophys Res Commun 351:438–442
7. Lin CH, Tallaksen-Greene S, Chien WM, Cearley JA, Jackson WS, Crouse AB, Ren S, Li XJ, Albin RL, Detloff PJ (2001) Neurological abnormalities in a knock-in mouse model of Huntington's disease. Hum Mol Genet 10:137–144
8. Rampoldi L, Dobson-Stone C, Rubio JP, Danek A, Chalmers RM, Wood NW, Verellen C, Ferrer X, Malandrini A, Fabrizi GM, Brown R, Vance J, Pericak-Vance M, Rudolf G, Carre S, Alonso E, Manfredi M, Nemeth AH, Monaco AP (2001) A conserved sorting-associated protein is mutant in chorea-acanthocytosis. Nat Genet 28:119–120
9. Stege JT, Laub MT, Loomis WF (1999) tip genes act in parallel pathways of early *Dictyostelium* development. Dev Genet 25:64–77
10. Tomemori Y, Ichiba M, Kusumoto A, Mizuno E, Sato D, Muroya S, Nakamura M, Kawaguchi H, Yoshida H, Ueno S, Nakao K, Nakamura K, Aiba A, Katsuki M, Sano A (2005) A gene-targeted mouse model for chorea-acanthocytosis. J Neurochem 92:759–766
11. Ueno S, Maruki Y, Nakamura M, Tomemori Y, Kamae K, Tanabe H, Yamashita Y, Matsuda S, Kaneko S, Sano A (2001) The gene encoding a newly discovered protein, chorein, is mutated in chorea-acanthocytosis. Nat Genet 28:121–122

Part IV
The Structural Basis of Brain Involvement in Neuroacanthocytosis

Functional Imaging in Neuroacanthocytosis

K.L. Leenders(✉) and H.H. Jung

Abstract The published functional neuroimaging results indicate that the major finding in neuroacanthocytosis (NA) is severe striatal hypometabolism, either with or without atrophy. There is some evidence that the right side is more affected than the left. Cortical abnormalities are either not present or subtle, but if present appear to be particularly located in the frontal cortex. As with NA, an intriguing variety of devastating movement disorders may arise when specific striatal lesions occur. Due to the low prevalence of NA only a few patients have been investigated to date, and further studies are needed.

1 Introduction

The neuroacanthocytosis (NA) syndromes which manifest with chorea consist of chorea-acanthocytosis (ChAc) and McLeod syndrome (MLS). The genetic background of both conditions has been clarified considerably and clearly distinguishes the two entities. Apart from ChAc and MLS there are a large number of other

K.L. Leenders
Department of Neurology, University Medical Centre Groningen (UMCG), University of Groningen, the Netherlands
k.l.leenders@neuro.umcg.nl

R.H. Walker et al. (eds.), *Neuroacanthocytosis Syndromes II.*

conditions such as Huntington's disease (HD), Huntington's disease-like 2 (HDL2), and pantothenate kinase-associated neurodegeneration (PKAN) which share hereditary chorea, dystonia or other movement disorders, accompanied by a spectrum of hematological, neurological and other sequelae. The genetic and clinical features of the NA syndromes are dealt with in detail elsewhere in this volume. Since chorea is a dominant clinical sign in these conditions, and the hallmark of the cerebral lesion – at least in the early stages – is atrophy and gliosis of the striatum, we review here the functional neuroimaging in NA (positron emission tomography [PET], single photon emission computed tomography [SPECT], magnetic resonance spectroscopy [MRS]) as documented in the literature. We also describe two further patients with ChAc. For details of structural imaging findings, see the chapter by Henkel et al.

2 Tracer Imaging in HD

The model disease in which the brunt of the pathological changes manifest, at least in the early stages of the disease, almost exclusively in the striatum, is HD. In this autosomal dominant disease the mutant gene leads to formation of an abnormal huntingtin protein resulting in malfunction and degeneration, in particular of the projection neurons of the striatum. Why this usually occurs after a delay of many years and why the disturbance mainly damages striatal neurons, is still a matter of speculation and investigation, but is undoubtedly of fundamental importance. HD can present clinically in various ways, but the most conspicuous neurological signs consist of movement disorders such as chorea, dystonia and bradykinesia [21]. In this sense HD is a typical example of a basal ganglia disorder. How exactly the abnormal movements characterized as chorea are generated by the altered regulation of cerebral motor networks remains to be elucidated. It appears that the altered final outflow pattern of the globus pallidus, due to the lesioned striatal projection neurons, projecting to the thalamus and cortex appears to be a necessary condition. It is intriguing, but as yet not clarified at a pathophysiological network level, that altered activity in the globus pallidus can lead to either chorea or to dystonia (and perhaps also to tics?).

The degeneration of the striatal neurons results at a certain stage in striatal atrophy, as has been shown many times in post mortem studies and during life in CT or MRI studies. When the disease has developed clinically, not only is structural atrophy detectable, but there is also markedly reduced local energy consumption and receptor loss, as shown, for example, by radiotracer PET studies. Changes in metabolism are demonstrated by regional glucose utilization studies using the tracer ^{18}[F]-fluoro-deoxyglucose (FDG) and PET (Fig. 1)[25]. Dopamine D2 receptor binding (using for example [^{11}C]-raclopride) indicates receptor binding on the cell membrane of striatal projection neurons [1, 5, 19].

Local biochemical alterations at the level of the striatum can be detected using the above-mentioned radiotracer methods many years before the disease becomes clinically manifest [2, 23]. This situation is currently being investigated in order to be able to predict the time course of the degenerative process. If local biochemical

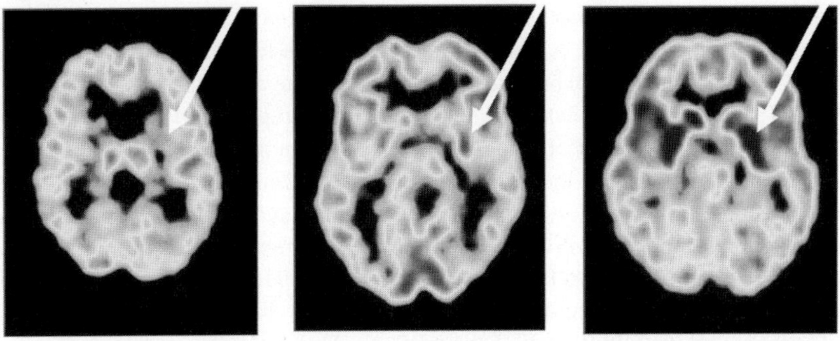

Fig. 1 Radiolabeled FDG ([^{18}F]fluoro-deoxyglucose) uptake in the brain of a patient with Huntington's disease (*left*), a healthy volunteer (*middle*) and a patient with Sydenham's chorea (*right*). One plane cutting through the striatal regions is shown. The top of the images is the front of the brain. The Huntington patient shows a marked decrease of striatal glucose utilisation related to the severity of the neurodegeneration, whereas the Sydenham's chorea patient shows a marked increase in striatal glucose metabolism

dysfunction is an early marker for disease progression then the appropriate patients could be selected to test potentially protective drugs.

It is interesting to realize that not only a loss of striatal projection neurons can cause chorea, but also selectively altered neuronal function, as apparently is the case in Sydenham's chorea (SC). Whereas in HD a loss of striatal synaptic activity results in lower local energy demands and altered functions, in SC a completely different pathophysiology results in local metabolic increases (Fig. 1), presumably due to aberrant immunologic activity towards striatal components in the context of a streptococcal infection. The disease process in SC apparently leads to almost identical network alterations as in HD as far as the generation of chorea is concerned. Usually SC is a self-limiting disease and the hypermetabolism disappears after some time [24]. This illustrates that chorea due to a striatal lesion must not automatically be associated with loss of local energy consumption.

3 Two Patients with Chorea-acanthocytosis (ChAc)

3.1 Patient 1

This male caucasian patient was born in 1963 of consanguineous parents. One younger sister had been diagnosed at age 28 as having NA. She died of suicide at age 39. At the time of her diagnosis, 2 of her 3 brothers, including patient 1, had mild signs on examination, but no symptoms. These two brothers had decreased tendon reflexes, elevated creatine phosphokinase (CPK) and acanthocytes in the blood. One of these two brothers died in an accident at sea under unclear circumstances. The other brother, patient 1, was seen for the first time in 1996 at age 33, for a first seizure for which he received carbamazepine and later also valproic acid. An MRI scan at the time was normal. Electromyography (EMG) showed a mild polyneuropathy. Electroencephalography (EEG) was normal except for slight bilateral irritative activity in the temporal lobes. Clinically he showed occasional orobuccolingual dyskinesias, vocalisations and other stereotypic movements of the face, neck and throat. There was almost complete areflexia and saccadic eye movements but no ataxia. The CPK was elevated at 1,059 U/L (normal < 200 U/L for males), and in the blood there were 10% echinocytes. Kell antigens showed a normal pattern. An FDG PET scan of the brain showed a reduction of striatal glucose utilisation (Fig. 2). Repeated neuropsychological testing showed a moderate decrease of executive functions and verbal fluency. During the next few years chorea gradually developed giving rise to increasing difficulty with walking and to dysarthria. Under antiepileptic coverage, risperidone was tried at a dose of 2 mg twice daily. This had a positive effect on the chorea, but was soon stopped because of increasing seizure frequency. A repeat MRI showed caudate nucleus atrophy. Serum CPK rose to high levels (between 5,000 and 7,000 U/L). In 2004 the blood of this patient was tested for mutations of the *VPS13A* gene and homozygous deletions were found of exons 8 and 9 [7]. Over the next few years his condition worsened gradually and in 2006 he died suddenly at age 43 without witnesses under unclear circumstances.

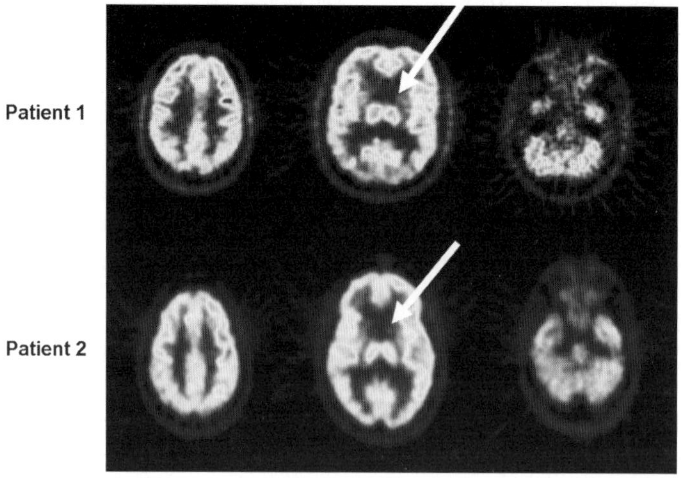

Fig. 2 Radiolabeled FDG ([^{18}F]fluoro-deoxyglucose) uptake in the brain of two patients with ChAc. An upper, middle and lower plane is shown. The top of the images is the front of the brain. A normal distribution of tracer uptake is seen except in striatal regions bilaterally where in both patients there is a marked loss of glucose utilisation. Compare with normal uptake in control in Fig. 1. Note that in these two patients the extent of striatal hypometabolism is more marked than in the HD patient of Fig. 1

3.2 Patient 2

This female patient of Asian descent was born in 1970 in Korea, but moved as a child to a foster family in Europe. Of her family nothing is known. At the age of 16 tics started to appear. These consisted initially of jerks of her head to the left, but gradually over the years more involuntary movements appeared, such as contractions of eyes, mouth and lips accompanied by vocalisations. In 1993 a diagnosis of Gilles de la Tourette syndrome was made, despite the findings of areflexia and a CPK of 700 U/L (normal < 170 U/L for females). As the symptoms worsened over the years psychiatric interventions became necessary. There was loss of concentration and increasingly chaotic behaviour. Haloperidol, tetrabenazine and pimozide were prescribed in high doses. These resulted in severe depression and somnolence, and an increase of orofacial dyskinesias was seen which was initially interpreted as tardive dyskinesia. Her situation worsened in that she showed self-mutilation with biting of her lips and tongue. A mechanical device partially prevented this. She increasingly experienced feeding problems with dysphagia and hypertrophy of her neck muscles, requiring placement of a percutaneous gastric tube. In 2002 she was admitted to our neurology ward and all neuroleptic medication was stopped. The patient's depression disappeared, but marked chorea, particularly during walking, came to light. The full blown clinical picture of NA (chorea, feeding dystonia, self-mutilation, areflexia, increased CPK, caudate atrophy on MRI and severe striatal glucose hypometabolism (Fig. 2), but without blood acanthocytes) became apparent. The patient's blood was sent for genetic testing [7] and a heterozygous mutation of the *VPS13A* gene (188-5 T > G) was found supporting the diagnosis.

4 Cerebral Glucose Utilization of Patients 1 and 2

FDG uptake in the brains of these two patients showed some remarkable features. As expected, on visual inspection the striatal regions showed a marked reduction of tracer uptake. However, visual inspection of the cortices did not reveal obvious differences with controls. This can be seen in Fig. 2 when compared to a normal control distribution as shown in Fig. 1.

In addition to visual inspection a semi-quantitative analysis was applied. This consisted of calculating the ratios of the average pixel counts in the target regions (i.e. putamen, caudate nucleus, and various cortical regions) vs. the average pixel counts of all brain regions in the patient's scan. Conspicuous differences were found as compared to healthy control values (the latter data not shown here). The Z-values (average normal value – patient value)/(standard deviation of normal values) for the various caudate regions reductions in both patients ranged from 3.3 to 7.5 and for the putamen from 2.3 to 6.3. In both patients the caudate nucleus showed only slightly greater reductions than the putamen. In both patients the right side was more affected than the left. No hypo- or hypermetabolism was seen in the cortical regions on visual inspection. However, looking at the semiquantitative values, metabolism in the lateral frontal cortex was relatively increased in both patients compared to healthy controls, and in both this was more pronounced on the left side (Z-values ranging from 2.5 to 3.5). In patient 1 the right lateral prefrontal cortex also showed an increase, in addition to an increase in some medial frontal regions (Z-values ranging from 2.0 to 3.0).

5 Review of the published imaging data in neuroacanthocytosis patients

5.1 *Structural Imaging*

CT and MRI studies demonstrate atrophy of caudate nucleus and putamen, particularly with advanced disease in both MLS and ChAc [6, 10, 13]. In MLS, striatal volumetry was performed in one study demonstrating a correlation between disease duration and reduction of striatal volumes [13]. In one report, white matter changes were described [17]. In ChAc, MRI may show increased signal on T2-weighted MRI in the caudate nucleus and putamen in addition to atrophy [22]. Voxel-based morphometry (VBM) revealed regional reduction of gray matter volumes symmetrically, in particularly of the head of the caudate nucleus [11] (and see chapter by Henkel et al.). Another study using diffusion tensor imaging (DTI) and apparent diffusion coefficient (ADC) maps demonstrated increased diffusibility in the putamen and caudate nucleus bilaterally, indicating disruption of tissue integrity in these regions [16].

5.2 *Tracer Imaging*

From the descriptions of NA patients in early publications, describing the results of functional imaging, it is not possible to make a distinction between ChAc and MLS. For convenience these are then listed under NA. In Table 1 the published patients are briefly summarized.

5.2.1 Neuroacanthocytosis

Early chapters report marked striatal glucose hypometabolism in a few patients with NA using FDG-PET scanning [3, 8, 12]. In one study concerning two brothers [8] prominent striatal glucose use reduction was present without evidence of atrophy of the striatum on MRI. Also in another study [3] the MRI showed only moderate caudate atrophy while glucose consumption was severely reduced. The degree of striatal loss of glucose utilization appeared to be associated with the severity of clinical signs [20], but the pattern of glucose consumption loss did not explain the differences in clinical presentation, specifically, it is unclear why in one patient the picture is dominated by tics and in another by chorea.

Table 1 Literature summary of PET/SPECT scanned neuroacanthocytosis patients

References		Diagnosis	Number of patients	Remarks	Type of scan
[8]	1989	NA	2	Brothers; no MRI atrophy	FDG PET
[3]	1998	ChAc	1		FDG PET
[22]	1998	ChAc	3	Frontal and temporal cortex reduced in addition to striatum	CBF and CMRO2 PET
[4]	1991	NA	6		
[5]	1994	MLS	1		D2 SPECT
[13]	2001	MLS	5		FDG PET
[18]	2001	MLS	2	In one no movement disorder No frontal lobe involvement	FDG PET
[20]	2004	ChAc	2	Some frontal hypometabolism Marked striatal hypometabolism, but different clinical severity	FDG PET
[9]	2006	MLS	5	Compared with five unaffected heterozygous females	MRS
[16]	2007	ChAc	2	Monozygotic twins	FDG PET Beta-CIT SPECT
Leenders (this chapter)	2007	ChAc	2	Marked frontal and temporal cortex hypometabolism	FDG PET

D2 dopamine D2 receptor, beta-CIT: dopamine transport site tracer, *CBF* cerebral blood flow, *CMRO2* cerebral metabolic rate of oxygen, *ChAc* chorea-acanthocytosis, *MLS* McLeod Syndrome, *FDG* fluoro-deoxyglucose, *NA* neuroacanthocytosis

A reduction in right striatal metabolism more than the left was reported in two monozygotic twins [16], as in our two patients described above. In addition widely distributed cortical increases were seen.

Not only was glucose use found to be affected, but also other indicators of neuronal energy metabolism were also involved, namely oxygen consumption and capillary perfusion in three patients with NA [22]. In that study oxygen consumption was also reduced in frontal and temporal brain regions. Reduction of frontal perfusion in three patients has also been reported [4].

Some studies used radiotracers related to the dopaminergic neurotransmitter system. In one patient with NA and parkinsonism, ^{18}[F]-fluoro-dopa (FDOPA) and PET showed a nigrostriatal deficit, correlating with the parkinsonism in that patient. [19]. Conversely, two monozygotic twins without parkinsonism showed essentially normal values of beta-CIT uptake into striatum, indicating normal nigrostriatal activity [16]. In six NA patients normal nigrostriatal dopaminergic activity was also found using FDOPA and PET [4], except in posterior putaminal regions where a reduction of 60% was seen. In that study loss of striatal dopamine receptors was confirmed using raclopride and PET, although only three patients had the latter investigation.

5.2.2 McLeod Syndrome

5.2.2.1 Positron Emission Tomography

FDG-PET demonstrated impaired striatal glucose metabolism in symptomatic as well as presymptomatic MLS patients [13, 18]. By using quantified FDG-PET, no impairment of glucose metabolism was found in the cerebral cortex [13]. However, this technique detected subtle alterations in the striatum of female heterozygotes, and demonstrated progressive impairment of FDG uptake in relation to disease duration in symptomatic MLS patients [13]. However, occasional male subjects carrying *XK* gene mutations will not show clinical symptoms, and in those subjects no apparent decrease in striatal energy consumption loss may be present [14].

One study used a SPECT dopamine D2 tracer in a single symptomatic patient and found reduced striatal receptor binding [5].

5.2.2.2 Magnetic Resonance Spectroscopy

In the only MRS study in NA, five McLeod patients, five asymptomatic heterozygous females of a Swiss McLeod family were compared with 10 age- and sex-matched healthy controls, using fast multiple spin-echo MRS [9]. Three McLeod patients with pronounced psychiatric or cognitive manifestations had pathological NAA/(Cr + Cho) ratios in frontal, temporal, and insular areas with an individual pattern. Two McLeod patients with a severe choreatic movement disorder had unilateral thalamic alterations. One McLeod patient with a moderate movement disorder and personality disorder had bilateral occipital alterations. Four female heterozygotes had normal findings. One female heterozygote had unilateral insular alterations, possibly indicating subclinical cerebral involvement. Although the prominent psychiatric and cognitive manifestations in McLeod patients suggest significant and widespread cortical abnormalities, previous neuroradiological and histopathological data have not revealed definite extrastriatal pathology. These MRS findings, demonstrating abnormalities in different brain regions of McLeod patients, might either reflect neuronal dysfunction due to impaired basal ganglia-thalamo-cortical circuits or subtle structural alterations in the particular cerebral areas [8].

6 Discussion

In MLS and ChAc, structural neuroradiological examinations demonstrate atrophy of caudate nucleus and putamen when the disease is manifest clinically. These alterations resemble those found in HD. The head of caudate nucleus might be most vulnerable for the neurodegenerative process in ChAc.

Functional neuroimaging studies have been reported in only a few studies and in each study only very few patients were included. Follow-up studies investigating functional neuroimaging results in relation to clinical progression have hardly been attempted. This is of course due to the low prevalence of these conditions.

FDG-PET in MLS and ChAc demonstrated impaired striatal glucose uptake even in presymptomatic patients and heterozygote females, and also sometimes in the absence of clear evidence of atrophy. Available data suggest progression of the metabolic impairments correlates with disease duration. Studies to date also suggest that the right striatum is more affected than the left, although in view of the few investigated subjects it is unclear whether this is coincidental or is truly due to a disease-specific process.

Radiotracers measuring aspects of the striatal dopaminergic neurotransmitter system have been applied only on a few occasions. Generally a loss of dopamine receptor binding was found. This is not surprising since these receptors are mainly located on the striatal projection neurons and thus will disappear with loss of these neurons as the disease progresses. Thus the dopamine receptor binding under these circumstances does not provide more information than the measurement of the striatal energy metabolism. The presynaptic striatal dopa metabolism (an expression of the dopaminergic activity of the nigrostriatal system) is generally unimpaired – as indeed is also seen in HD – except if ChAc is accompanied by clinical parkinsonism.

One intriguing phenomenon is the dominant pathological process at the level of the striatum by the time the disease clinically manifests. In post mortem studies marked changes are demonstrated in those regions, but also in some other nuclei connected with the striatum. Cortical atrophy is present only seldom, or perhaps is more prominent at later stages of the disease or under special conditions. This is also reflected in the metabolic tracer studies which show subtle alterations mainly in the frontal cortex if at all present. In one study MRS demonstrated subtle extrastriatal alterations corresponding to the neuropsychiatric symptoms without evidence for specific alterations.

One question is how it can be explained that so many and complex clinical phenomena are caused by absent or malfunctioning striatal regions? The various well-known movement disorders such as tic syndromes, chorea, dystonia and bradykinesia are all clearly expressed either separately or simultaneously. Behavioural abnormalities, such as self-mutilation and obsessive-compulsive phenomena, in addition to psychiatric manifestions such as depression and schizophrenia, and cognitive decline are often seen. As the cortex is relatively normal, but the striatum shows marked degeneraton, we postulate that these symptoms are the result of dysfunction of striato-cortical neuronal circuits. Therefore, these diseases could form model conditions to investigate the fundamental brain mechanisms underlying the above-mentioned movement disorders and neuropsychiatric problems. To this end it would be conceivable that a small number of carefully selected subjects, with genetically confirmed disease, but in whom no or only few clinical signs are demonstrable, could undergo detailed longitudinal investigation applying clinical, neuropsychological, neurophysiological and neuroimaging methods. This might yield insights into abnormal brain function in general, beyond the context of NA. Naturally it is also mandatory to continue research into the mechanisms leading to neuronal malfunction and death in this sad disease, hopefully ultimately leading to a preventive or curative therapy.

References

1. Antonini A, Leenders KL, Spiegel R, Meier D, Vontobel P, Weigell-Weber M, Sanchez-Pernaute R, de Yebenez JG, Boesiger P, Weindl A, Maguire RP (1996) Striatal glucose metabolism and dopamine D2 receptor binding in asymptomatic gene carriers and patients with Huntington's disease. Brain 119:2085–2095
2. Antonini A, Leenders KL, Eidelberg D (1998) [11C]raclopride-PET studies of the Huntington's disease rate of progression: relevance of the trinucleotide repeat length. Ann Neurol 43:253–255
3. Bohlega S, Riley W, Powe J, Baynton R, Roberts G (1998) Neuroacanthocytosis and aprebetalipoproteinemia. Neurology 50:1912–1914
4. Brooks DJ, Ibanez V, Playford ED, Sawle GV, Leigh PN, Kocen RS, Harding AE, Marsden CD (1991) Presynaptic and postsynaptic striatal dopaminergic function in neuroacanthocytosis: a positron emission tomographic study.Ann Neurol 30:166–171
5. Danek A, Uttner I, Vogl T, Tatsch K, Witt TN (1994) Cerebral involvement in McLeod syndrome. Neurology 44:117–120
6. Danek A, Rubio JP, Rampoldi L, Ho M, Dobson-Stone C, Tison F, Symmans WA, Oechsner M, Kalckreuth W, Watt JM, Corbett AJ, Hamdalla HHM, Marshall AG, Sutton I, Dotti MT, Malandrini A, Walker RH, Daniels G, Monaco AP (2001) McLeod neuroacanthocytosis: genotype and phenotype. Ann Neurol 50:755–764
7. Dobson-Stone C, Velayos-Baeza A, Filippone LA, Westbury S, Storch A, Erdmann T, Wroe SJ, Leenders KL, Lang AE, Dotti MT, Federico A, Mohiddin SA, Fananapazir L, Daniels G, Danek A, Monaco AP (2004) Chorein detection for the diagnosis of chorea-acanthocytosis. Ann Neurol 56:299–302
8. Dubinsky RM, Hallett M, Levey R, Di chiro G (1989) Regional brain glucose metabolism in neuroacanthocytosis. Neurology 39:1253–1255
9. Dydak U, Mueller S, Sandor PS, Meier D, Boesiger P, Jung HH (2006) Cerebral metabolic alterations in McLeod syndrome. Eur Neurol 56:17–23
10. Hardie RJ, Pullon HW, Harding AE, Owen JS, Pires M, Daniels GL, Imai Y, Misra VP, King RH, Jacobs JM (1991) Neuroacanthocytosis. A clinical, haematological and pathological study of 19 cases. Brain 114:13–49
11. Henkel K, Danek A, Grafman J, Butman J, Kassubek J (2006) Head of the caudate nucleus is most vulnerable in chorea-acanthocytosis: a voxel-based morphometry study. Mov Disord 21:1728–1731
12. Hosokawa S, Ichiya Y, Kuwabara Y, Ayabe Z, Mitsuo K, Goto I, Kato M (1987) Positron emission tomography in cases of chorea with different underlying diseases. J Neurol Neurosurg Psychiatry 50:1284–1287
13. Jung HH, Hergersberg M, Kneifel S, Alkadhi H, Schiess R, Weigell-Weber M, Daniels G, Kollias S, Hess K (2001) McLeod syndrome: a novel mutation, predominant psychiatric manifestations, and distinct striatal imaging findings. Ann Neurol 49:384–392
14. Jung HH, Hergersberg M, Vogt M, Pahnke J, Treyer V, Röthlisberger B, Kollias SS, Russo D, Frey BM (2003) McLeod phenotype associated with a XK missense mutation without hematologic, neuromuscular, or cerebral involvement. Transfusion 43:928–938
15. Leenders KL, Frackowiak RS, Quinn N, Marsden CD (1986) Brain energy metabolism and dopaminergic function in Huntington's disease measured in vivo using positron emission tomography. Mov Disord 1:69–77
16. Muller-Vahl KR, Berding G, Emrich HM, Peschel T (2007) Chorea-acanthocytosis in monozygotic twins: clinical findings and neuropathological changes as detected by diffusion tensor imaging, FDG-PET and (123)I-beta-CIT-SPECT. J Neurol Feb 8; [Epub ahead of print]

17. Nicholl DJ, Sutton I, Dotti MT, Supple SG, Danek A, Lawden M (2004) White matter abnormalities on MRI in neuroacanthocytosis. J Neurol Neurosurg Psychiatry 75:1200–1201
18. Oechsner M, Buchert R, Beyer W, Danek A (2001) Reduction of striatal glucose metabolism in McLeod choreoacanthocytosis. J Neurol Neurosurg Psychiatry 70:517–520
19. Pavese N, Andrews TC, Brooks DJ, Ho AK, Rosser AE, Barker RA, Robbins TW, Sahakian BJ, Dunnett SB, Piccini P (2003) Progressive striatal and cortical dopamine receptor dysfunction in Huntington's disease: a PET study. Brain 126:1127–1135
20. Saiki S, Hirose G, Sakai K, Matsunari I, Higashi K, Saiki M, Kataoka S, Hori A, Shimazaki K (2004) Chorea-acanthocytosis associated with tourettism. Mov Disord 19:833–836
21. Sanchez-Pernaute R, Kunig G, del Barrio Alba A, de Yebenes JG, Vontobel P, Leenders KL (2000) Bradykinesia in early Huntington's disease. Neurology 54:119–125
22. Tanaka M, Hirai S, Kondo S, Sun X, Nakagawa T, Tanaka S, Hayashi K, Okamoto K (1998) Cerebral hypoperfusion and hypometabolism with altered striatal signal intensity in chorea-acanthocytosis: a combined PET and MRI study [see comments]. Mov Disord 13:100–107
23. van Oostrom JC, Maguire RP, Verschuuren-Bemelmans CC, Veenma-van der Duin L, Pruim J, Roos RA, Leenders KL (2005) Striatal dopamine D2 receptors, metabolism, and volume in preclinical Huntington disease. Neurology 65:941–943
24. Weindl A, Kuwert T, Leenders KL, Poremba M, Grafin von Einsiedel H, Antonini A, Herzog H, Scholz D, Feinendegen LE, Conrad B (1993) Increased striatal glucose consumption in Sydenham's chorea. Mov Disord 8:437–444
25. Young AB, Penney JB, Starosta-Rubinstein S, Markel DS, Berent S, Giordani B, Ehrenkaufer R, Jewett D, Hichwa R (1986) PET scan investigations of Huntington's disease: cerebral metabolic correlates of neurological features and functional decline. Ann Neurol 20:296–303

Volumetric Neuroimaging in Neuroacanthocytosis

K. Henkel, M. Walterfang(✉), D. Velakoulis, A. Danek, and J. Kassubek

Abstract Neuroimaging provides an in vivo method of understanding the changes to brain structure and function that occur with CNS disorders, and have provided important insights into a range of neurodegenerative disorders. The rarity of chorea-acanthocytosis has thus far limited the application of neuroimaging methodologies to the disease. Increasing recognition of the disorder and improved diagnosis with antibody screening has allowed for initial study of structural brain changes. Voxel-based morphometry takes a whole-brain, hypothesis-neutral approach and determines where maximal differences occur in brain regions between groups. When six patients were compared to controls, maximal reductions were found bilaterally in the caudate head. Manual volumetric approaches allow for more sophisticated between-group analyses. When the caudate nuclei of ten patients were delineated, reductions of approximately 80% of total volume were shown in chorea-acanthocytosis patients, again maximal in the caudate head. Future studies are required that follow brain changes longitudinally, and compare patients with this rare disorder to related disorders such as Huntington's disease.

1 Introduction

A variety of neuroimaging approaches offer the possibility to assess brain volumes in vivo. Beside the determination of a general volume loss there is a special interest in localizing and quantifying regional brain atrophy in neurodegenerative disorders that may correlate to distinct dysfunction of neuronal tissues.

Mark Walterfang
Melbourne Neuropsychiatry Centre, University of Melbourne, and Neuropsychiatry Unit, Royal Melbourne Hospital, Melbourne, Australia
mark.walterfang@mh.org.au

R.H. Walker et al. (eds.), *Neuroacanthocytosis Syndromes II.*

Early studies measured ventricular width as a correlate of global brain atrophy, but were not able to define regional changes due to neurodegeneration. Other volumetric imaging studies have examined predefined regions of interest (ROI) using formulas for an approximation to spatial shapes or the summing up of manually-traced imaging slices. Newer techniques, like the brain parenchymal fraction method and voxel based morphometry (VBM) use automated segmentation procedures and standardized parametric statistics. These allow a more operator-independent quantitative determination of structural alterations in neurological disorders at the group level as well as comparisons to a normal data base.

Neuroacanthocytosis syndromes are uncommon and as a result only a handful of imaging studies exploring structural brain changes have been undertaken. Damage to the basal ganglia has been described in a number of cases [8, 10]. In McLeod syndrome (MLS) two patients were multimodally investigated with different imaging techniques [14] and a volume reduction of the caudate nucleus and putamen was found, with volume loss correlating with disease duration.

The recent developments of a mutation test of the *VPS13A* gene and of chorein detection in red cell membranes [19, 25] have allowed for the definitive diagnosis of one further neuroacanthocytosis type: autosomal recessive chorea-acanthocytosis (ChAc, *OMIM 200150*). This enables the identification of a neurobiologically relatively homogeneous group of ChAc patients for future studies. Two approaches are described below.

2 Quantitative Studies in Chorea-acanthocytosis

2.1 Voxel-Based Morphometry in Chorea-acanthocytosis

In contrast to predefined ROI analysis, voxel based morphometry (VBM) is a whole brain-based statistical approach to compare volumes of the complete imaging data set on a voxel basis as described by Ashburner and Friston [3]. VBM offers the possibility of exploring regional brain changes unconstrained by a-priori hypotheses about affected structures by comparing patient brains with normal controls.

Volumetric data were analyzed from six ChAc patients (five male and one female, median age: 37, range: 26–44 years) with a genetically confirmed diagnosis [12]. Disease duration, defined as time since onset of first motor symptoms, was 13 (range: 6–18) years. As an a priori hypothesis, changes were expected to be localized in the striatum, in accordance with the results of former imaging studies. Three-dimensional T1-weighted high-resolution volume-rendering MRI scans were acquired at the Clinical Center, National Institutes of Health (Bethesda, MD, USA), using a 1.5 Tesla MRI scanner and a 3D magnetization prepared rapid gradient echo pulse sequence (MP-RAGE). As the standard tool, the Statistical Parametric Mapping software (SPM Wellcome Department of Imaging Neuroscience Group, London UK.; http//www.fil.ion.ucl.uk/spm) was used. The processing of the imaging data included normalization to the 3D-stereotaxic Montreal Neurological Institute (MNI) standard space, automated segmentation into the compartments of grey matter, white matter and cerebrospinal fluid and smoothing with a 6 mm isotropic Gaussian kernel. The grey matter maps of the patients and 15 age-matched healthy controls were statistically compared voxel-by-voxel in a parametric group analysis to detect differences of grey matter density. Data were thresholded at $p < 0.001$, and a correction for small volumes was performed.

The relative global brain volume was estimated by calculation of the brain parenchymal fraction (BPF) as described by Kassubek et al. [6, 15]. Different tissue densities were automatically separated into grey matter (GM), white matter (WM) and cerebrospinal fluid (CSF). The BPF is defined as the ratio of brain parenchymal volume (GM and WM) to total brain volume. In the group-comparison of the mean BPF of the six ChAc patients with age-matched controls, no significant reductions of brain volume was found (0.817 vs. 0.837). However, in two of the patients a reduction of more than two standard deviations was detected. Their mean disease duration did not differ significantly from the other four patients (12 vs. 16.7). In summary, this argues against a gross general brain atrophy in the first years of the disease.

In the VBM component of this study, regional atrophy was found almost exclusively in the head of the caudate nucleus (Figs. 1 and 2). The global maxima were

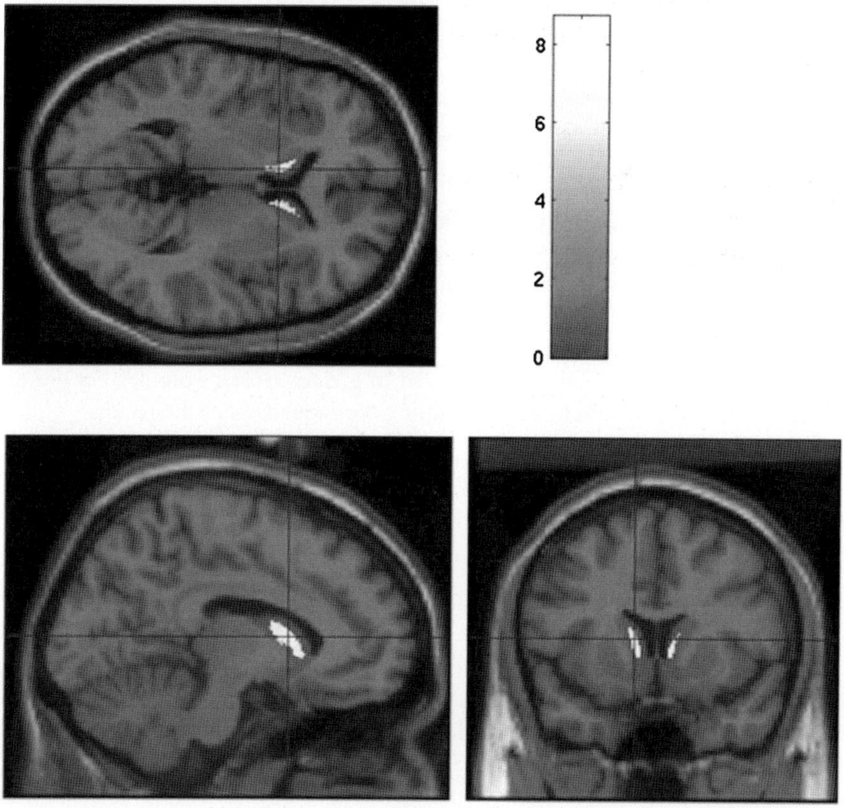

Fig. 1 Overlay of significant atrophic areas of the group analysis on a patient's brain demonstrating alterations in the caudate nucleus bilaterally (crosshair indicating the global maximum. Z-score is indexed by the greyscale)

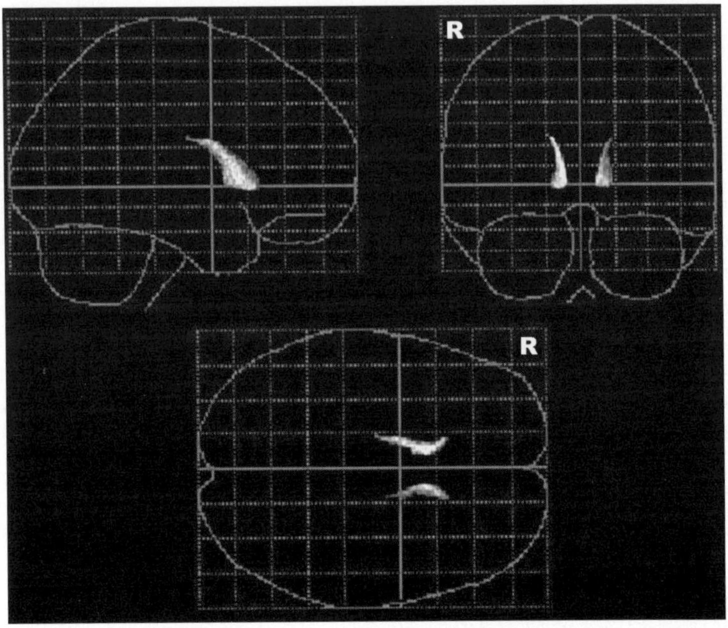

Fig. 2 Maximum intensity projection of all areas with significantly decreased grey matter in the SPM group analysis, overlain on the SPM glass brain template

localized nearly symmetrically in the head of the caudate nucleus at the Talairach coordinates *x*,*y*,*z*: -8,12,4 and 9,12,4. Z-scores of the global maxima were 5.46 and 5.37, respectively, showing a robust result. Symmetry was shown by the limitation of side difference of the size of the significant voxel clusters to less than 10%. No areas of increased grey matter density were found. The use of a less significant threshold with an uncorrected $P < 0.05$ led to greater cluster volumes but did not include other cerebral structures, underlining the robustness of the results.

2.2 Region of Interest Approach in Chorea-acanthocytosis

Manual volumetry was performed on ten individuals (seven male and three female) with ChAc confirmed genetically or with the chorein Western blot.

Region-of-interest (ROI) volumetry analyses remain the "gold standard" for determining volumetric change in morphometry studies, as they are not susceptible to a number of the statistical methodological difficulties that plague VBM research, such as the introduction of shape differences through misregistration during normalization [7]; the displacement of true differences through choice of smoothing kernel, which is often arbitrarily chosen without reference to the size of expected anatomical change [13]; and the inability of the method to take into account local sulcal/gyral variability [7]. VBM is probably not well-suited to detection of subtle change, but its inherent methodological problems are not likely to affect results where there is gross focal atrophy, such as in ChAc. VBM also can only provide one-dimensional information about where differences between two sets of images are greatest, but cannot describe shape differences or differential changes between disparate brain regions, or their associations. As a result, an approach that combines both VBM and ROI methodologies is widely felt to be most powerful.

The ROI method was an adaptation of a method previously described [4, 20]. T1-weighted images sliced coronally were used to measure the volume of the caudate nucleus on contiguous slices in the coronal plane. The medial boundary was the lateral border of the lateral ventricle; the lateral boundary was the medial aspect of the internal capsule; the superior boundary was the most superior aspect of the caudate nucleus visualized; the anterior boundary was generally considered to be the subcallosal fasciculus; and the most posterior extension of the tail used as the posterior boundary when visible (Fig. 3). Images were manually traced using Analyze 7.0 software (Biomedical Image Resource, May Clinic, USA) on 30–60 slices, and the volumes from each slice were summed in millimetre cube, taking into account slice thickness of each coronal image.

Ideally, in this type of volumetric study, all images should be placed into the same stereotaxic space using a 12-parameter affine transformation as described previously [20], which obviates the need to covary for whole-brain or intracranial volume, and also allows for direct region-to-region comparison of caudate sections through (repeated measures) analysis of variance to detect shape as well as size

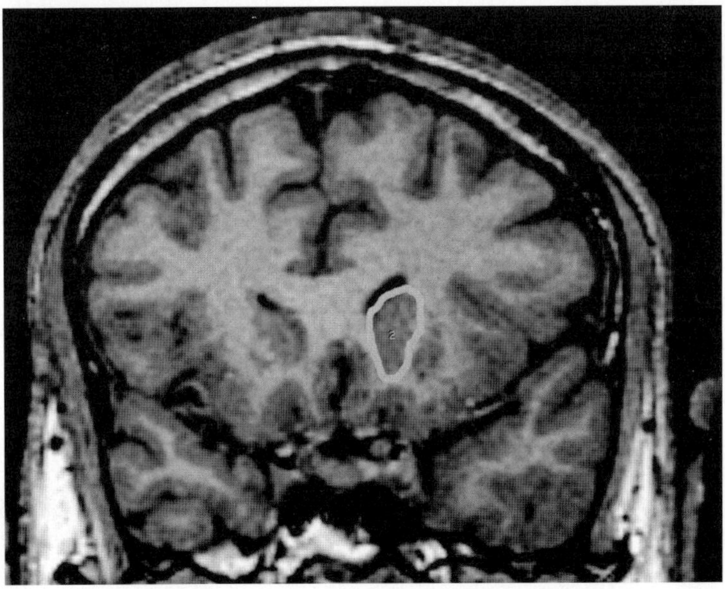

Fig. 3 Coronal slices showing tracing around left caudate head for caudate volumetry using Analyze 7.0 software

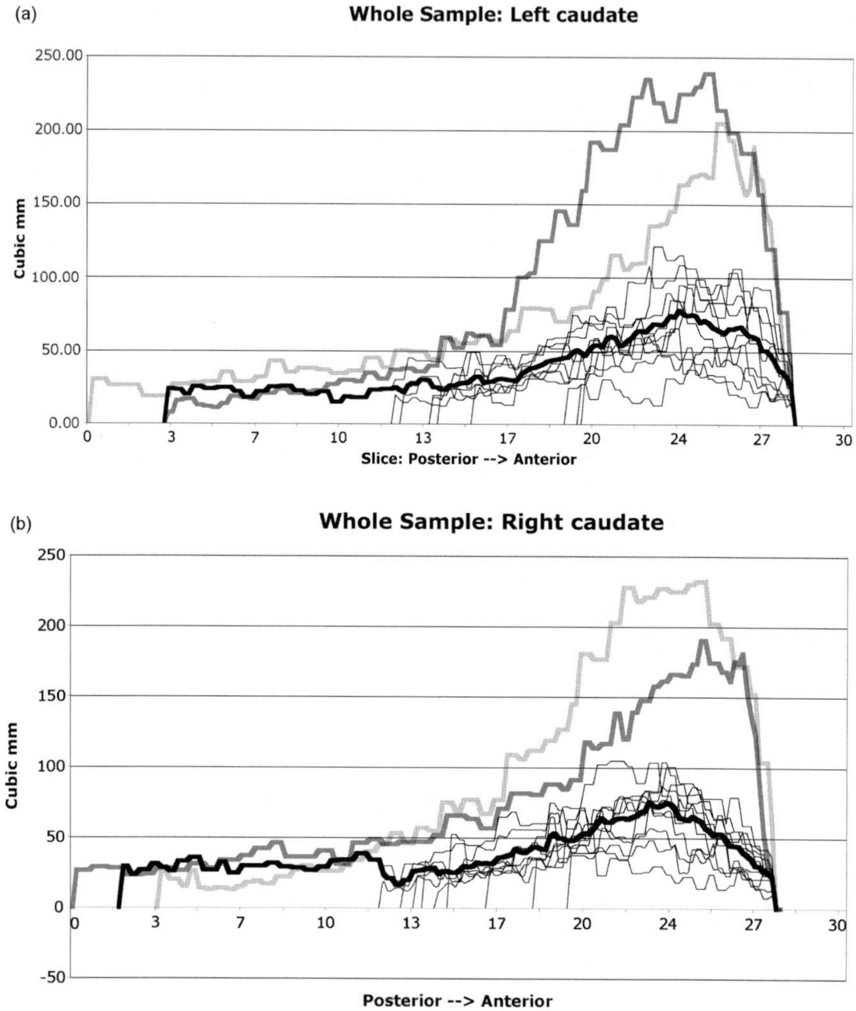

Fig. 4 Caudate shape [left (**a**) and right (**b**)] of 10 ChAc patients (*thin lines*) plotted against caudate shape, with measurements from male and female controls (*thicker grey lines*). The means of the patients' caudate volumes can be seen in the *thicker black line*. Gross volumetric reduction of the caudate is seen, particularly in the head

differences. This method has been well-described for shape analysis of the corpus callosum previously [9].

What this method has thus far allowed for is a quantification of total volume of the left and right caudate nuclei, and for visual inspection of shape changes. The mean volume of the left caudate in the patient group was 1,567(±556) mm^3, and the right caudate 1,624(±551) mm^3. Using the same methodology, the volumes of matched male and female controls were 5,308 mm^3 (left), 5,074 mm^3 (right) and 4,872 mm^3 (left), 4,872 mm^3 (right) respectively. These volumes suggest that caudate volumes in ChAc are reduced by 50–80% compared to similarly aged controls, a reduction not seen in any other disorder.

The shape plot from posterior caudate (body and tail) to anterior caudate (head), plotted against the volumes from healthy controls shows that the caudate is globally and significantly reduced in volume by 50–80% in most ChAc patients, with much of this loss coming from the caudate head (Fig. 4; to the right of each plot). This regional reduction matches the co-ordinates of the previously described VBM work, and the small number of neuropathological studies undertaken in ChAc.

3 Discussion

From previous neuropathology and imaging reports of ChAc atrophy of the striatum and the caudate nucleus/putamen was known [8, 10]. The VBM analysis reported above supports the restriction of volume changes to striatal tissues, with a peculiar predilection for the head of the caudate nucleus, but failed to show significant atrophy of the putamen, even when reduced thresholds for significance were used. The ROI analysis supports the VBM findings of special caudate vulnerability. These results are ideally the precursor to imaging studies of a larger number of patients, applying robust analysis methodologies.

The finding of a marked volumetric reduction of the head of the caudate nucleus appears to be in good agreement with the clinical picture seen in ChAc, of disturbances to frontostriatal function. These manifest as cognitive (particularly executive) impairments, personality changes, compulsive motor behaviour and the development of chorea [28]. The particular vulnerability of the head of the caudate nucleus, where the peak area of volume loss was found, correlates with the finding of cognitive and behavioural disturbances which are thought to be related to dysfunction of cortico-subcortical loops that connect the anterior cingulate and orbito-frontal cortex with the caudate head [1]. Dysfunction in this loop is described in obsessive-compulsive disorder [5], which is perhaps the most common major mental disorder seen in ChAc. The presentation in adolescence and early adulthood could be related to interruption of the developmental trajectory of these fronto-striatal connections which begin to come "on line" in late adolescence [22]. This may result in impaired selective motor response inhibition and the characteristic motor compulsions seen in the disorder such as tongue/lip biting, trichotillomania or complex motor rituals.

4 Comparison to Neuropathology and to Other Basal Ganglia Disorders

Marked striatal atrophy has been demonstrated post mortem repeatedly [11] and was most prominent in the head of the caudate nucleus, in putamen and pallidum. The substantia nigra was also affected. This loss of tissue appears to be neurodegenerative

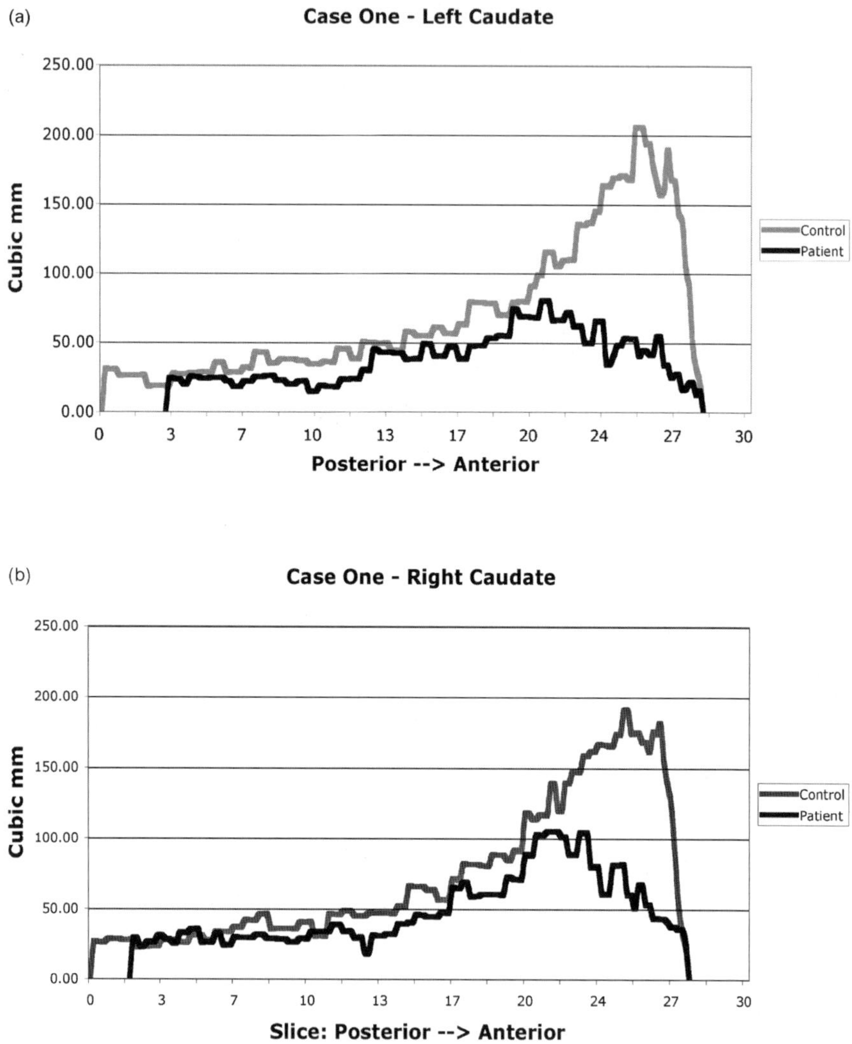

Fig. 5 Caudate volume/shape in two previously described male and female ChAc patients [18, 21] (*black lines*), compared with age-matched controls (*grey lines*); (**a**) left and (**b**) right caudate volumes of a 33-year-old man with OCD onset at the age of 10, who developed seizures at 21 and chorea at 25

in origin, with severe neuronal loss and gliosis in the caudate and, to a lesser degree, the putamen and pallidum [26]. Neuropathologic investigations using a stereological technique for cell counting [2] support this, again demonstrating marked striatal atrophy with a predilection for the caudate nucleus. Additionally, in contrast to former studies, the group of Arzberger also found a more distinct and diffuse cortical cell loss suggesting more diffuse cerebral involvement of the neurodegenerative process. Thus, we would expect future neuroimaging studies to show selective

(c)

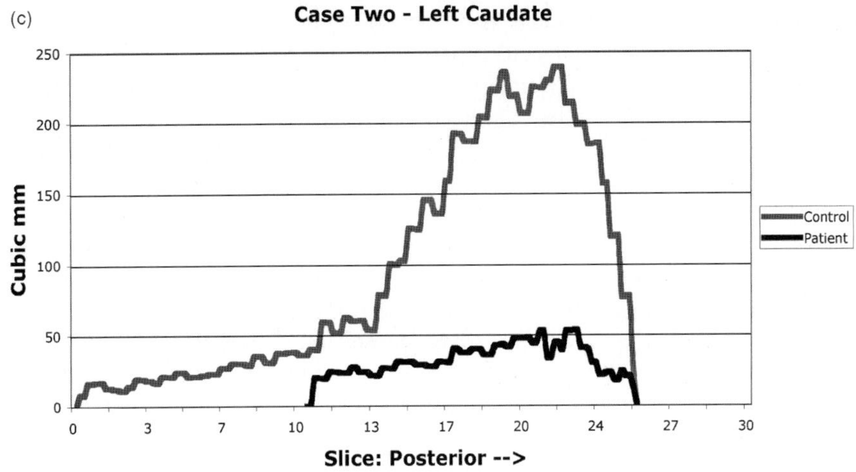

(d)

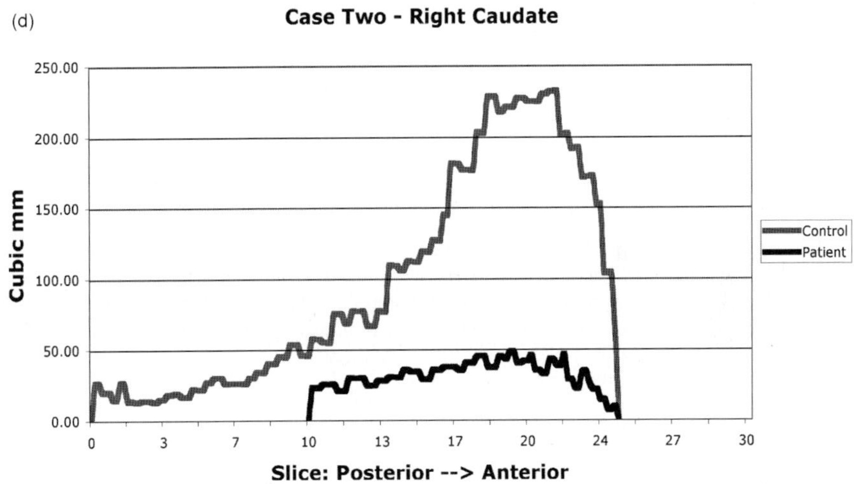

Fig. 5 (continued) (**c**) left and (**d**) right caudate volumes of a 38-year-old woman who developed OCD at 16, her first seizure at 21 and choreiform movements at 27

regional loss in the striatum, but also to show cortical volume loss, possibly with a predilection for frontal cortical regions, as might be expected given the pattern of neuropsychological impairments and behavioural disturbances.

In Huntington's disease (HD), global brain parenchyma reduction seems to be an early feature. In VBM studies brain parenchymal fractions in early HD patients (stage I and II according to Shoulson et al. [23]) were significantly reduced compared to age-matched controls [16]. Regional volume loss was found bilaterally in striatal areas as well as in the hypothalamus and the opercular cortex, and unilaterally in the right paracentral lobule [17] as well in dorsal midbrain and bilateral intra-parietal sulcus [24]. The topography of striatal changes in HD correspond to the dorso-medial to ventro-lateral gradient of neuronal loss that was found in neuropathological studies [27]. Additionally a correlation between cerebral grey matter loss with clinical severity and CAG repeat length was found in this study.

In comparison to HD, the small number of available ChAc patients limits extended morphometric studies. The present data suggest a greater total reduction of striatal volume with differential involvement of striatal subregions in comparison to HD. The ventrocaudal to dorsolateral gradient described in HD, has not been found. Additionally the caudate nucleus seems to be more involved than the putamen, whereas in HD the whole striatum shows marked pathology. Adequately powered studies are needed in the future to directly compare these two groups, as well as other relevant patient groups (e.g. McLeod syndrome, Huntington disease like-2, and obsessive-compulsive disorder), to reveal their differential effects on striatal and extra-striatal regions. Eventually, genotype–phenotype and structure–function correlations may be further clarified. Neuroimaging studies in general, and volumetric studies in particular, are able to at least partly provide "missing links." Understanding the progression of pathology from genome to proteome, through microstructure and macrostructure, to clinical phenotype and illness course, will ultimately allow for the nature of the illness to be more fully understood, and for future treatments to be tailored and introduced appropriately to ChAc patients.

References

1. Alexander GE, Crutcher MD (1990) Functional architecture of basal ganglia circuits: neural substrates of parallel processing. Trends Neurosci 13:266–271
2. Arzberger T, Heinsen H, Buresch N, Dobson-Stone C, Danek A, Kretzschmar H (2005) The neuropathology of chorea-acanthocytosis: from stereology to an immunohistochemical detection of chorein in: 2005 organizing committee. Anderman E, Danek A, Irvine G, Jung HH, Rampoldi L, Tison F, Walker RH. Second international neuroacanthocytosis symposium: expanding the spectrum of choreatic syndromes: Montreal neurological hospital and institute. Mov Disord 20: 1673–1684
3. Ashburner J, Friston KJ (2000) Voxel-based morphometry – the methods. Neuroimage 11:805–821
4. Bridle N, Pantelis C, Wood S, Coppola R, Velakoulis D, McStephen M, Tierney P, Le T, Torrey E, Weinberger D (2002) Thalamic and caudate volumes in monozygotic twins discordant for schizophrenia. Aust N Z J Psychiatry 36:347–354
5. Chamberlain S, Blackwell A, Fineberg N, Robbins T, Sahakian B (2005) The neuropsychology of obsessive-compulsive disorder: the importance of failures in cognitive and behavioural inhibition as candidate endophenotypic markers. Neurosci Biobehav Rev 29:399–419
6. Chard DT, Parker GJ, Griffin CM, Thompson AJ, Miller DH (2002) The reproducibility and sensitivity of brain tissue volume measurements derived from an SPM-based segmentation methodology. J Magn Reson Imaging 15:259–267
7. Crum W, Griffin L, Hill D, Hawkes D (2003) Zen and the art of medical image registration: correspondence, homology, and quality: Neuroimage 20:1425–1437
8. Danek A, Sheesley L, Tierney M, Uttner I, Grafman J. (2004) Cognitive and neuropsychiatric findings in McLeod syndrome and in chorea-acanthocytosis. In: Danek A (ed) Neuroacanthocytosis syndromes. Springer, Dordrecht, The Netherlands, pp 95–115
9. Downhill JE, Buchsbaum MS, Wei T et al. (2000) Shape and size of the corpus callosum in schizophrenia and schizotypal personality disorder. Schizophr Res 42:193–208

10. Gradstein L, Danek A, Grafman J, Fitzgibbon EJ (2005) Eye movements in chorea-acanthocytosis. Invest Ophthalmol Vis Sci 46:1979–1987
11. Hardie RJ, Pullon HW, Harding AE, Owen JS, Pires M, Daniels GL, Imai Y, Misra VP, King RH, Jacobs JM, et al. (1991) Neuroacanthocytosis. A clinical, haematological and pathological study of 19 cases. Brain 114:13–49
12. Henkel K, Danek A, Grafman J, Butman J, Kassubek J (2006) Head of the caudate nucleus is most vulnerable in choreo-acanthocytosis: a voxel-based morphometry study. Mov Disord 21:1728–1731
13. Jones D, Symms M, Cercignani M, Howard R (2005) The effect of filter size on VBM analyses of DTI-MRI data. Neuroimage 26:546–564
14. Jung HH, Hergersberg M, Kneifel S, Alkadhi H, Schiess R, Weigell-Weber M, Daniels G, Kollias S, Hess K (2001) McLeod syndrome: a novel mutation, predominant psychiatric manifestations, and distinct striatal imaging findings. Ann Neurol 49:384–392
15. Kassubek J, Tumani H, Ecker D, Kurt A, Ludolph AC, Juengling FD (2003) Age-related brain parenchymal fraction is significantly decreased in young multiple sclerosis patients: a quantitative MRI study. Neuroreport 14:427–430
16. Kassubek J, Bernhard Landwehrmeyer G, Ecker D, Juengling FD, Muche R, Schuller S, Weindl A, Peinemann A (2004) Global cerebral atrophy in early stages of Huntington's disease: quantitative MRI study. Neuroreport 15:363–365
17. Kassubek J, Juengling FD, Kioschies T, Henkel K, Karitzky J, Kramer B, Ecker D, Andrich J, Saft C, Kraus P, Aschoff AJ, Ludolph AC, Landwehrmeyer GB (2004) Topography of cerebral atrophy in early Huntington's disease: a voxel based morphometric MRI study. J Neurol Neurosurg Psychiatry 75:213–220
18. Lim S (2006) Orofacial dyskinesias and obsessive compulsive disorder. J Clin Neurosci 13:1018
19. Rampoldi L, Dobson-Stone C, Rubio JP, Danek A, Chalmers RM, Wood NW, Verellen C, Ferrer X, Malandrini A, Fabrizi GM, Brown R, Vance J, Pericak-Vance M, Rudolf G, Carre S, Alonso E, Manfredi M, Nemeth AH, Monaco AP (2001) A conserved sorting-associated protein is mutant in chorea-acanthocytosis. Nat Genet 28:119–120
20. Riffkin J, Yucel M, Maruff P, Wood S, Soulsby B, Olver J, Kyrios M, Velakoulis D, Pantelis C (2005) A manual and automated MRI study of anterior cingulate and orbitofrontal cortices, and caudate nucleus in obsessive-compulsive disorder: comparison with healthy controls and patients with schizophrenia. Psychiatry Res Neuroimaging 138:99–113
21. Robertson B, Evans A, Walterfang M, Ng A, Velakoulis D (2008) Epilepsy, progressive movement disorder and cognitive decline. J Clin Neurosci (in press)
22. Rosenberg D, Keshavan M (1998) Toward a neurodevelopmental model of obsessive-compulsive disorder. Biol Psychiatry 43:623–640
23. Shoulson I, Odoroff C, Oakes D, Behr J, Goldblatt D, Caine E, Kennedy J, Miller C, Bamford K, Rubin A et al. (1989) A controlled clinical trial of baclofen as protective therapy in early Huntington's disease. Ann Neurol 25:252–259
24. Thieben MJ, Duggins AJ, Good CD, Gomes L, Mahant N, Richards F, McCusker E, Frackowiak RS (2002) The distribution of structural neuropathology in pre-clinical Huntington's disease. Brain 125:1815–1828
25. Ueno S, Maruki Y, Nakamura M, Tomemori Y, Kamae K, Tanabe H, Yamashita Y, Matsuda S, Kaneko S, Sano A (2001) The gene encoding a newly discovered protein, chorein, is mutated in chorea-acanthocytosis. Nat Genet 28:121–122
26. Vital A, Bouillot S, Burbaud P, Ferrer X, Vital C (2002) Chorea-acanthocytosis: neuropathology of brain and peripheral nerve. Clin Neuropathol 21:77–81
27. Vonsattel JP, DiFiglia M (1998) Huntington disease. J Neuropathol Exp Neurol 57:369–384
28. Walker RH, Danek A, Dobson-Stone C, Guerrini R, Jung HH, Lafontaine AL, Rampoldi L, Tison F, Andermann E (2006) Developments in neuroacanthocytosis: expanding the spectrum of choreatic syndromes. Mov Disord 21:1794–1805

Neuropathology of Chorea-Acanthocytosis

B. Bader(✉), T. Arzberger, H. Heinsen, C. Dobson-Stone, H.A. Kretzschmar, and A. Danek

Abstract General neuropathology in chorea-acanthocytosis (ChAc) is marked by a striking decrease in neurons, predominantly in the striatum, accompanied by a strong reactive gliosis and microglial activation, even more distinctive than that found in Huntington's disease patients. These findings correlate with an impressive macroscopic atrophy of cortical and subcortical structures in ChAc. In Western blot analysis of unaffected brain tissue, chorein expression is ubiquitous. Interestingly, additional bands corresponding to proteins of approximately 160 kDa and 94 kDa comprising chorein N-terminal structures appear in brain as well as in several peripheral tissues. Furthermore, chorein is found at a high degree in testis and erythrocytes and at lower levels in muscle.

B. Bader
Ludwig-Maximilians-Universität, Zentrum für Neuropathologie und Prionforschung, Feodor-Lynen-Straße 23, 81377 München, Germany
benedikt.bader@med.uni-muenchen.de

R.H. Walker et al. (eds.), *Neuroacanthocytosis Syndromes II.*

1 Introduction

Chorea-acanthocytosis (ChAc) is marked by the presence of acanthocytes in blood and neurodegeneration causing a choreiform movement disorder. The connection between acanthocytes in blood and neurodegeneration in brain is poorly understood. Of the two major neuroacanthocytosis syndromes, McLeod syndrome is linked to mutations in the *XK* gene coding for a membrane transport protein which carries the Kx antigen [19], and chorea-acanthocytosis (ChAc) is connected with mutations of the *VPS13A* gene coding for chorein, a protein of as yet unknown function [7, 18].

In contrast to most inherited neurodegenerative diseases, in which we see pathological accumulations of proteins which incorporate the mutated protein (e.g. huntingtin in HD etc), in chorea-acanthocytosis the absence of chorein apparently results in the formation of acanthocytes and neurodegeneration without protein deposits of any sort.

The knowledge of pathological mechanisms is essential in understanding the disease and developing therapeutic strategies. We have used histopathologic and stereologic techniques as well as Western blot in order to further characterize the affected brain regions, the degree of neurodegeneration and to learn more about normal and pathological appearance of chorein due to qualify neurodegeneration in ChAc.

2 Materials and Methods

2.1 *Stereology*

After fixation in 4% formaldehyde one hemisphere of the cerebrum was completely cut into 440 μm thick sections. Every second section was mounted, digitally photographed and stained with gallocyanin for stereological investigations. Stereology was performed according to the protocol described previously [12]. The antibody against chorein has been reported previously [5].

2.2 *Histology*

Selected areas of the non-stained sections were cut out and embedded in paraffin for histological and immunohistochemical investigations. Histologic slices were stained using primary antibodies against glial fibrillary acidic protein (GFAP) (Dako, Germany) and CD68 (Dako, Germany) respectively. Secondary antibody staining and alkaline phosphatase-anti-alkaline phosphatase (APAAP) reaction was carried out on a Ventana Benchmark system (Ventana, Germany).

2.3 *Western Blot*

Ten percent brain homogenate was prepared on ice by mechanical disruption of the tissue using a S24 glass homogenizer (Schütt, Germany) and brain lysis buffer (100 mM NaCl, 10 mM Tris, 10 mM EDTA, 0,5% Nonident P-40 detergent, 0,5% desoxycholate DOC, pH 7,6). Brain homogenate was cleared for 5 min at 800 rpm and an equal sample amount was denatured with NuPAGE LDS buffer (Invitrogen, Germany) and sample reducing buffer (Invitrogen, Germany). Gel electrophoresis, polyvinylidene difluoride (PVDF) membrane transfer, erythrocyte membranes preparation and processing, and primary antibody detection, were carried out as previously described [5].

3 Basic Neuropathology

In vivo, neurodegeneration can be observed by MRI scans, which show a special vulnerability of the head of the caudate nucleus [13] in ChAc. FDG-PET scans reveal a marked decrease in neuronal glucose metabolism in striatal structures [17, 20], whereas the cortex show regular glucose uptake. Postmortem macroscopic findings in ChAc reveal a significant degeneration of the caudate nucleus, the putamen and minor atrophy of the globus pallidus and the substantia nigra [1, 3, 10, 22]. These findings are very similar to those found in Huntington's disease (HD). On histological and immunohistochemical examination, there was loss of 90% of striatal neurons, accompanied by a striking astroglial and oligodendroglial proliferation as well as microglial activation as shown in Fig. 1 [1, 3, 22]. This results in an increased striatal glial index (number of striatal astroglial and oligodendroglial cells divided by the number of striatal neurons) in ChAc of 46.9 (two cases) compared to 22.9 in HD (five cases) and 3.4 in non-affected controls (five controls). Gliosis was also found in

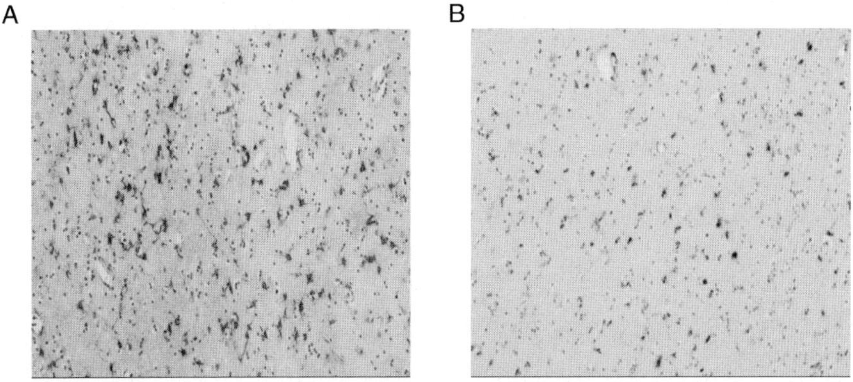

Fig. 1 Immunohistochemistry of the putamen in ChAc (×200 enlargement). (**a**) Staining for glial fibrillary acidic protein (GFAP) shows numerous reactive astrocytes, (**b**) staining for CD68 shows many activated microglial cells or macrophages

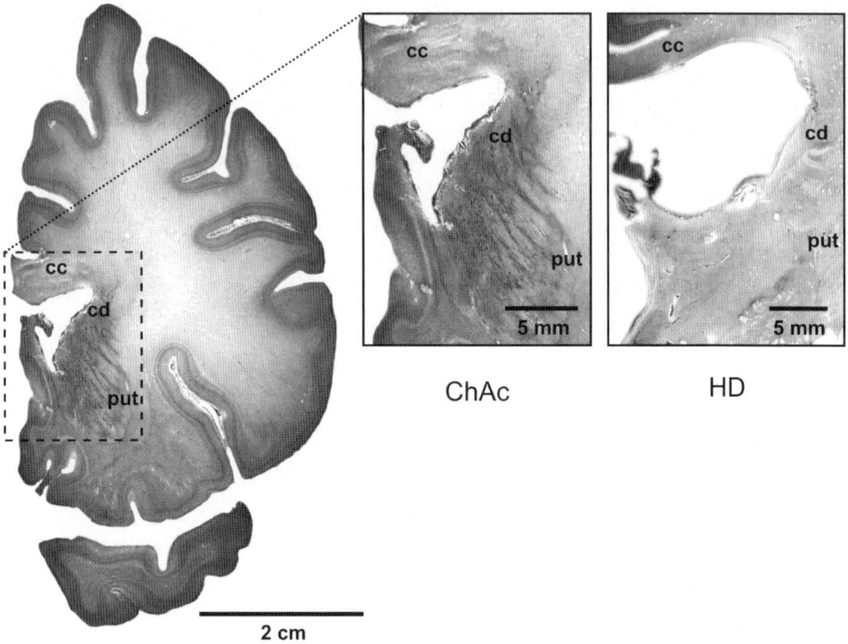

Fig. 2 Postmortem brain section of a 35-year-old male ChAc patient at the level of the anterior striatum. Gallocyanin staining was applied to the tissue to highlight neuronal areas. The cutout in the big picture is shown in the left enlargement. For comparison, an enlargement of approximately the same region of a HD-affected brain is shown in the right enlargement. *cc*, corpus callosum; *cd*, caudate nucleus; *put*, putamen

thalamic regions [22], together with moderate atrophy of the anterior and centromedian nuclei [1]. Compared to HD, pathology of the corpus callosum in ChAc was only mild at best.

Figure 2 shows a post mortem brain section of a 35-year-old male ChAc patient at the level of the anterior striatum. An impressive reduction of the mediolateral diameter of the caudate nucleus to 4 mm can be observed. Brains were embedded in celloidin, cut and stained with gallocyanin [12]. In contrast to HD, there was no apparent atrophy of the corpus callosum (Fig. 1). Three-dimensional reconstructions of striatal structures in HD and ChAc from postmortem sections also demonstrate a dramatic atrophy in both neurodegenerative diseases compared to healthy controls (Fig. 3).

Stereological investigations [3] estimated the total number of astroglial, oligodendroglial and neuronal cells within the striatum of two post mortem ChAc brains compared to HD brains and non-affected control brains (Fig. 3). In ChAc, the number of glial cells (astroglial and oligodendroglial cells) is markedly increased relative to the decrease of neurons. This effect is even more striking than

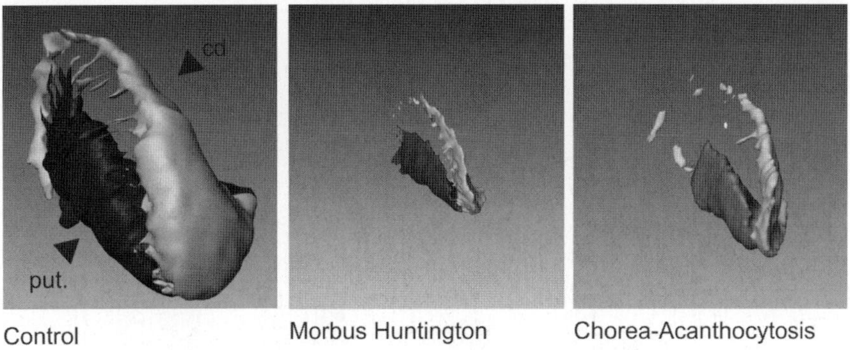

Fig. 3 3D reconstructions of the striatum from digitalized consecutive 440 μm coronal brain sections of healthy controls (control), *HD* (Morbus Huntington) and ChAc. *cd*, caudate nucleus, light grey; put, putamen, dark grey

in five HD patients. Hence, the striatal glia index in ChAc is 2.1 times higher than in HD, and 14 times higher than in non-affected controls (Fig. 4).

4 Chorein Expression In Brain Regions

In non-affected control brains we found chorein expressed at approximately the same level in all examined brain regions (Fig. 5). A similar pattern was found also in substantia nigra, globus pallidus, occipital cortex and cerebellum (data not shown). Full length chorein is represented by a band at approx. 370 kDa (consistent with the predicted molecular weight of chorein of 360 kDa). This band was absent in tissue from ChAc cases (Fig. 5).

In control brain tissue, we found at least two clear bands at approximately 160 kDa and 94 kDa that were not present in ChAc tissue. The origin of these bands is unclear. They represent proteins of distinct size matching chorein epitopes within the amino acids 27–326 of full length chorein due to the synthetic peptide used for the antibody production. This comprises approximately the first 10% of chorein's N-terminus [5]. It is unlikely that they result from unspecific degradation products since these bands do not vary in size and are present in all examined regions. These proteins may represent N-terminal fragments of 25% (94 kDa) and 43% (160 kDa), respectively, of full length chorein e.g. due to alternative splicing or clear defined post-translational modifications.

Bands at approximately 310 kDa and 82 kDa were present in brain tissue from ChAc patients as well as from controls, and therefore correspond most likely to nonspecific cross reactions of the polyclonal chorein antiserum used for detection in Western blot.

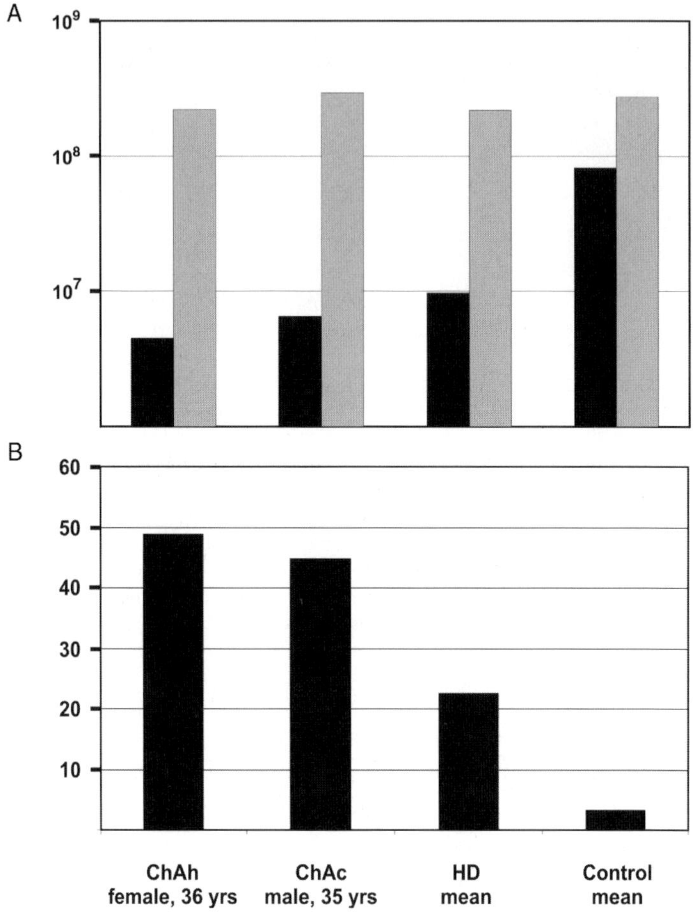

Fig. 4 Stereological data show an increased proportion of glial cells within the striatum in ChAc patients. (**a**) Cell numbers of astroglial and oligodendroglial cells (grey) and neurons (black) estimated [12] for the striatum of two ChAc cases, the mean of five HD patients and the mean of five non-affected controls. (**b**) The striatal glial index was calculated

5 Chorein Expression In Peripheral Tissues

Kurano et al. reported that in mice, chorein is highly expressed in brain, testis, kidney, spleen and muscle [15]. In human, chorein is found in brain and blood, and at a high level in testis (Fig. 6). Minor expression can be observed in muscle tissue. So far, chorein was not detected in peripheral nerve, liver and kidney (Fig. 6) or spleen, small intestine and colon (data not shown). In most peripheral tissues except peripheral nerve, a band was found at 82 kDa, which was also seen in brain of

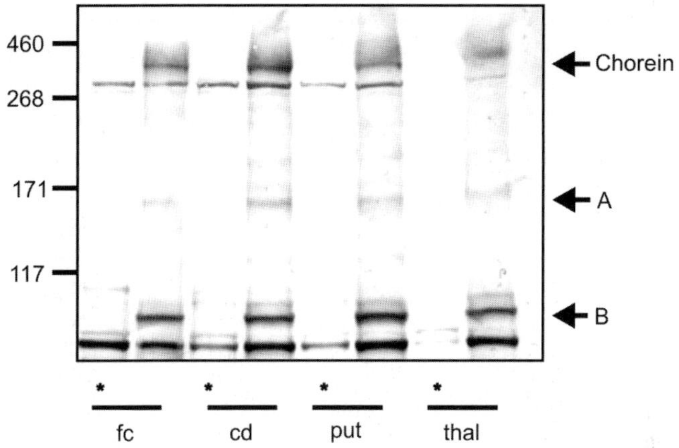

Fig. 5 Western blot for chorein in different brain regions from an individual affected by ChAc (*) and a non-affected control subject. In control, tissue a band corresponding to chorein is seen at 370 kDa. In ChAc tissue, chorein is absent. The two bands visible in these lanes (69 kDa and 306 kDa) are due to nonspecific cross reactions. In brain homogenate of the control, two further bands are seen (A, 160 kDa; B, 94 kDa), which may be due to alternative splicing or post-translational modifications of chorein. *fc*, frontal cortex; *cd*, caudate nucleus; *put*, putamen; *thal*, thalamus. Numbers at the left side indicate molecular weight calibration in kDa

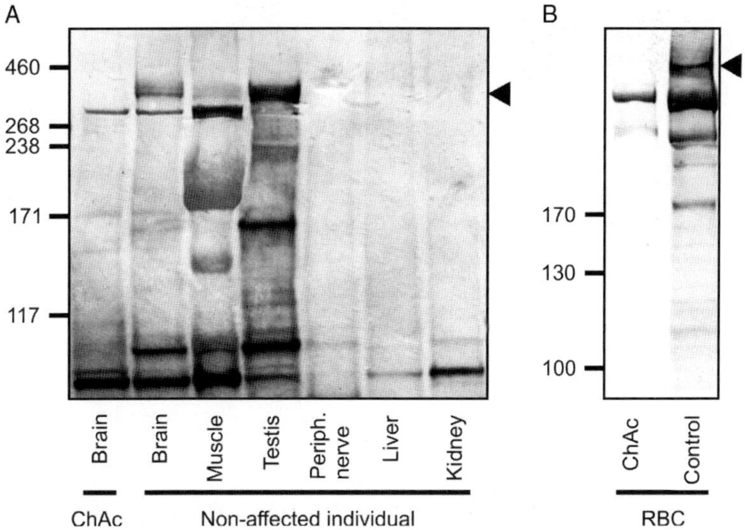

Fig. 6 (**a**) Western blot of peripheral tissues of a non-affected individual (42-year-old male) compared to brain tissue of an affected ChAc patient (35-year-old male). (**b**) Western blot of erythrocyte membrane preparations (RBC, red blood cells) of a ChAc patient and a non-affected control patient. Numbers at the left side indicate molecular weight calibration in kDa. Arrowheads indicate height of chorein band

affected and non-affected patients and therefore likely to be due to nonspecific cross-reactivity. A strong band at 317 kDa was also very likely to be attributable to nonspecific antibody cross-reactivity. In brain, as well as in muscle and especially in testis, a clear band appeared in the range of 94–97 kDa. This band was not present in ChAc brain and for that reason was possibly a product of alternative splicing of *VPS13A* or specific post-translational proteolytic modifications of chorein. A strong band at about 160 kDa was found both in testis and in unaffected brain tissue. Additionally, a large band appears in unaffected muscle tissue with its maximum at 199 kDa. ChAc muscle tissue has not yet been examined. There are several minor bands of as yet unknown origin showing up in testis. Further experiments have to be carried out to check whether these bands are chorein specific or not.

6 Discussion

Loss of neurons is a hallmark of neurodegenerative diseases. Due to the generally non-regenerative nature of neuronal cells in the adult central nervous system, this results in an irreversible and fatal process of neurodegeneration. Whereas necrotic cell death does not require active participation of the cell itself, non-necrotic processes are usually regulated by autonomous processes of the affected cell. Among several types of non-necrotic cell death, apoptosis is one of the most common forms of tissue degeneration [4, 14]. In neurodegenerative diseases, cell death as a result of apoptotic degeneration has been reported for Alzheimer's, Parkinson's and Huntington's diseases [2, 6, 21]. However, several studies have failed to prove the hypothesis of apoptosis being directly responsible for cell death in neurodegenerative diseases [9]. In ChAc, the mechanism of cell death is still a mystery. Necrotic residues can not be observed, but loss of neurons is remarkable.

In recent years the theory of neuronal cell death caused by proteins other than the classical apoptosis-related proteins, Bcl-2 and caspase families, has emerged. Among these, the amyloid precursor protein, alpha-synuclein, presenilins and huntingtin seem to play important roles, strongly supported by the finding of mutations of the genes producing these proteins in familial forms of the major neurodegenerative diseases [8, 11, 16]. However, apoptosis can not be ruled out as a subsequent effect of alterations of other proteins. It is possible that chorein plays a regulatory role in the cellular cascade leading towards apoptosis which is abolished by missing chorein. Unlike the protein aggregations observed in various neurodegenerative diseases like Alzheimer's disease, Huntington's disease and Parkinson's disease, we do not observe an accumulation of protein in ChAc. It is therefore possible, that the pathological phenomena in ChAc occur as a result of loss of function due to the loss of chorein rather than a gain of function. Future research concerning the physiological function of chorein will help to uncover the pathological pathways leading to neurodegeneration in ChAc.

References

1. Alonso ME, Teixeira F, Jimenez G, Escobar A (1989) Chorea-acanthocytosis: report of a family and neuropathological study of two cases. Can J Neurol Sci 16:426–431
2. Andersen JK (2001) Does neuronal loss in Parkinson's disease involve programmed cell death? Bioessays 23:640–646
3. Arzberger T, Heinsen H, Buresch N, Dobson-Stone C, Danek A, Kretzschmar H (2005) The neuropathology of chorea-acanthocytosis: from stereology to an immunohistochemical detection of chorein. Mov Disord 20:1679
4. Clarke PG (1990) Developmental cell death: morphological diversity and multiple mechanisms. Anat Embryol (Berl) 181:195–213
5. Dobson-Stone C, Velayos-Baeza A, Filippone LA, Westbury S et al. (2004) Chorein detection for the diagnosis of chorea-acanthocytosis. Ann Neurol 56:299–302
6. Evert BO, Wullner U, Klockgether T (2000) Cell death in polyglutamine diseases. Cell Tissue Res 301:189–204
7. Fuller R (2005) Soi1/VPS13 function in TGN-endosomal cycling in yeast: a paradigm for CHAC function? Mov Disord 20:1682
8. Goedert M, Spillantini MG (2006) A century of Alzheimer's disease. Science 314:777–781
9. Graeber MB, Moran LB (2002) Mechanisms of cell death in neurodegenerative diseases: fashion, fiction, and facts. Brain Pathol 12:385–390
10. Hardie R (1998) Cerebral hypoperfusion and hypometabolism in chorea-acanthocytosis. Mov Disord 13:853–854
11. Hardy J, Cai H, Cookson MR, Gwinn-Hardy K et al. (2006) Genetics of Parkinson's disease and parkinsonism. Ann Neurol 60:389–398
12. Heinsen H, Arzberger T, Schmitz C (2000) Celloidin mounting (embedding without infiltration) – a new, simple and reliable method for producing serial sections of high thickness through complete human brains and its application to stereological and immunohistochemical investigations. J Chem Neuroanat 20:49–59
13. Henkel K, Danek A, Grafman J, Butman J et al. (2006) Head of the caudate nucleus is most vulnerable in chorea-acanthocytosis: a voxel-based morphometry study. Mov Disord 21:1728–1731
14. Kerr JF, Wyllie AH, Currie AR (1972) Apoptosis: a basic biological phenomenon with wide-ranging implications in tissue kinetics. Br J Cancer 26:239–257
15. Kurano Y, Nakamura M, Ichiba M, Matsuda M et al. (2006) In vivo distribution and localization of chorein. Biochem Biophys Res Commun 352:431
16. Lee ST, Kim M (2006) Aging and neurodegeneration. Molecular mechanisms of neuronal loss in Huntington's disease. Mech Ageing Dev 127:432–435
17. Leenders K (2005) Tracer imaging in basal ganglia degeneration. Mov Disord 20:1679
18. Levecque C, Dobson-Stone C, Velayos-Baeza A, Monaco AP (2005) Potential interaction partners of VPS13 proteins. Mov Disord 20:1683
19. Russo DC, Lee S, Reid ME, Redman CM (2002) Point mutations causing the McLeod phenotype. Transfusion 42:287–293
20. Saiki S, Sakai K, Kitagawa Y, Saiki M et al. (2003) Mutation in the CHAC gene in a family of autosomal dominant chorea-acanthocytosis. Neurology 61:1614–1616
21. Su JH, Anderson AJ, Cummings BJ, Cotman CW (1994) Immunohistochemical evidence for apoptosis in Alzheimer's disease. Neuroreport 5:2529–2533
22. Vital A, Bouillot S, Burbaud P, Ferrer X et al. (2002) Chorea-acanthocytosis: neuropathology of brain and peripheral nerve. Clin Neuropathol 21:77–81

The Neuropathology of McLeod Syndrome

F. Geser, M. Tolnay, and H.H. Jung(✉)

Abstract McLeod syndrome (MLS) belongs to the heterogeneous group of neuroacanthocytosis (NA) syndromes that are characterized by an involvement of the hematological and nervous systems. Central nervous system symptoms of MLS resemble Huntington's disease (HD) or choreoacanthocytosis (ChAc) and include a choreatic movement disorder, psychiatric abnormalities, cognitive decline, and generalized seizures. In MLS, rather non-specific pathological changes are present in the caudate nucleus, putamen and pallidum, which are characterized by neuronal loss and astrogliosis. ChAc may show an additional involvement of the substantia nigra and thalamus, and HD features more widespread pathology and the presence of distinctive intranuclear inclusions. Cortical pathology predominantly occurs in HD, is less pronounced in ChAc, and most likely present to an only minor extent in MLS. However, the nature of cortical, subcortical, and basal ganglia pathology in MLS remains to be investigated in more detail in larger autopsy series.

H.H. Jung
Department of Neurology, University Hospital Zürich, 8091 Zürich, Switzerland
hans.jung@usz.ch

R.H. Walker et al. (eds.), *Neuroacanthocytosis Syndromes II.*

1 Introduction

McLeod syndrome (MLS) belongs to the heterogeneous group of neuroacanthocytosis (NA) syndromes that are characterized by involvement of both the hematological and nervous system [8, 9, 16, 28, 29]. Central nervous system symptoms of MLS are similar to Huntington's disease (HD) and comprise a choreatic movement disorder, psychiatric abnormalities, cognitive decline, and generalized seizures [9, 16]. About 150 MLS patients are reported worldwide, but data on brain pathology consist of only three cases published in either the form of a book chapter, abstracts or small case series [4, 5, 13, 22, 23]. Here we summarize the neuropathological features of these cases as reported in the literature and compare them with choreoacanthocytosis (ChAc), which is another NA syndrome of differential diagnostic interest, and HD.

2 Case Reports

2.1 *Case 1 (From Literature)*

In 1993, Brin and colleagues reported on a male patient from a New Zealand family [3] with MLS who died at the age of 50 of unknown cause after a disease duration of at least 9 years [4, 5] (This case is discussed in more detail in the chapter by Danek et al.). Clinical features included generalized chorea, dysarthria, dysphagia, wide-based gait, proximal myopathy with fasciculations, and absent deep tendon reflexes. Sensation was normal except for mild distal vibration loss. Gross examination of the brain showed marked atrophy of the caudate nucleus. Microscopic evaluation revealed neuronal loss and astrogliosis, which was severe and widespread throughout the caudate nucleus and moderate in the putamen. Only mild diffuse astrogliosis was described in the pallidum. Astrogliosis was also found in a small region of the substantia nigra, which, however, did not show overt neuronal loss. Other brain regions, including the cortex, the brainstem and the cerebellum, were reported to be normal and white matter alterations were not mentioned [4, 5].

2.2 *Case 2 (From Literature)*

In 1994, Rinne and colleagues published two papers on the neuropathology of three cases with "neuroacanthocytosis", one of whom had MLS [22, 23]. Of note, this was a manifesting 51 years old female heterozygote patient published a few years earlier by Hardie et al. ([14], case 5; also discussed in chapters by Danek et al. and Gandhi et al.). The woman had the prototypic clinical features of MLS including progressive chorea and cognitive decline [14]. Neuropathological evaluation revealed a normal brain weight [23]. On coronal sections, a symmetric dilatation of the lateral ventricles and marked atrophy of the striatum, in particular in its anterior areas, were seen. The brainstem was described to be slightly reduced in bulk. On microscopic evaluation, the caudate showed severe neuronal loss and astrogliosis, mainly in the posterior regions. Considerable neuronal loss and astrocytic gliosis was also found in the putamen, albeit to a lesser degree than the caudate. Both segments of the pallidum were also gliotic. In the neocortex no "obvious nerve cell loss or gliosis" [23] was present. The white matter showed slight myelin pallor in the frontal, temporal, and occipital lobes. The substantia nigra was described as normal with a neuronal density at the lower limit of the control range [22]. Some extraneuronal pigment was noted, most likely related to aging.

2.3 *Case 3*

Recently, we described the clinical features, the neuroradiological findings, and the neuropathology of a patient with MLS [13, 15–17]. The clinical presentation was dominated by recurrent psychotic episodes, subsequent generalized chorea, and moderate cognitive decline. In addition, neurological examination showed absent deep tendon reflexes and moderate generalized muscular atrophy. The patient died, probably due to cardiac arrhythmia, at the age of 55 after a disease duration of 25 years. On gross examination, the external surface of the brain was normal; in particular no atrophy of the gyri was found (Fig. 1a). On coronal sections, a pronounced symmetric dilatation of the lateral ventricles was evident, especially of the frontal horns and central parts. The caudate nucleus and putamen were severely atrophic (Fig. 1b), and the globus pallidus was almost as severely

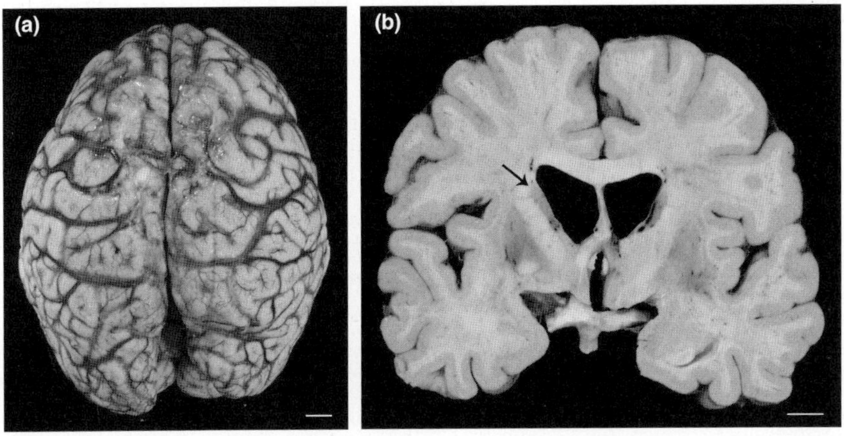

Fig. 1 Gross appearance (bars = 1cm). **(a)** External surface of the brain with no abnormalities **(b)** Coronal section showing enlargement of the lateral ventricles and atrophy of the basalganglia, in particular of the caudate nucleus (arrow)

reduced in size as the striatum. The substantia nigra and locus coeruleus were normally pigmented. Microscopic evaluation demonstrated an almost complete neuronal loss in the striatum. The remaining tissue showed a spongiform appearance with a marked reactive astrogliosis (Fig. 2a and 2b). Pronounced neuronal cell loss and astrogliosis were also found in the globus pallidus – albeit to a slightly lesser degree than in the striatum. No disease-defining intraneuronal or intranuclear inclusions were identified by routine histochemical and immunohistochemical staining. Brain areas without pathology included thalamus, subthalamic nucleus, cerebellum, midbrain, pons, and medulla. The substantia nigra as well as the locus coeruleus demonstrated a normal density of pigmented cells and no extracellular pigment was detected. In addition to the basal ganglia pathology, immunohistochemical staining with an antibody directed against glial fibrillary acidic protein revealed moderate focal subcortical white matter and subtle cortical astrogliosis, in particular in frontal areas (Fig. 3a and 3b).

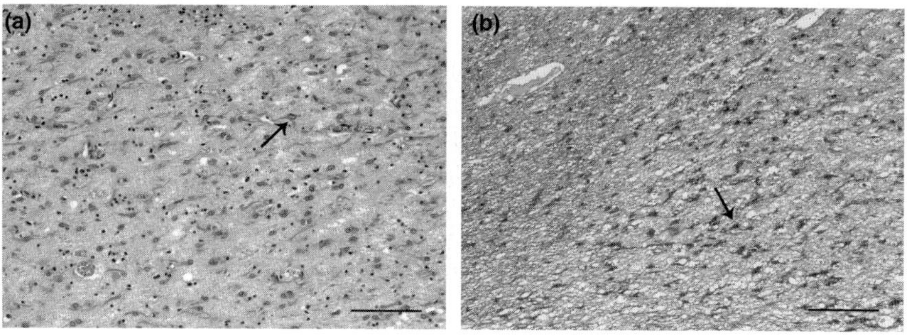

Fig. 2 Microphotograph of the striatum. Severe striatal neuronal loss and astrogliosis (e.g., arrows) (**a**) Hematoxylin and eosin stain (bar = 100 μm), (**b**) Anti-glial fibrillary acidic protein immunohistochemistry (bar = 200 μm)

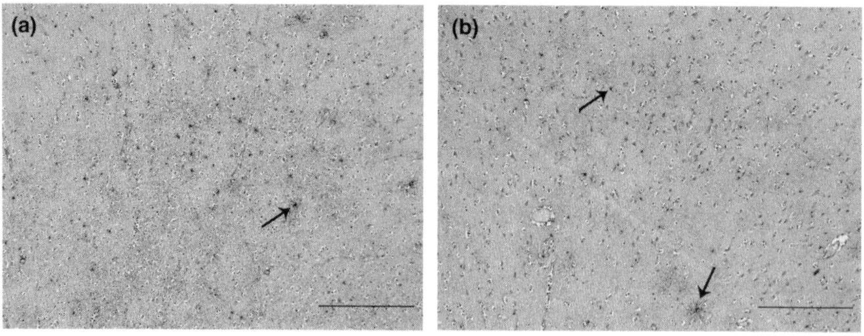

Fig. 3 Microphotograph of the temporal lobe. (anti-glial fibrillary acidic protein immunohistochemistry, bars = 200 μm). Moderate subcortical (**a**) and subtle cortical astrogliosis (**b**) (e.g., arrows)

3 Discussion

The common neuropathological features of the three published McLeod cases include neuronal loss and reactive astrogliosis predominantly in the basal ganglia, especially in the caudate nucleus, without disease-defining intraneuronal or intranuclear inclusions. Cortical or subcortical white matter abnormalities including subtle gliosis or myelin pallor were not consistently reported.

The minimal cortical pathology of our own case (case 3) is in line with previous magnetic resonance spectroscopic data demonstrating only mild extrastriatal changes in this disease [11]. These changes might reflect a subtle histopathological correlate of a functional disruption basal ganglia-cortical circuits, resulting not only in a motor disorder, but also in the prominent psychiatric and cognitive symptoms.

Neuropathological studies of ChAc showed many similarities but also revealed differing features. As in MLS, predominant atrophy of the striatum, and, to a lesser extent, of the globus pallidus were found [1, 2, 6, 12, 14, 19, 23, 24, 25, 28]. In some cases, however, white matter [1, 19], or cortical pathology was also described [2, 19]. Also in contrast to MLS, pathological changes in the thalamus [1, 6, 23, 24, 28] and nigral neuronal loss coupled with astrogliosis of variable degrees have been reported [1, 12, 14, 19, 20, 22, 23]. Correspondingly, almost one third of ChAc patients have parkinsonian features during life [8]. No definite nigral involvement is found in MLS and parkinsonian features are found in less than 20% of McLeod patients [8, 9].

HD has many clinical similarities to MLS and ChAc. Pathologically, HD is characterized by neuronal loss and astrogliosis of the striatum. The degenerative process displays a caudo-rostral, dorso-ventral, and medio-lateral direction which is reflected in a grading system of striatal atrophy ranging from grade 1 to 4 [26]. Although there is only a small number of MLS cases with reported neuropathology available there seems to be no neostriatal pathology gradient like that found in HD [27]. In higher HD grades, the pallidum as well as neo- and allocortex, thalamus, hypothalamus, subthalamic nucleus, white matter, pons, or cerebellum may also be involved [18, 27]. The pars reticulata of the substantia nigra may show neuronal loss while the pars compacta – albeit reported to be thinner – has a normal number of neurons [7, 21, 27]. The pathology of HD is characterized by an accumulation of mutant huntingtin protein fragments within intranuclear inclusion bodies or neurites which have been shown to be widely distributed throughout the neocortex [10, 18]. Despite the clinical similarities between HD and MLS, the different and more restricted pattern of degeneration as well as the lack of a specific inclusion pathology in the latter allows a clear neuropathological distinction between these two disorders.

4 Conclusion

In MLS, rather non-specific pathological changes are present in the caudate nucleus, putamen and pallidum, characterized by neuronal loss and astrogliosis. ChAc may show an additional involvement of the substantia nigra and thalamus. HD is characterized by a more widespread pathology and the presence of distinctive intranuclear inclusions. Cortical pathology predominantly occurs in HD, is less pronounced in ChAc, and most likely present to an only minor extent in MLS. However, the degree and topographical distribution of cortical, subcortical, and basal ganglia pathology in MLS remains to be elucidated in larger autopsy series [27].

References

1. Alonso ME, Teixeira F, Jimenez G, Escobar A (1989) Chorea-acanthocytosis: report of a family and neuropathological study of two cases. Can J Neurol Sci 16(4):426–431
2. Arzberger T, Heinsen H, Buresch N, Dobson-Stone C, Danek A, Kretzschmar H (2005) The neuropathology of chorea-acanthocytosis: from stereology to an immunohistochemical detection of chorein. Mov Disord 20(12):1679
3. Bertelson CJ, Pogo AO, Chaudhuri A et al. (1988) Localization of the McLeod locus (XK) within Xp21 by deletion analysis. Am J Hum Genet 42(5):703–711
4. Brin MF (1993) Acanthocytosis. In: Goetz CG, Tanner CM, Aminoff MJ (eds) Handbook of clinical neurology: systemic diseases, Part I. Elsevier, Amsterdam, pp 271–299
5. Brin MF, Hays A, Symmans WA, Marsh WL, Rowland LP (1993) Neuropathology of McLeod phenotype is like chorea-acanthocytosis (CA). Can J Neurol Sci 20(Suppl):234
6. Burbaud P, Vital A, Rougier A et al. (2002) Minimal tissue damage after stimulation of the motor thalamus in a case of chorea-acanthocytosis. Neurology 59(12):1982–1984
7. Campbell AM, Corner B, Norman RM, Urich H (1961) The rigid form of Huntington's disease. J Neurol Neurosurg Psychiatry 24:71–77
8. Danek A, Jung HH, Melone MA, Rampoldi L, Broccoli V, Walker RH (2005) Neuroacanthocytosis: new developments in a neglected group of dementing disorders. J Neurol Sci 229–230:171–186
9. Danek A, Rubio JP, Rampoldi L et al. (2001) McLeod neuroacanthocytosis: genotype and phenotype. Ann Neurol 50(6):755–764
10. Dietrich P, Dragatsis I (2005) Knock-in and knock-out models of Huntington disease. In: LeDoux M (ed) Animal models of movement disorders. Elsevier, Oxford, pp 317–328
11. Dydak U, Mueller S, Sandor PS, Meier D, Boesiger P, Jung HH (2006) Cerebral metabolic alterations in McLeod syndrome. Eur Neurol 56(1):17–23
12. Galatioto S, Serra S, Batolo D, Marafioti T (1993) Amyotrophic choreo-acanthocytosis: a neuropathological and immunocytochemical study. Ital J Neurol Sci 14(1):49–54
13. Geser F, Prokop S, Glatzel M, Tolnay M, Jung H (2006) The neuropathology of McLeod syndrome: a case study. Mov Disord 21(Suppl 5):357
14. Hardie RJ, Pullon HW, Harding AE et al. (1991) Neuroacanthocytosis. A clinical, haematological and pathological study of 19 cases. Brain 114 (Pt 1A):13–49
15. Jung HH, Haker H (2004) Schizophrenia as a manifestation of X-linked Mcleod-Neuroacanthocytosis syndrome. J Clin Psychiatry 65(5):722–723
16. Jung HH, Hergersberg M, Kneifel S et al. (2001) McLeod syndrome: a novel mutation, predominant psychiatric manifestations, and distinct striatal imaging findings. Ann Neurol 49(3):384–392
17. Jung HH, Russo D, Redman C, Brandner S (2001) Kell and XK immunohistochemistry in McLeod myopathy. Muscle Nerve 24(10):1346–1351
18. Lowe JS, Leigh N (2002) Disorders of movement and system degenerations. In: Graham DI, Lantos PL (eds) Disorders of movement and system degenerations, 7th edn. Arnold, London, New York, New Delhi, pp 325–430
19. Rafalowska J, Drac H, Jamrozik Z (1996) Neuroacanthocytosis. Review of literature and case report. Folia Neuropathol 34(4):178–183

20. Rampoldi L, Danek A, Monaco AP (2002) Clinical features and molecular bases of neuroacanthocytosis. J Mol Med 80(8):475–491
21. Richardson EPJ (1990) Third Dorothy S. Russell memorial lecture. Huntington's disease: some recent neuropathological studies. Neuropathol Appl Neurobiol 16(6):451–460
22. Rinne JO, Daniel SE, Scaravilli F, Harding AE, Marsden CD (1994) Nigral degeneration in neuroacanthocytosis. Neurology 44(9):1629–1632
23. Rinne JO, Daniel SE, Scaravilli F, Pires M, Harding AE, Marsden CD (1994) The neuropathological features of neuroacanthocytosis. Mov Disord 9(3):297–304
24. Stevenson VL, Hardie RJ (2001) Acanthocytosis and neurological disorders. J Neurol 248(2):87–94
25. Vital A, Bouillot S, Burbaud P, Ferrer X, Vital C (2002) Chorea-acanthocytosis: neuropathology of brain and peripheral nerve. Clin Neuropathol 21(2):77–81
26. Vonsattel JP, Myers RH, Stevens TJ, Ferrante RJ, Bird ED, Richardson EP Jr (1985) Neuropathological classification of Huntington's disease. J Neuropathol Exp Neurol 44(6):559–577
27. Vonsattel JP, Lianski M (2004) Huntington's disease. In: Esiri MM, Lee VMY, Trojanowksi JQ (eds) The neuropathology of dementia, 2nd edn. Cambridge University Press, CA, pp 376–401
28. Walker RH, Danek A, Dobson-Stone C et al. (2006) Developments in neuroacanthocytosis: expanding the spectrum of choreatic syndromes. Mov Disord 21(11):1794–1805
29. Walker RH, Jung HH, Dobson-Stone C et al. (2007) Neurologic phenotypes associated with acanthocytosis. Neurology 68(2):92–98

Cerebral Involvement in McLeod Syndrome: The First Autopsy Revisited

A. Danek(✉), M. Neumann, M.F. Brin, W.A. Symmans+, and A.P. Hays

Abstract Patient WA was a member of the New Zealand "S" family with an *XK* gene deletion, and was among the first patients with McLeod syndrome (MLS) comprehensively assessed by Laurence Marsh and collaborators in the 1970s. He displayed some hyperkinetic movements from his third decade of life and within 20 years developed a full picture of MLS: pathognomonic Kell phenotype, hepatosplenomegaly, cardiomyopathy, creatine kinase (CK) and liver enzyme elevation, muscle weakness and atrophy, neuropathy, dysarthria, dysphagia, chorea, and neuropsychiatric abnormalities. He died at the age of 50 years of unknown cause. The post mortem findings have not been fully reported up to now, although the autopsy study was the very first known of MLS. The neuropathological changes were largely limited to the striatum and consisted of bilateral neuronal loss and astrocytosis, more severe in the caudate nucleus than in the putamen. Only two other autopsy reports on MLS are available to date. These showed a similar picture to case WA and also resembled the findings in post mortem cases with a molecular diagnosis of chorea-acanthocytosis (ChAc).

A role for endothelins has been previously the subject of speculation as a potential mechanism for pathophysiology in MLS. However, this now seems implausible on the basis of recent expression studies of the endothelin-cleaving protein Kell and the McLeod protein XK in the brains of experimental animals and of humans. In brain only XK was found, whereas in red cell membranes and other peripheral tissues XK colocalises with Kell, to which it is covalently linked.

A. Danek
Neurologische Klinik, Ludwig-Maximilians-Universität, D-81366 München, Germany
danek@lmu.de

R.H. Walker et al. (eds.), *Neuroacanthocytosis Syndromes II.*

At present, study of the neuropathology of both conditions, MLS and ChAc, is still in its infancy. Increasing clinical recognition of neuroacanthocytosis syndromes and increasing availability of autopsy material will hopefully soon clarify some of the many open issues.

1 Introduction

For a considerable time the finding of weak Kell antigens in the dental student Hugh McLeod had remained an oddity only known to immunohematologists, such as the groups of F.H. Allen in Boston and of W.L. Marsh in New York [1]. The occurrence of his peculiar "McLeod" erythrocyte phenotype in boys with chronic granulomatous disease (CGD) was not noted until after a decade had passed [21, 34] and neither was the shape-change of the antigen-deficient red cells because of acanthocytic deformation [19, 49].

The McLeod syndrome (MLS) with X-linkage and a variety of features apart from red cell involvement was delineated by W.L. Marsh only in the 1970s. On a lecture trip through New Zealand, local physicians asked him to consult concerning a family with "acanthocytosis associated with haemolytic anaemia" (Fig. 1) . After his return, Marsh on November 25, 1975 wrote a letter to Dr. McLeod who was in regular contact with the New York Blood Center as a blood donor for CGD patients: "I thought you would like to know that you are no longer unique for we have now tested a large family sent from New Zealand and two of the male

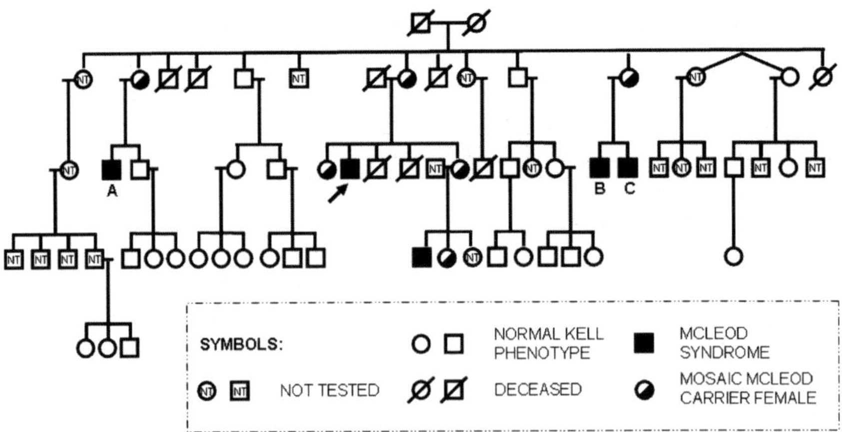

Fig. 1 Pedigree of the "S" family with McLeod syndrome due to an *XK* deletion [33, 45]. The index patient of the present communication, case WA, is marked by an *arrow* and corresponds to patient III-7 of [33] (as well as to II-4 of [45] and case 8 of [13]). Cases A, B, and C in [33] correspond to III-2 (also case 7 of [13]), III-17 (also case 2 of [6] and case 5 of [13]), and III-16 (also case 1 of [6] and case 6 of [13]), respectively. Please note that the two brothers B and C had erroneously been described as cousins or single case ("S/H") in the past [6, 24, 26]

members have the red cell McLeod phenotype. ... We have been able to show that some female members in the New Zealand family are carriers of the gene."

The New Zealand "S" family [45] was studied repeatedly for the multi-system features that were first properly noted in its members [32, 33, 42] and was also important in the first attempts to define the syndrome's molecular basis [6, 24, 26]. The first autopsy of a male (WA) with MLS is from this family, briefly mentioned in reviews of MLS [13, 33, 45] and individually reported with little detail in a book chapter [8] and an abstract [9]. Here we provide the first comprehensive account of the clinical and pathological findings of this patient.

2 Case Report

This white man from the "S" family (II-4 in the pedigree of [45], see also Fig. 1), was originally a grocer's shop assistant but retired in his forties because of increasing neurological impairment. He had displayed minimal signs of choreatic movements from his mid twenties. WA had lived by himself and was discovered dead in his bed at the age of 50 years.

He had first presented to his local hospital at the age of 34 years with tiredness and intermittent left-sided upper abdominal pain. Both remitted after an enlarged spleen was removed. Mild hypertension (elevated to 160/100 mmHg) was treated medically. At the ages of 43, 44, and 47 years, chest x-rays showed mainly left ventricular enlargement of the heart, increased in comparison to previous films. Electrocardiogram showed sinus rhythm with marked ectopic activity. Cardiac ultrasound at age 47 disclosed no mitral or aortic valve abnormalities, but showed thickened walls of the left ventricle and of the interventricular septum (both 20 mm) as well as enlargement of the left atrium and left ventricle and generalized hypokinesia.

Laboratory assessments between the ages of 34 and 50 disclosed elevated creatine kinase (CK) levels that ranged from 788 to 3240 U/l (normal < 200 U/l) as well as slightly elevated levels of lactate dehydrogenase (LDH) (254–367 U/l, normal < 250), Alanine transminase (ALT) (43–80 U/l, normal < 55 U/l) and aspartate aminotransferase (AST) (60–100 U/l, normal < 48 U/l). Gamma-glutamyl transpeptidase (γGT), in contrast, was within normal limits as were hepatitis markers, bilirubin, and alkaline phosphatase as well as serum cholesterol, triglycerides and lipoprotein electrophoresis, and haptoglobin. Hemoglobin was 14.0–16.0 g/dl, MCV 87fl, and reticulocyte count 1.8–2.0%. Serum iron was 18 μmol/l, iron binding capacity 72 μmol/l, and ferritin 144 μg/l. Blood films showed 30–40% acanthocytes.

His Kell erythrocyte typing results have been reported previously and confirmed the McLeod red cell phenotype [45]. Although this was not directly assessed, a major deletion of the McLeod gene *XK*, affecting the promoter and exon 1 with resultant absence of the XK protein is likely as this deletion was found in his cousins, B and C (see Fig. 1, also known as S/H [6, 24, 26]).

From his early forties the choreatic movements became progressively more frequent and intense and he developed dysarthria. Examination at the age of 47 years showed distortion of speech by extra movements of his lips and of the tongue that moved in his mouth in a restless manner. There was also a fine tongue tremor. Facial expression was described as "slightly lop sided with jaw forward". Muscle power was moderately reduced and muscle bulk was globally diminished in this man of body mass index 21 (183 cm; 70 kg), disproportionately affecting proximal limb muscles (triceps, biceps, and deltoid in particular). Muscle tone was globally reduced and fasciculations were present, e.g. in the triceps and biceps bilaterally, increasing on gentle tapping. Generalized areflexia had been noted several years previously. Plantar responses were flexor. Sensation was normal except for absent vibration sense at the ankles. There was no ataxia on finger-nose-testing. His gait was wide-based and bent forward. His choreatic movements were of small amplitude, and irregularly and unpredictably affected his face, arms, abdominal wall, thigh and feet. These movements waxed and waned over periods of several weeks, and increased during tension. They were felt to respond somewhat to tetrabenazine 25 mg q.i.d. but over the course of 2 years had clearly deteriorated, severely affecting everyday activities and making dressing, walking, and cycling difficult. He was able to cycle for at least 30 km, but was reminiscent of an inebriate because of his erratic and unsteady course. Feeding was impaired because he spilled and jerked food around through sudden involuntary movements. However, he continued to live by himself, with little outside help.

Although somewhat socially withdrawn he had belonged to a darts club and had appeared quite proficient at it. There were some reports of unusual behavior with sexual deviancy and documented exhibitionism. His mental state seemed to deteriorate but was difficult to assess because of dysarthria. He had a fair knowledge of world affairs and recent events and was able to speedily and accurately solve easy mental arithmetic such as the serial 7s test.

When seen shortly before his death, WA continued to have marked choreoathetoid movements of trunk and limbs. Involuntary movements also involved the head and tongue. His speech had further deteriorated, having become quite slurred. Tongue movements were slow and there was occasional drooling. The gag reflex was prominent. His stance was stable, but his walking was more stooped and unstable and the head more flexed. There was prominent wrinkling of his brow, a rather flat facial affect with paucity of expression. Most prominent was the increasing dysphagia for both solids and liquids but especially for swallowing soft food.

Computerized tomography of the brain had shown some hydrocephalus with a high density lesion corresponding to an anterior communicating artery aneurysm. His death was not related to this but an exact cause could not be given.

On autopsy, the heart was found moderately enlarged (650 g) and left and right ventricles were dilated (left ventricular wall thickness 2.5 cm). Also the liver was enlarged (1800 g). Lymph nodes and bone marrow were unremarkable. The spleen removed in vivo had weighed 450 g and histologically was reported to show retention of the normal splenic architecture with an increase in red pulp, congestion of pulp sinusoids that were lined by mildly hyperplastic endothelial cells. These findings had been thought to resemble those of hereditary spherocytosis.

3 Neuropathological Findings

The unfixed brain weighed 1350 g. On macroscopic inspection of the cerebral hemi spheres, gyri, sulci and leptomeninges and leptomeningeal blood vessels were normal as were the cranial nerve roots, the brainstem and the cerebellum. An aneurysm of 3 mm diameter was found arising from the anterior communicating artery on the right into the interhemispheric fissure. No herniations were present. Coronal sections showed normal cortex and white matter of the cerebral hemispheres. Except for the dilatation of the lateral ventricles and the findings in the caudate nucleus and putamen (see below), the gray matter of other basal ganglia structures, thalamus, and hypothalamus was unremarkable. Sections of the cerebellum and brainstem showed normal cerebellar folia and white matter, normal dentate nucleus and well-pigmented substantia nigra and locus coeruleus. The spinal cord was not examined.

The striatum showed marked diminution in the size of the caudate nuclei bilaterally, with the nuclei appearing as thin bands of tissue adjacent to the lateral ventricles. The putamen also appeared smaller than normal bilaterally (Fig. 2) .

For histological examination, representative brain regions were fixed in 4% formalin, embedded in paraffin and processed for routine histology. Sections were stained with hematoxylin and eosin, Prussian blue, Luxol-fast blue, Luxol fast blue-PAS, phosphotungstic acid hematoxylin (PTAH), Mallory iron, and Gallyas and Bielschowsky silver impregnation methods. The following primary antibodies

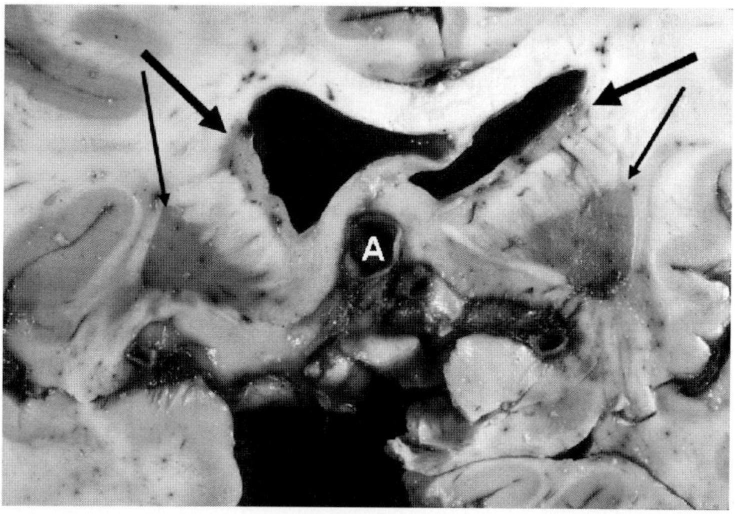

Fig. 2 Macroscopic appearance of a frontal section of the brain of patient WA with McLeod syndrome at the level of the anterior commissure demonstrates the unruptured aneurysm of the anterior communicating artery ("A") and the atrophy of the striatum. In both hemispheres, the small size of the putamen is obvious (*small arrows*) as well as the shrinkage of the caudate nucleus that is reduced to bands of tissue lining the lateral ventricles (*large arrows*)

were used for immunohistochemistry: mouse anti-glial fibrillary acidic protein (GFAP) monoclonal antibody (mAb) (DAKO, Glostrup, Denmark), rat anti-α-synuclein mAb (clone 15G7, Connex, Munich, Germany), rabbit anti-ubiquitin antiserum (DAKO) and mouse anti-tau mAb (clone AT8, Innogenetics, Ghent, Belgium). For immunohistochemistry 4 μm sections were deparaffinized in xylene and rehydrated using graded alcohols. To enhance immunoreactivity for α-synuclein sections were boiled in 0.01 M citrate buffer (pH 6.0) in a microwave oven. Sections were incubated in PBS with 2% BSA and 0.01% Triton X-100 at room temperature (RT) for 30 minutes. Incubation with primary antibody was performed at RT for 1 hour. Detection of antibody binding was performed with the alkaline phosphatase anti-alkaline phosphatase system (DAKO) according to the manufacturer's instructions using neufuchsin as chromogen.

Histological examination revealed extensive and marked atrophy of the head of the caudate nucleus, which was shrunken and pale (Fig. 3a), and to a lesser degree of the body of the caudate and the putamen. These affected regions showed severe spongiosis, neuronal loss and astrocytic gliosis (Figs. 3b, c). Within bundles of white matter in the caudate nucleus there was focal loss of axons (Bielschowsky) and of myelin (Luxol fast blue-PAS). PTAH disclosed mild proliferation of glial fibers. The head of the caudate nucleus had a few blue, iron-positive deposits (Mallory). These appeared as a fine line along the surface of erythrocytes within the lumen of blood vessels or on the surface of neural tissue components. The significance of this small amount of iron is not known, but the finding may be an artifact. No macrophages, no cytoplasmic or nuclear inclusions in nerve cells, no abnormal PAS-positive material were observed. Gallyas and Bielschowsky silver stains revealed no abnormal argyrophilic inclusions. In addition, no abnormal labelling was obtained in immunostains for ubiquitin, phosphorylated tau and α-synuclein. Changes in the putamen resembled those in caudate nucleus, but they were less severe and patchy in distribution. Bundles of white matter in the putamen were more severely depleted of myelinated fibers than those in caudate nucleus and

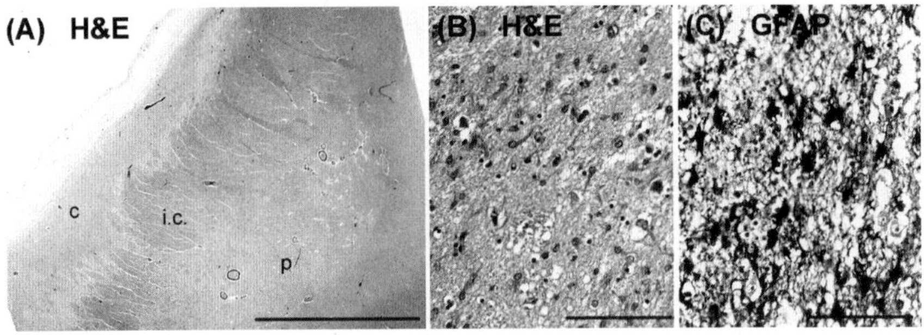

Fig. 3 Histological findings in patient WA with McLeod syndrome (**a**) Severe atrophy of the head of the caudate nucleus (frontal section; *c*: caudate nucleus; *i.c.*: internal capsule; *p*: putamen), (**b**) Higher magnification of the caudate with marked astrocytic gliosis and neuronal loss, (**c**) Anti-GFAP immunostaining showing numerous reactive astrocytes. Scale bars: a, 5 mm, b, and c, 100 μm

showed a moderately severe loss of axons. The globus pallidus was hypercellular and had a mild, diffuse proliferation of glia, but neurons were not clearly reduced in number. There was a small focus of hypercellularity at the lateral aspect of the right substantia nigra (near the subthalamic nucleus) with small, round nuclei and no visible cytoplasm of the proliferated cells that presumably were astrocytes. A similar but smaller focus of hypercellularity appeared present in the middle of the right thalamus and on the medial aspect of the left thalamus. In the medulla two pink bodies were seen in a nucleus of the tegmentum just superior to the inferior olivary nucleus on one side, possibly corresponding to degenerating neurons.

4 Discussion

In summary, this man with the characteristic weak erythrocyte Kell phenotype, from a family with a corresponding *XK* gene deletion, with hepatosplenomegaly, elevation of CK levels and of liver enzymes had displayed some hyperkinetic movements since his twenties and developed areflexia, myopathy and cardiomyopathy as well as chorea with some (possibly drug-related) Parkinsonian features in his forties. Before death at age 50 he also showed progressive dysarthria, dysphagia and neuropsychiatric abnormalities. His clinical presentation is typical for McLeod syndrome which was correctly recognized by his local physicians in collaboration with Laurence Marsh.

A cause for the patient's death, probably in sleep, was not identified on post mortem examination. His neuropathological findings were quite distinct and are focused on the striatum. There was severe and widespread neuronal loss and astrocytosis in the caudate nuclei and moderately severe and focal neuronal loss and astrocytosis in the putamen. Minor alterations were found in the globus pallidus (mild diffuse astrocytosis), the thalamus and substantia nigra (small foci of astrocytosis).

The earlier abstract publication on WA [9] had emphasised the similarity of his findings and those observed in ChAc but at that time a definite distinction between the two neuroacanthocytosis syndromes, MLS and ChAc, was impossible. Reliable conclusions therefore cannot be made from the approximately ten early post mortem reports of neuroacanthocytosis [7, 17, 18, 22, 28, 31, 37, 39–41, 43, 44].

After the genetic discoveries of 1994 and 2001, respectively [24, 38, 46], the question of neuropathological similarity between MLS and ChAc can now be examined. At present, there are neuropathological reports of ten molecularly-diagnosed cases of ChAc [2–5, 10, 15, 16, 22, 27, 36, 39, 40, 48], yet neuropathological data on MLS are only available for two other confirmed cases.

The first of these is very unusual, being one of the very few females known to express MLS clinically (case 5 of [22]; also discussed in chapters by Gandhi et al. and Geser et al.). Her carrier status of an *XK* deletion (exon 2, 350delT) [25] became known only after her autopsy findings had been published (case 1 of [40]) and this observation is therefore commonly overlooked. Practically identical to case WA, coronal slices showed symmetrically dilated lateral ventricles and shrunken striata bilaterally. The caudate nucleus showed most severe neuronal loss and gliosis, with the putamen and globus pallidus being somewhat less affected. Other structures, including the substantia nigra (reported as case 1 of [39]) appeared unaffected.

The second patient, a 55-year-old male with a V299X point mutation, again had severe involvement of the striatum and globus pallidus with neuronal loss and astrogliosis (discussed in detail as case 3 in the chapter by Geser et al.). In distinction to WA and the previous report there was cortical gliosis, especially in the frontal lobes [20]. Comparing the findings to those in ChAc, a general similarity can be confirmed. In both conditions recent reports have emphasized some cortical involvement in addition to the predominant striatopallidal damage [3, 20].

Very little is known about preferential involvement of cell subpopulations in MLS and ChAc. In ChAc, it appears that striatal neuronal loss affects mainly small and medium size neurons and in contrast to the findings in MLS the substantia nigra is involved to a variable degree [39]. In ChAc measurements of dopamine in the striatum and of GABA and substance P have been performed [35], however nothing is yet known about neurotransmitter levels in MLS.

No inclusions were found in our present case of MLS, in contrast to the ubiquitin-positive nuclear inclusions found in surviving striatal and cortical neurons of patients with Huntington's disease (HD). In HD the severity of striatal changes has a regional gradient, with more severe neuropathological changes in the tail of the caudate nucleus than in the body, and in the body more so than in the head [35]. Recent voxel-based approaches suggest a specific vulnerability of the caudate head in ChAc [23], but no such data are available for MLS.

Concerning the mechanism of loss of striatal neurons in MLS one may speculate that functional XK protein is essential for neuronal survival, as proposed in our initial report of cerebral involvement in this condition [14]. However, it was not until 2007 that XK expression was demonstrated in the brain of mouse, rat and humans [11, 30]. XK was found to be highly expressed in the striatum, in thalamic and hypothalamic nuclei, the mammillary bodies, the hippocampal fields CA1, CA2 and CA3 (pyramidal cells), in the dentate gyrus (granular cells) and in all layers of the neocortex. The expression of XK appears restricted to neurons, in particular to the cytoplasmatic compartment of the cell soma, colocalizing with calreticulin, a marker of the endoplasmatic reticulum, and 58-K, a marker of the Golgi apparatus. XK was also found in dendritic processes, but not within synaptic vesicles or nerve terminals. XK was absent from glia [11]. Specifically in humans, XK was found expressed in the spinal cord, medulla, cerebellum, hippocampus, amygdala, striatum, thalamus, cerebral cortex and frontal, temporal and occipital lobes. It was proposed that neurological findings in MLS may derive from prolonged dysfunction caused by absence of XK in many different types of neurons, and not particularly from pure basal ganglia dysfunction. It is possible that the late onset of neurological symptoms may be due to compensatory expression of the related protein XPLAC [30].

In the red cell, XK is coupled to the Kell protein, however, expression of Kell was not detected in the central nervous system [11, 30]. This finding argues against the previous speculation [12, 47] of an involvement of endothelins (that are enzymatically processed by Kell [29]) in the neuropathology of MLS.

More detailed studies are needed to delineate the course and pattern of neuronal damage in both MLS and ChAc. These studies should aim at understanding the pathways that lead from mutations in the *XK* and *VPS13A* genes to clinically and morphologically similar pictures. It is tempting to speculate that the similarities between these two neurodegenerative conditions indicate that these pathways share some of their intermediate steps.

Acknowledgements We thank John T. Blois, Jordan Grafman, Laurence Marsh†, Colvin Redman, Lewis P. Rowland, and Alexander Vortmeyer for their contributions during the many stages of this intercontinental investigation.

References

1. Allen FH, Krabbe SMR, Corcoran PA (1961) A new phenotype (McLeod) in the Kell blood-group system. Vox Sang 6:555–560
2. Alonso ME, Teixeira F, Jimenez G et al. (1989) Chorea-acanthocytosis: report of a family and neuropathological study of two cases. Can J Neurol Sci 16:426–431
3. Arzberger T, Heinsen H, Buresch N et al. (2006) The neuropathology of chorea-acanthocytosis: from stereology to an immunohistochemical detection of chorein. Mov Disord 20:1679
4. Bader B (2007) Neuropathology of chorea-acanthocytosis. Mov Disord 22:VI
5. Bader B, Arzberger T, Dobson-Stone C et al. (2007) Detection of chorein in postmortem brain tissue of chorea-acanthocytosis patients and nonaffected controls by Western blot. Mov Disord 22:VI
6. Bertelson CJ, Pogo AO, Chaudhuri A et al. (1988) Localization of the McLeod locus (*XK*) within Xp21 by deletion analysis. Am J Hum Genet 42:703–711
7. Bird TD, Cederbaum S, Valpey RW et al. (1978) Familial degeneration of the basal ganglia with acanthocytosis: a clinical, neuropathological, and neurochemical study. Ann Neurol 3:253–258
8. Brin MF (1993) Acanthocytosis. In: Goetz CG, Tanner CM, Aminoff MJ (eds) Handbook of clinical neurology, vol 19(63): Systemic diseases, Part I. Elsevier, Amsterdam, pp 271–299
9. Brin MF, Hays A, Symmans WA et al. (1993) Neuropathology of McLeod phenotype is like choreaacanthocytosis (CA). Can J Neurol Sci 20 (Suppl):S234
10. Burbaud P, Vital A, Rougier A et al. (2002) Minimal tissue damage after stimulation of the motor thalamus in a case of chorea-acanthocytosis. Neurology 59:1982–1984
11. Claperon A, Hattab C, Armand V et al. (2007) The Kell and XK proteins of the Kell blood group are not co-expressed in the central nervous system. Brain Res 1147:12–24
12. Danek A (2004) Neuroacanthocytosis syndromes: what links red blood cells and neurons? In: Danek A (ed) Neuroacanthocytosis syndromes. Springer, Dordrecht, The Netherlands, pp 1–14
13. Danek A, Rubio JP, Rampoldi L et al. (2001) McLeod neuroacanthocytosis: genotype and phenotype. Ann Neurol 50:755–764
14. Danek A, Uttner I, Vogl T et al. (1994) Cerebral involvement in McLeod syndrome. Neurology 44:117–120
15. de Yébenes JG, Brin MF, Mena MA et al. (1988) Neurochemical findings in neuroacanthocytosis. Mov Disord 3:300–312
16. de Yébenes JG, Vazquez A, Martínez A et al. (1988) Biochemical findings in symptomatic dystonias. Adv Neurol 50:167–175
17. Feinberg TE, Cianci CD, Morrow JS et al. (1991) Diagnostic tests for choreoacanthocytosis. Neurology 41:1000–1006
18. Galatioto S, Serra S, Batolo D et al. (1993) Amyotrophic choreo-acanthocytosis: a neuropathological and immunocytochemical study. Ital J Neurol Sci 14:49–54

19. Galey WR, Evan AP, Van Nice PS et al. (1978) Morphology and physiology of the McLeod erythrocyte. Vox Sang 34:152–161
20. Geser F, Prokop S, Glatzel M et al. (2007) The neuropathology of McLeod syndrome: a case study. Mov Disord 22:VI
21. Giblett ER, Klebanoff SJ, Pincus SH et al. (1971) Kell phenotypes in chronic granulomatous disease: a potential transfusion hazard. Lancet I:1235–1236
22. Hardie RJ, Pullon HWH, Harding AE et al. (1991) Neuroacanthocytosis: a clinical, haematological and pathological study of 19 cases. Brain 114:13–49
23. Henkel K, Danek A, Grafman J et al. (2006) Head of the caudate nucleus is most vulnerable in chorea-acanthocytosis: a voxel-based morphometry study. Mov Disord 21:1728–1731
24. Ho M, Chelly J, Carter N et al. (1994) Isolation of the gene for McLeod syndrome that encodes a novel membrane transport protein. Cell 77:869–880
25. Ho MF, Chalmers RM, Davis MB et al. (1996) A novel point mutation in the McLeod syndrome gene in neuroacanthocytosis. Ann Neurol 39:672–675
26. Ho MF, Monaco AP, Blonden LAJ et al. (1992) Fine mapping of the McLeod locus (XK) to a 150–380-kb region in Xp21. Am J Hum Genet 50:317–330
27. Ishida C, Makifuchi T, Saiki S et al. (2007) An autopsied case of chorea-acanthocytosis with a single detected mutation of *VPS13A*. Mov Disord 22:VI
28. Iwata M, Fuse S, Sakuta M et al. (1984) Neuropathological study of chorea-acanthocytosis. Jpn J Med 23:118–122
29. Lee S, Lin M, Mele A et al. (1999) Proteolytic processing of big endothelin-3 by the Kell blood group protein. Blood 94:1440–1450
30. Lee S, Sha Q, Wu X et al. (2007) Expression profiles of mouse Kell, XK, and XPLAC mRNA. J Histochem Cytochem 55:365–374
31. Malandrini A, Fabrizi GM, Palmeri S et al. (1993) Choreo-acanthocytosis like phenotype without acanthocytes: clinicopathological case report. Acta Neuropathol (Berl) 86:651–658
32. Marsh NJ, Marsh WL, Johnson CL et al. (1980) Evidence of a muscular defect in subjects with McLeod syndrome. Transfusion 20:618–619
33. Marsh WL, Marsh NJ, Moore A et al. (1981) Elevated serum creatine phosphokinase in subjects with McLeod syndrome. Vox Sang 40:403–411
34. Marsh WL, Oyen R, Nichols ME et al. (1975) Chronic granulomatous disease and the Kell blood groups. Br J Haematol 29:247–262
35. Martínez A, Mena MA, Jamrozik Z et al. (2004) Pathology of neuroacanthocytosis and of Huntington's disease. In: Danek A (ed) Neuroacanthocytosis syndromes. Springer, Dordrecht, pp 87–94
36. Ochiai J, Takeuchi Y, Mabuchi C et al. (2007) A case of chorea-acanthocytosis: clinico-pathological presentation. Mov Disord 22:VI
37. Rafalowska J, Drac H, Jamrozik S (1996) Neuroacanthocytosis. Review of literature and case report. Folia Neuropathol 34:178–183
38. Rampoldi L, Dobson-Stone C, Rubio JP et al. (2001) A conserved sorting-associated protein is mutant in chorea-acanthocytosis. Nat Genet 28:119–120
39. Rinne JO, Daniel SE, Scaravilli F et al. (1994) Nigral degeneration in neuroacanthocytosis. Neurology 44:1629–1632
40. Rinne JO, Daniel SE, Scaravilli F et al. (1994) The neuropathological features of neuroacanthocytosis. Mov Disord 9:297–304
41. Sato Y, Ohnishi A, Tateishi J et al. (1984) An autopsy case of chorea-acanthocytosis. Special reference to the histopathological and biochemical findings of basal ganglia. Brain Nerve (Tokyo) 36:105–111
42. Schwartz SA, Marsh WL, Symmans A et al. (1982) "New" clinical features of McLeod syndrome. Transfusion 22:404
43. Sobue G, Mukai E, Fujii K et al. (1986) Peripheral nerve involvement in familial chorea-acanthocytosis. J Neurol Sci 76:347–356
44. Spencer SE, Walker FO, Moore SA (1987) Chorea-amyotrophy with chronic hemolytic anemia: a variant of chorea-amyotrophy with acanthocytosis. Neurology 37:645–649

45. Symmans WA, Shepherd CS, Marsh WL et al. (1979) Hereditary acanthocytosis associated with the McLeod phenotype of the Kell blood group system. Br J Haematol 42:575–583
46. Ueno S-I, Maruki Y, Nakamura M et al. (2001) The gene encoding a newly discovered protein, chorein, is mutated in chorea-acanthocytosis. Nat Genet 28:121–122
47. van den Buuse M (2004) Endothelins as basal ganglia transmitters. In: Danek A (ed) Neuroacanthocytosis syndromes. Springer, Dordrecht, The Netherlands, pp 205–212
48. Vital A, Bouillot S, Burbaud P et al. (2002) Chorea-acanthocytosis: neuropathology of brain and peripheral nerve. Clin Neuropathol 21:77–81
49. Wimer BM, Marsh WL, Taswell HF et al. (1977) Haematological changes associated with the McLeod phenotype of the Kell blood group system. Br J Haematol 36:219–224

Part V
Clinical Aspects and Therapy

Psychiatric Morbidity in Neuroacanthocytosis

Akira Sano

Abstract Chorea-acanthocytosis and McLeod syndrome are the core of neuroacanthocytosis syndromes in which there is primarily neurodegeneration in the striatum. In addition to chorea, neuropsychiatric features such as cognitive impairment, personality change, depression, obsessive-compulsive symptoms, epilepsy-related symptoms and psychosis are frequently seen in these diseases. These conditions are attributed to some extent to dysfunction of the fronto-subcortical circuit.

1 Early Case Reports

In the late 1960s, Levine et al. [7] and Critchley et al. [2] first described neuroacanthocytosis (NA) in two independent studies of American families. Members of these families had choreic involuntary movements accompanied by acanthocytosis with normal serum lipoproteins. Levine et al. reported dementia, paranoid ideation, and negativism, and the proband's brother had previously been reported as a case with acanthocytosis associated with schizophrenia. Critchley et al. also reported forgetfulness in the family members of their pedigree, in which three of the proband's parent's full siblings had "spells," had become psychotic, and died in mental institutions. Since these originally reported cases, a variety of neuropsychiatric symptoms have been noted in NA syndromes.

Akira Sano
Department of Psychiatry, Kagoshima University Graduate School of Medical and Dental Sciences, Japan.
sano@m3.kufm.kagoshima-u.ac.jp

R.H. Walker et al. (eds.), *Neuroacanthocytosis Syndromes II.*

In descriptions by psychiatrists of Japanese chorea-acanthocytosis (ChAc) cases, psychiatric features are described in detail. Takahashi et al. described a 26-year-old male patient with moderate dementia, compulsive acts, personality change, depersonalization, stupor, and twilight state, in addition to generalized tonic-clonic and complex partial seizures [13]. Naito et al. [10] described a 45-year-old female patient with delirium, depression, suicide attempts, euphoria, personality change, persecutory delusions, visual and somatosensory hallucinations, self-mutilation, and progressive subcortical-type dementia, and also generalized tonic-clonic and psychomotor seizures. The authors concluded that the neuropsychiatric symptoms could be summarized as organic brain syndrome as in the DSM III. Shiraishi et al. [12] reported a 31-year-old male case who had a systematized love delusion. He suffered from a twilight state, self-mutilation, dysphoria, personality change, depersonalization, depression, dementia, and schizophrenia-like symptoms. He also had generalized tonic-clonic and complex partial seizures. Mizukami et al. [8] reported a female patient with onset at age 17 of a schizophrenia-like psychosis with auditory hallucinations, thought-broadcasting and -hearing, persecutory delusions of reference and of poisoning, and xenopathic experience. Three years later, oral dyskinesias, personality change and dementia became marked.

Self-mutilation is one of the most characteristic features of ChAc. Takahashi et al. [13] proposed that self-mutilation does not result from oral dyskinesias alone but is related to psychiatric symptoms such as unstable mood, anxiety, phobia, obsessive-compulsive traits, and episodic psychotic symptoms. Walker et al. [16] similarly proposed that the head self-excoriations manifested by their ChAc patient were related to an obsessive – compulsive-type behavioral disorder, and thus that the orolingual self-mutilation seen in other ChAc patients is also due to psychiatric pathology rather than the movement disorder.

2 Genetically Confirmed Cases

Recently the molecular genetic defects have been identified for McLeod syndrome (MLS) [6] and ChAc [11, 14]. *XK* is the gene responsible for MLS, and *VPS13A* for ChAc. Based on the clinical data of molecular-genetically diagnosed MLS and ChAc cases, Danek et al. [3] reported the occurrence of various symptoms (Table 1). They calculated the frequencies of neuropsychiatric symptoms and cognitive changes, with each of the symptoms being seen in over half of the cases.

In molecularly diagnosed ChAc, half of the patients show some impairment of higher cerebral function at presentation, with the earliest manifestations below the age of 10. Of two siblings, the girl had developed behavioral problems in school, including ritualistic touching and kissing of objects. At age 24, her IQ was low. Her brother was diagnosed as dyslexic, with poor visuospatial skills, writing and spelling. His educational attainments were below average when he left school aged 16

Table 1 Clinical comparison of chorea-acanthocytosis (ChAc) and McLeod syndrome (MLS) (cited from Danek et al. [6])

Finding	% Affected in ChAc	% Affected in MLS
Seizures	42	50
Neuropsychiatric	60	83
Cognitive changes	73	54
Involuntary vocalisations	62	58
Tongue and lip biting	40	8

years [5]. Another young man was diagnosed with schizophrenia in adolescence, and neurological signs were noted only many years later [1]. Recently, Müller-Vahl et al. [9] reported monozygotic twins with ChAc. Twin 1 had been hospitalised with an acute episode of paranoid schizophrenia manifesting with paranoid delusions, affective lability, disorganized thoughts and suicidal ideation. In both twins, psychiatric examination revealed similar changes with obsessive-compulsive symptoms, infantile behavior, inappropriate cheerfulness, self-neglect and dementia. Unlike his brother twin 2 showed slightly less cognitive impairment (IQ 70), no significant depression and no psychosis.

Although the inheritance of ChAc has been recognized as autosomal recessive, single heterozygous *VPS13A* mutations have occasionally been found in patients with typical disease manifestations. Dobson-Stone and associates reported 57 different *VPS13A* mutations in 43 ChAc patients, with 7 of 43 patients possessing only a single heterozygous *VPS13A* mutation [4]. In our original family study in which we identified the gene responsible for ChAc, we found that heterozygous carriers of the disease-causing mutation sometimes showed neuropsychiatric symptoms such as emotional lability or cognitive impairment [15].

3 Summary of Psychiatric Symptoms

The neuropsychiatric symptoms of ChAc can be divided into two broad categories as summarized in Table 2. One is epilepsy-related symptoms like psychomotor seizures, seizure-related delirium or twilight state, and seizure-related intermittent dysphoria. Other symptoms can be grouped together as an organic brain syndrome (DSM III) or organic mental disorder (F0, ICD-10), which includes delirium, cognitive disturbances, personality change, obsessive-compulsive symptoms or stereotypies, schizophrenia-like symptoms, depression and anxiety. Cognitive disturbances that consist typically of slowness in thinking and behavior, and forgetfulness, and neuropsychiatric symptoms such as personality change and obsessive-compulsive symptoms or stereotypies, can be attributable to dysfunction of frontosubcortical circuits due to degeneration of the striatum.

Table 2 Neuropsychiatric symptoms of chorea-acanthocytosis (ChAc)

Epilepsy-related Symptoms
Psychomotor seizure
Delirium or twilight state
Intermittent dysphoria
Organic Brain Syndrome
Delirium
Frontosubcortical Dementia
Cognitive Disturbance
Slowness in thinking and behavior, forgetfulness
Personality Change
Childish attitude, lack of concentration, irritability, apathy, Euphoria, impulsiveness, emotional lability, disinhibition
Obsessive-compulsive symptoms or stereotypies
Schizophrenia-like symptoms
Delusions, hallucinations
Depression, anxiety

References

1. Bruneau MA, Lesperance P, Chouinard S (2003) Schizophrenia-like presentation of neuroacanthocytosis. J Neuropsychiatry Clin Neurosci 15:378–380
2. Critchley EM, Clark DB, Wikler A (1968) Acanthocytosis and neurological disorder without betalipoproteinemia. Arch Neurol 18:134–140
3. Danek A, Jung HH, Melone MA, Rampoldi L, Broccoli V, Walker RH (2005) Neuroacanthocytosis: new developments in a neglected group of dementing disorders: J Neurol Sci 229–230:171–186
4. Dobson-Stone C, Danek A, Rampoldi L, Hardie RJ, Chalmers RM, Wood NW, Bohlega S, Dotti MT, Federico A, Shizuka M, Tanaka M, Watanabe M, Ikeda Y, Brin M, Goldfarb LG, Karp BI, Mohiddin S, Fananapazir L, Storch A, Fryer AE, Maddison P, Sibon I, Trevisol-Bittencourt PC, Singer C, Caballero IR, Aasly JO, Schmierer K, Dengler R, Hiersemenzel LP, Zeviani M, Meiner V, Lossos A, Johnson S, Mercado FC, Sorrentino G, Dupre N, Rouleau GA, Volkmann J, Arpa J, Lees A, Geraud G, Chouinard S, Nemeth A, Monaco AP (2002) Mutational spectrum of the CHAC gene in patients with chorea-acanthocytosis. Eur J Hum Genet 10:773–781
5. Hardie RJ, Pullon HWH, Harding AE, Owen JS, Pires M, Daniels GL, Imai Y, Misra VP, King RHM, Jacobs JM, Tippett P, Duchen LW, Thomas PK, Marsden CD (1991) Neuroacanthocytosis: a clinical, haematological and pathological study of 19 cases. Brain 114:13–49
6. Ho M, Chelly J, Carter N, Danek A, Crocker P, Monaco AP (1994) Isolation of the gene for McLeod syndrome that encodes a novel membrane transport protein. Cell 77:869–880
7. Levine IM, Estes JW, Looney JM (1968) Hereditary neurological disease with acanthocytosis. A new syndrome. Arch Neurol 19:403–409
8. Mizukami K, Kawanishi Y, Shiraishi H, Suzuki E (1995) A case of chorea-acanthocytosis with various psychiatric symptoms. Clin Psychiatry (Seishinigaku) 37:743–749 (in Japanese)
9. Müller-Vahl KR, Berding G, Emrich HM, Peschel T (2007) Chorea-acanthocytosis in monozygotic twins: clinical findings and neuropathological changes as detected by diffusion tensor imaging, FDG-PET and ^{123}I-β-CIT-SPECT. J Neurol published online: 8 February

10. Naito A, Hasegawa M, Kaji S (1986) A case of chorea-acanthocytosis with epileptic seizures and various psychotic symptoms. Clin Psychiatry (Seishinigaku) 28:897–906 (in Japanese)
11. Rampoldi L, Dobson-Stone C, Rubio JP, Danek A, Chalmers RM, Wood NW, Verellen C, Ferrer X, Malandrini A, Fabrizi GM, Brown R, Vance J, Pericak-Vance M, Rudolf G, Carre S, Alonso E, Manfredi M, Nemeth AH, Monaco AP (2001) A conserved sorting-associated protein is mutant in chorea-acanthocytosis. Nat Genet 8:119–120
12. Shiraishi K, Takahashi Y, Okubo Y, Mochizuki A, Fukuzawa H, Kariya T (1987) A follow-up of a case of chorea-acanthocytosis on development of neuro-psychiatric symptoms and clinical data. Clin Psychiatry (Seishinigaku) 29:1345–1347 (in Japanese)
13. Takahashi Y, Kojima T, Atsumi Y, Okubo Y, Shimazono Y (1983) A case of chorea-acanthocytosis with various psychotic symptoms. (in Japanese) Psych Neurol Japonica 85:457–472
14. Ueno S, Maruki Y, Nakamura M, Tomemori Y, Kamae K, Tanabe H, Yamashita Y, Matsuda S, Kaneko S, Sano A (2001) The gene encoding a newly discovered protein, chorein, is mutated in chorea-acanthocytosis. Nat Genet 28:121–122
15. Ueno S, Kamae K, Yamashita Y, Maruki Y, Tomemori Y, Nakamura M, Ikeda M, Tanabe H, Sano A (2005) Chorea-acanthocytosis with the ehime-deletion mutation. In: Danek A (ed) Neuroacanthocytosis Syndromes, Springer, Dordrecht, Netherlands, pp 39–43
16. Walker RH, Liu Q, Ichiba M, Muroya S, Nakamura M, Sano A, Kennedy CA (2006) Self-mutilation in chorea-acanthocytosis: manifestation of movement disorder or psychopathology. Mov Disord 21:2268–2269

Muscular Aspects of Chorea-Acanthocytosis

S. Saiki(✉) and Y. Tamura

Abstract We report new insights into muscular aspects in chorea-acanthocytosis (ChAc), a hereditary disease characterized by involuntary movements and amyotrophy with elevation of serum creatine kinase (sCK). In addition, we review the literature regarding muscular aspects of genetically-confirmed ChAc cases. All ChAc cases, except for one, showed sCK elevation, while clinical neuromuscular symptoms are variously reported in most. Conventional muscle stains, such as hematoxylin and eosin (H&E), nicotinamide adenine dinucleotide-tetrazolium reductase (NADH-TR), and ATPase, showed mild neurogenic changes and/or mild myopathic changes. Recently, evidence has accumulated of a primary myopathy in ChAc. Increase of tTGase-derived Nε-(-γ-glutamyl)lysine isopeptide cross-links was shown in skeletal muscle as well as in erythrocytes [19]. It was also reported that nemaline rods were found within myofibres in a patient with ChAc [27]. Moreover, immunohistochemical staining with two specific antibodies against

S. Saiki
Department of Medical Genetics, Cambridge Institute for Medical Genetics, Wellcome Trust/MRC Building, Addenbrooke's Hospital, Hills Road, Cambridge, CB2 0XY, UK
ss644@cam.ac.uk

R.H. Walker et al. (eds.), *Neuroacanthocytosis Syndromes II*.

chorein showed distribution changes of the protein in three heterozygous ChAc cases. Although the precise mechanism of the myopathic changes is still unclear, these findings suggest that skeletal muscles as well as peripheral nerves are primarily involved in the pathophysiology of this disorder.

1 Introduction

Neuroacanthocytosis syndromes, such as chorea-acanthocytosis (ChAc; MIM200150), McLeod syndrome (MLS; MIM314850), and Huntington's disease-like 2 (HDL2; MIM606438), share many clinical features, most notably peripheral acanthocytosis and involuntary movements. ChAc is a hereditary neurodegenerative disease with a world-wide distribution, particularly common in Japan. It usually presents in the second, third or fourth decades with involuntary movements, psychiatric symptoms, amyotrophy, and erythrocyte acanthocytosis [5, 6, 21]. Clinical neuromuscular symptoms such as muscle weakness/atrophy, reduced or absent deep tendon reflexes, and hypotonia are common hallmarks and account for the early definition of the disease as "familial amyotrophic acanthocytosis" [10, 14]. Amyotrophy with elevation of serum creatine kinase (sCK) is an important characteristic of ChAc, and is not observed in HDL-2 or Huntington's disease (HD) [1, 11, 20, 21, 25, 30]. Amyotrophy with elevation of sCK in ChAc has been usually regarded as secondary to neurogenic muscular atrophy. Nerve biopsy findings and findings of nerve conduction velocity studies (NCV) and electromyography (EMG) have been indicative of axonal degeneration [25]. A secondary inflammatory response during the progression of the disease has been described and a ChAc case with anti-GM1 antibodies has been reported [12, 16]. However, recent studies associated with muscular aspects have shown some evidence of primary muscular involvement [19, 24, 27]. Here, we describe primary myopathic findings detected by modified Gomori trichrome staining and immunostaining with anti-chorein (VPS13A protein) antibodies in our ChAc cases and provide a review of the muscular aspects of published ChAc cases with genetic diagnoses (Fig. 1).

2 Muscular Aspects of ChAc

2.1 Clinical Neuromuscular Symptoms

Table 1 summarises the clinical aspects of previously reported genetically-diagnosed ChAc cases [2, 3, 7, 23, 24, 26–28, 31]. Neurological examinations revealed muscle weakness and/or atrophy in eight cases (Cases 1–3, 8–12). The severity of neuromuscular symptoms observed in all cases is mild-moderate. Except for three cases showing hyper-reflexia with extensor plantar responses [2], decreased or absent deep tendon reflexes (DTRs) were noted.

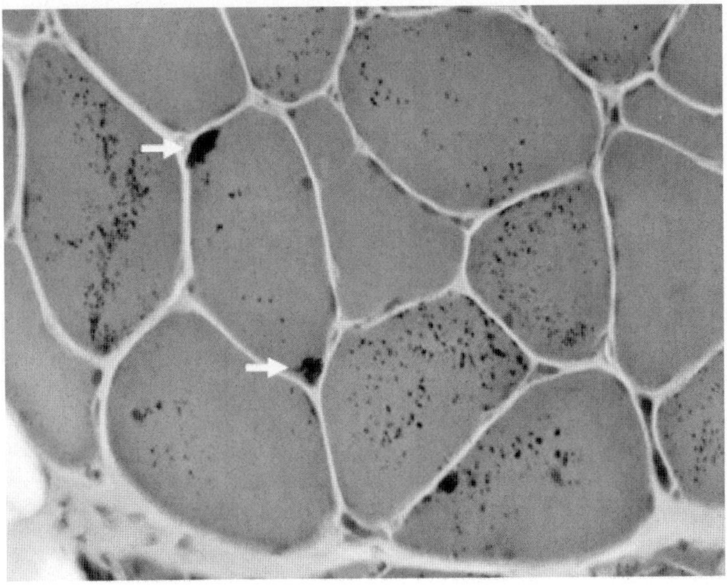

Fig. 1 Nemaline rods in ChAc Case 15[27]. Modified Gomori trichrome staining transverse section of the biopsied muscle demonstrated nemaline rods clustering preferentially in a subsarcolemmal location (*arrows*). Original magnification: ×400

In early clinical stages obvious neuromuscular involvement is rare, but gradually develops with disease evolution. Dysphagia and inability to eat are usually common and probably represent a mixture of impairments by involuntary movements (oromandibular and lingual dystonia and/or chorea and skeletal muscle weakness [4]. Nine ChAc cases (Cases 1–5, 10, 13, 14, 16) showed impairment of postural reflexes and/or gait difficulty. Although dystonia and choreiform movements of the trunk and limbs are associated with increasing risk of falls, as in HD, skeletal muscle weakness/atrophy may contribute to some extent [9].

2.2 *Elevation of Serum Creatine Kinase (sCK)*

All ChAc cases except for Case 1 showed elevation of sCK of variable degrees, typically reaching at least 2- to 3-fold higher than the normal upper limit (Table 1). sCK elevation was thought to be derived from secondary degeneration of skeletal muscle tissue and/or violent choreiform movements of the limbs [25]. However, marked elevation of sCK was noted in Case 12 who did not have neurogenic changes on electrophysiological or conventional pathological studies, in Case 8 without muscular atrophy, and in Case 7, who presented with akinesia, without involuntary movements.

Table 1 Summary of the clinical features of ChAc cases

Author	Case No.	Age at onset	Postural instability/ decreased postural reflexes	Involuntary movements	Muscle weak-ness/ atrophy	Anatomical distribution of muscle weakness/atrophy	Deep tendon reflexes	Serum CK (<upper limits) IU/l
Tanaka et al. [28]	1	23	+	Vocal, facial, limb, and truncal tics, chorea	+	Distal limbs	Hypoactive	WNL
	2	37	+	Vocal tics, chorea	–	–	Hypoactive	2369
	3	33	+	Orofacial dyskinesia, Chorea	+	Limbs	Hypoactive or absent	307
Sorrentino G et al. [19, 26]	4	20	+	Orofaciolingual dyskinesia, chorea	+	All limbs (especially in interossei muscles)	Absent	2355
Wihl G et al. [31]	5	17	+	Dystonia, chorea, phonic tics	+	Lower limbs	Absent	334 (<80)
Burbaud P et al. [3]	6	31	ND	Chorea	ND	ND	ND	ND
Bohlega S et al. [2]	7	30	ND	Oral dyskinesia, facial tics, chorea	ND	ND	Hyperactive	3360 (<195)
	8	20	ND	Dystonia, chorea	ND	ND	Hyperactive	224 (<195)
	9	20	ND	–	ND	ND	ND	1393 (<195)
	10	26	+	Dystonia, chorea	+	Masseters, deltoids, other shoulder girdle muscles	Hyperactive	7175 (<195)
	11	30	ND	Orofacial dyskinesia, chorea	+	Temporalis, distal limbs	Absent	388 (<195)
Saiki S et al. [23, 24]	12	35	–	Chorea	+	Distal lower limbs	Hypoactive	851 (<212)
	13	25	+	Chorea, tourettism	+	Lower limbs	Hypoactive	1179 (<212)
	14	29	+	Orolingual dyskinesia, chorea	+	Lower limbs	Hypoactive	1517 (<212)
Tamura Y et al. [27]	15	32	+	Chorea, vocal tics	–	–	Hypoactive or absent	975 (<200)
Dotti MT et al. [8]	17–23 (7 cases)	25.1 (mean)	ND	Orofacial dyskinesia, chorea, dystonia	71% of cases	ND	ND	Increased

Abbreviations: *CK* creatine kinase, *ChAc* chorea-acanthocytosis, *ND* not described, *NE* not examined, *WNL* within normal limits

3 Neurophysiology

Table 2 summarises the findings of the NCV and EMG studies. Two cases showed a mild decrease of CVs and amplitudes of compound muscle action potentials. Sensory nerve action potentials were reduced in two cases. EMG showed evidence of neurogenic changes in three cases. Mild myopathic changes were noted only in one case.

4 Neuropathology

4.1 Muscle Biopsy

4.1.1 Traditional Studies (H&E, NADH-TR, and ATPase Staining)

Muscle biopsy findings are summarised in Table 2. Conventional staining showed neurogenic changes (neurogenic atrophy, grouped atrophy, and small angular fibres) in three cases, and myopathic changes (increase of central nuclei and variable fibre diameters) in three cases. No abnormal findings were noted in one case. NADH-TR staining and ATPase staining did not show any abnormalities in Cases 10–12. Compared with the previous reports, the severity of these changes may be milder, but may be attributable to the clinical stage at which muscle biopsy was performed.

4.1.2 Nemaline Rods

We have reported a homozygous ChAc case (Case 15 in Table 1) with abnormal histopathological findings in addition to typical clinical features of ChAc [27]. Briefly, a 37-year-old woman developed mild dysarthria and a gait disturbance in addition to the typical hyperkinetic involuntary movements (choreiform movements of the limbs and tics). Although hypotonia with hyporeflexia was noted, neither muscle weakness nor atrophy was noted. Laboratory and radiological findings were compatible with ChAc. H&E staining revealed mild neurogenic changes (grouped atrophy, small angular fibres, and fibres with some internal nuclei), while modified Gomori trichrome staining demonstrated a moderate number of collections of nemaline rods (Fig. 2). The frequency of nemaline rods was quite high and similar to that of adult-onset nemaline myopathy. There were approximately ten rod-containing myofibres per fascicle. Electron microscopy showed the presence of rods in the subsarcolemmal location within myofibres.

Table 2 Summary of the neurophysiological and pathological findings of ChAc cases

Case	NCV	EMG	Muscle biopsy: Conventional staining (H&E, ATPase, NADH-TR)	Muscle biopsy: Gomori-trichrome, immunohistochemistry, etc	Sural nerve biopsy
1	WNL	WNL	ND	ND	ND
2	WNL	High amplitude MUPs, prolonged duration	ND	ND	ND
3	WNL	WNL	ND	ND	ND
4	WNL	NE	Neurogenic atrophy with occasional necrotic fibres, internal nuclei	Increase of anti-tTGase immune activity, amorphous, dense, egg-shaped elements (EM) within muscle fibres	Slight loss of myelinated fibres, many onion bulbs and fibres in regeneration clusters, segmental demyelination
5	Mild axonal neuropathy	ND	ND	ND	ND
6	ND	Neurogenic findings	Neurogenic atrophy	ND	ND
7	ND	ND	ND	ND	ND
8	ND	ND	ND	ND	ND
9	Mild sensory-motor neuropathy	ND	ND	ND	ND
10	WNL	Spontaneous MUPs, denervation potentials	ND	ND	ND
11	ND	ND	ND	ND	ND
12	Slight prolongation of CV in PNs	WNL	Mild increase of central nuclei and variable fibre diameter	Uneven and discontinuous CI along the sarcolemmas, increase of CI in the sarcolemma	Mild decrease of large myelinated fibres

13	Mild prolongation of CV in PNs	Short duration MUPs	Mild increase of central nuclei and variable fibre diameter	Uneven and discontinuous CI along the sarcolemma	NE
14	WNL	WNL	WNL	Uneven and discontinuous CI along the sarcolemma	NE
15	Low CMAP and SNAP with normal CVs	High amplitude MUPs, reduced inteference	Variation in size of muscle fibres, grouped atrophy, small angular fibres, internal nuclei	Nemaline rods located under sarcolemma	Decrease of large myelinated fibres, a few myelin-ovoids
17–23	Sensory-motor-neuropathy (1/6), axonal neuropathy (3/6), normal (2/6)	Grouped atrophy with necrotic fibres and clumps of vesicular nuclei, increase of central nuclei (2/2)	NE	Loss of myelinated fibres (6/6), accumulation of membrano-lamellar profiles in the axoplasm (5/6)	

NCV nerve conduction velocity, *EMG* electromyography, *ND* not described, *NE* not examined, *CV* conduction velocity, *PNs* peroneal nerves, *WNL* within normal limit, *MUPs* motor unit potentials, *CI* chorein immunoreactivity

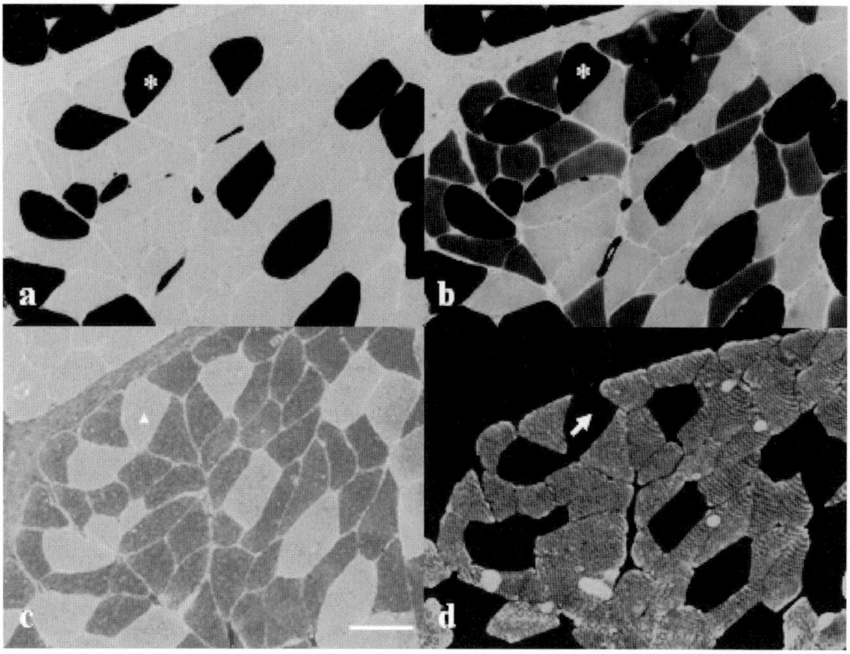

Fig. 2 Chorein immunostaining in normal control (**c**) and comparison with ATPase staining (pH = 4.3: **a**; pH = 4.6: **b**) and myosin heavy chain fast (MHC-F) immunostaining (**d**). Chorein was present at the sarcolemma and sarcoplasma. Chorein positive fibres (*arrowhead* in **c**) were predominantly present in type I fibres and positive for ATPase (*asterisks* in **a**, **b**) and MHC-F negative fibres (*arrow* in **d**). Bar, 50 μm

4.1.3 tTGases

Using immunohistochemical, high pressure liquid chromatography, and Western blot analysis, Melone et al. found increased amounts of tTGase-derived Nε-(-γ-glutamyl)lysine isopeptide cross-links in erythrocytes and skeletal muscles from their two ChAc cases, which also showed internal nuclei and some necrotic fibres, suggesting primary myopathic changes [5, 19]. Electron microscopy demonstrated amorphous elements with osmiophilic microfibrils in degenerating myofibres and at the perinuclear level. Combined with these findings, they suggested that tTGase is involved in the production of these inclusion bodies in ChAc muscles.

4.1.4 Immunochemical Staining

We have reported a pedigree with a heterozygous mutation of *VPS13A* showing apparent autosomal dominant transmission [23]. Using rabbit polyclonal anti-chorein antibodies reactive to two synthetic peptides (amino acids 109–122 and

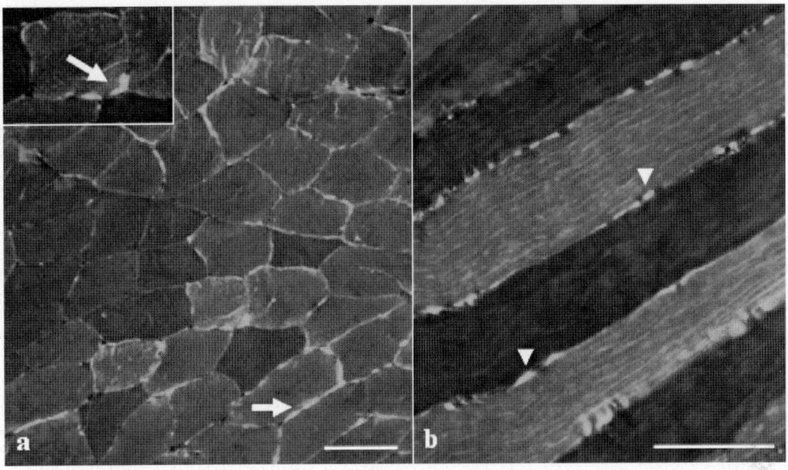

Fig. 3 Immunostaining for chorein in chorea-acanthocytosis. Accumulation of chorein (*arrows* in a and upper inset) was localized along the sarcolemma and at the sarcoplasma in ChAc Case 13. In a longitudinal section from Case 13, it was partially weakened and the surface of the myofibres was significantly irregular (*arrowheads* in **b**). ChAc: chorea-acanthocytosis; Bar, 50 μm

2579–2592), immunohistochemical studies of skeletal muscle were performed. In normal muscle from controls, chorein immunoreactivity was found linearly along the sarcolemma and was intense in some fibres or present at a lesser degree in others (Fig. 2c). Compared with the fibre type distribution of NADH-TR and ATPase (pH = 4.6, 4.3) staining and immunohistochemistry with anti-myosin heavy chain fast antibody, chorein labeling was predominantly present in type I fibres and mainly localized in the sarcolemma (Fig. 2a, b, d). These findings suggest that the chorein function may be associated with mitochondrial activity. Muscle tissues from three heterozygous ChAc cases (Cases 10–12) showed uneven and discontinuous chorein immunoreactivity along the sarcolemma, and in some fibres an increase in chorein immunoreactivity in the sarcoplasma was also noted (Fig. 3). Those chorein accumulations along the sarcolemma were negative for ubiquitin immunostaining. These findings are not seen in typical MLS or genetically-confirmed HD (data not shown).

4.2 *Nerve Biopsy Findings*

According to the report of Dotti et al., common findings are (1) mild reduction in large myelinated fibres, (2) some axonal degeneration, (3) slight remyelination signs and (4) cytoskeletal abnormalities suggesting distal axonopathy. Electron microscopy showed accumulation of membrano-lamellar profiles in the axoplasm [8]. As summarised in Table 1, most other reported cases show similar findings consistent with axonal neuropathy. There has been no immunohistochemical study focusing on the association between the neuropathy and chorein localisation using anti-chorein antibodies of biopsied peripheral nerves in ChAc.

5 Discussion

Previous neurophysiological and neuropathological studies have shown peripheral neuropathy in ChAc diagnosed by clinical features, laboratory and radiological tests [17, 20, 25]. According to the review by Dotti et al., six molecularly-confirmed, homozygous ChAc cases showed axonal neuropathy of various severities [8] . In the present review, on NCV and EMG studies, one case showed only mild myopathic changes, whereas other cases revealed mild/moderate neurogenic changes. Moreover, mild/moderate neurogenic changes identified by conventional muscular staining are noted in most ChAc cases. Although how chorein abnormalities in neurons and Schwann cells contribute to the pathophysiology of neuropathy is unclear, these results establish the fact that peripheral neuropathy is definitely one of the prominent features of ChAc. Further studies are needed to delineate the differential effects of heterozygous versus homozygous mutations of *VPS13A*, specifically the effects of reduced as compared with absent chorein. Dotti et al. have hypothesized that the impairment of chorein may have an influence on axonal transport because of the pathological findings such as neuroaxonal dystrophy, also observed in the spinal cord in autopsied cases [18]. Recently, using transgenic mice models of ChAc, Sano and colleagues have shown high chorein expression levels in microsomal and synaptosomal fractions as well as chorein immunoreactivity observed in the neuronal perinuclear region, cytoplasm and fibres [15]. Those findings may support the hypothesis of Dotti et al.

In ChAc, skeletal muscle involvement with elevation of sCK has usually been regarded as secondary to peripheral neuropathy. However, when there is moderate/severe neurogenic atrophy in pathological studies, concomitant primary myopathic changes may be obscured and can be misdiagnosed. Myopathic changes in four ChAc cases without genetic diagnoses have been reported [17]. Ishikawa et al. reported that in ChAc patients, muscle CT revealed selective symmetric muscular atrophy of the lower legs, which is not consistent with the peripheral nerve distribution [13]. Furthermore, in the present review, most of the ChAc cases showed marked sCK elevation regardless of neuromuscular symptoms and/or involuntary movements of the limbs. In either early or advanced stages of HD, elevation of sCK is unusual despite choreiform movements or parkinsonian symptoms [9]. Therefore, we conclude that elevation of sCK, either with or without amyotrophy, is one of the important features of ChAc.

Melone et al. have reported abnormal accumulation of tTGase products in both skeletal muscles and erythrocytes in genetically-diagnosed ChAc cases [19]. They also demonstrated that homogenous amorphous elements existed in some myofibres on electron microscopic analysis. Based on the muscular findings, they have suggested that a sarcolemmal alteration in the myofibres and primary instability of muscle structure may be associated with myopathic changes in ChAc. We have reported the presence of nemaline rods in addition to myogenic histopathological findings such as internal nuclei, necrotic and regenerating myofibres on H&E staining [27]. Nemaline rods may occur in various pathological conditions and the precise mechanisms of rod formation are still unclear. However, in the context of the gradual evolution of this multisystem disease, one suggestion is that ChAc involves a myopathy which results in rod formation. The identification of the responsible gene *VPS13A* (formerly *CHAC*) opened a window of opportunity to examine the pathophysiology of this disorder [22, 29]. The expression of chorein is high in the skeletal muscles as well as the central nervous system [22, 29]. In our previous report and this review, chorein immunoreactivities of the ChAc skeletal muscles were uneven and discontinuous along the sarcolemma and increased in the sarcoplasma especially in type I fibres in heterozygous ChAc cases [24]. Combined with these findings, it is suggested that in addition to the central and peripheral nervous systems, the skeletal muscles are also involved in ChAc.

Based on the homologue analysis of chorein, a central role of the protein is thought to be involved in protein trafficking at the trans-Golgi network and maintenance of the plasma membrane. However, the association of chorein with myopathy is unclear. Although Melone et al. have reported the inclusion elements with osmiophilic microfibrils in the myofibres, it is unknown whether or not those inclusion bodies positive for tTGase-derived Nε-(-γ-glutamyl)lysine isopeptide in ChAc muscles are associated with protein degradation pathways like the ubiquitin-proteasome system. Our immunohistochemical studies have shown that chorein accumulations are negative for ubiquitin and that staining pattern of beta-spectrin, which is a cytoskeletal protein mainly expressed in the skeletal muscles and erythrocytes, is similar to that of normal control. Although chorein accumulations may be pathologic, incidental, or a beneficial coping response, we can speculate that chorein may be associated with other proteolysis pathways like the macroautophagy-lysosome system and/or have an effect on the maintenance of other cytoskeletal proteins in myofibres.

In summary, both primary muscular involvement and peripheral neuropathy are commonly noted in ChAc, although it may be difficult to detect the muscular involvement by conventional methods in some cases, and to differentiate the effect of muscular involvement from that of peripheral neuropathy. Furthermore, it remains to be determined if decreased or absent chorein contributes to the pathogenesis of both peripheral neuropathy and myopathy. Identification of the function of chorein and potential effector (or binding) proteins will be important to clarify the pathophysiological events involved in ChAc.

Acknowledgements We would like to express our deep thanks to Drs. Genjiro Hirose, Kiyoharu Inoue, Koichiro Sakai, and Ken-ya Murata for their advice and supervision.

References

1. Aminoff MJ (1972) Acanthocytosis and neurological disease. Brain 95:749–760
2. Bohlega S, Al-Jishi A, Dobson-Stone C et al. (2003) Chorea-acanthocytosis: clinical and genetic findings in three families from the Arabian peninsula. Mov Disord 18:403–407
3. Burbaud P, Rougier A, Ferrer X et al. (2002) Improvement of severe trunk spasms by bilateral high-frequency stimulation of the motor thalamus in a patient with chorea-acanthocytosis. Mov Disord 17:204–207
4. Chahin N, Selcen D, Engel AG (2005) Sporadic late onset nemaline myopathy. Neurology 65:1158–1164
5. Danek A, Jung HH, Melone MA et al. (2005) Neuroacanthocytosis: new developments in a neglected group of dementing disorders. J Neurol Sci 229–230:171–186
6. Danek A, Walker RH (2005) Neuroacanthocytosis. Curr Opin Neurol 18:386–392
7. Dobson-Stone C, Velayos-Baeza A, Jansen A et al. (2005) Identification of a VPS13A founder mutation in French Canadian families with chorea-acanthocytosis. Neurogenetics 6:151–158
8. Dotti MT, Malandrini A, Federico A (2004) Neuromuscular findings in eight Italian families with neuroacanthocytosis. In: Danek A (ed) Neuroacanthocytosis syndrome. Springer, Dordrecht, pp 127–138
9. Durr A (2006) Huntington's disease. In: Jankovic J, Tolosa E (ed) Parkinson's disease and movement disorders. Lippincott Williams & Wilkins, Philadelphia
10. Gross KB, Skrivanek JA, Carlson KC, Kaufman DM (1985) Familial amyotrophic chorea with acanthocytosis. New clinical and laboratory investigations. Arch Neurol 42:753–756
11. Hardie RJ, Pullon HW, Harding AE et al. (1991) Neuroacanthocytosis. A clinical, haematological and pathological study of 19 cases. Brain 114 (Pt 1A):13–49
12. Hirayama M, Hamano T, Shiratori M et al. (1997) Chorea-acanthocytosis with polyclonal antibodies to ganglioside GM1. J Neurol Sci 151:23–24
13. Ishikawa S, Tachibana N, Tabata KI et al. (2000) Muscle CT scan findings in McLeod syndrome and chorea-acanthocytosis. Muscle Nerve 23:1113–1116
14. Kito S, Itoga E, Hiroshige Y, Matsumoto N, Miwa S (1980) A pedigree of amyotrophic chorea with acanthocytosis. Arch Neurol 37:514–517
15. Kurano Y, Nakamura M, Ichiba M et al. (2007) In vivo distribution and localization of chorein. Biochem Biophys Res Commun 353:431–435
16. Lagreze HL, Kornguth SE, Brooks BR, Levine RL (1988) Delayed immune response in chorea-amyotrophy with spherocytosis. Neurology 38:1642–1643
17. Limos LC, Ohnishi A, Sakai T et al. (1982) "Myopathic" changes in chorea-acanthocytosis. Clinical and histopathological studies. J Neurol Sci 55:49–58
18. Malandrini A, Fabrizi GM, Palmeri S et al. (1993) Choreo-acanthocytosis like phenotype without acanthocytes: clinicopathological case report. A contribution to the knowledge of the functional pathology of the caudate nucleus. Acta Neuropathol (Berl) 86:651–658
19. Melone MA, Di Fede G, Peluso G et al. (2002) Abnormal accumulation of tTGase products in muscle and erythrocytes of chorea-acanthocytosis patients. J Neuropathol Exp Neurol 61:841–848
20. Ohnishi A, Sato Y, Nagara H et al. (1981) Neurogenic muscular atrophy and low density of large myelinated fibres of sural nerve in chorea-acanthocytosis. J Neurol Neurosurg Psychiatry 44:645–648
21. Rampoldi L, Danek A, Monaco AP (2002) Clinical features and molecular bases of neuroacanthocytosis. J Mol Med 80:475–491
22. Rampoldi L, Dobson-Stone C, Rubio JP et al. (2001) A conserved sorting-associated protein is mutant in chorea-acanthocytosis. Nat Genet 28:119–120
23. Saiki S, Sakai K, Kitagawa Y et al. (2003) Mutation in the CHAC gene in a family of autosomal dominant chorea-acanthocytosis. Neurology 61:1614–1616
24. Saiki S, Sakai K, Murata K et al. (2007) Primary skeletal muscle involvement in chorea-acanthocytosis. Mov Disord 22:848–852

25. Sobue G, Mukai E, Fujii K, Mitsuma T, Takahashi A (1986) Peripheral nerve involvement in familial chorea-acanthocytosis. J Neurol Sci 76:347–356
26. Sorrentino G, De Renzo A, Miniello S, Nori O, Bonavita V (1999) Late appearance of acanthocytes during the course of chorea-acanthocytosis. J Neurol Sci 163:175–178
27. Tamura Y, Matsui K, Yaguchi H, Hashimoto M, Inoue K (2005) Nemaline rods in chorea-acanthocytosis. Muscle Nerve 31:516–519
28. Tanaka M, Hirai S, Kondo S et al. (1998) Cerebral hypoperfusion and hypometabolism with altered striatal signal intensity in chorea-acanthocytosis: a combined PET and MRI study. Mov Disord 13:100–107
29. Ueno S, Maruki Y, Nakamura M et al. (2001) The gene encoding a newly discovered protein, chorein, is mutated in chorea-acanthocytosis. Nat Genet 28:121–122
30. Walker RH, Jankovic J, O'Hearn E, Margolis RL (2003) Phenotypic features of Huntington's disease-like 2. Mov Disord 18:1527–1530
31. Wihl G, Volkmann J, Allert N et al. (2001) Deep brain stimulation of the internal pallidum did not improve chorea in a patient with neuro-acanthocytosis. Mov Disord 16:572–575

Autonomic Dysfunction in Neuroacanthocytosis and Causes of Sudden Death: Analysis of a Case of Chorea-Acanthocytosis with Dysautonomia

M. Kihara(✉), Y. Kawamura, and J.D. Schmelzer

Abstract We reported a 25-year-old woman with chorea-acanthocytosis (ChAc), with typical features of lip- and tongue-biting, choreiform movements and increased levels of serum creatine phosphokinase. She presented with orthostatic hypotension, and on evaluation had a moderately decreased heart rate response to deep breathing and mildly abnormal QSART in the lower extremities. These results indicated that autonomic dysfunction in our patient was mainly due to a postganglionic lesion. Although a few previous reports have described autonomic dysfunction associated with ChAc and have indicated that impairments are primarily sympathetic, our case study suggests that there is also parasympathetic dysfunction. We hypothesize that

---

M. Kihara
Suzuki Clinic, Department of Neurology, 3-10-6, Toyama, Sinjuku-ku, Tokyo, 162–0052, Japan,
mikikihara@aol.com

R.H. Walker et al. (eds.), *Neuroacanthocytosis Syndromes II*.

ChAc, and possibly other neuroacanthocytosis syndromes, have post-ganglionic sympathetic neuropathy in the early stages, which results in denervation hypersensitivity to norepinephrine in heart muscle. This denervation hypersensitivity results in an increase in norepinephrine receptor density and affinity on cardiac muscle and these abnormalities may induce fatal arrhythmia.

1 Introduction

Chorea acanthocytosis (ChAc) is one of the neuroacanthocytosis (NA) syndromes characterized by neurological disturbances such as orofacial dyskinesia, limb chorea, lip and tongue biting, distal muscle wasting, and absent or diminished tendon reflexes with adult onset [13, 18]. The pathogenesis of ChAc is still unknown, but there could be a general membrane abnormality [19]. Although autonomic dysfunction is described in ChAc, information is limited since reports are few and studies have not been comprehensive [16, 19].

It is well known that autonomic cardiovascular indices correlate with function. As the cardiovascular autonomic neuropathy worsens, cardiovascular performance and systemic peripheral resistance response become more abnormal so that autonomic neuropathy may contribute to exercise intolerance [17]. The most important reason for performing an autonomic evaluation is that patients with autonomic failure show an increase in mortality and sudden death [1–3]. It has been reported that quantitative autonomic studies and the use of a composite score is helpful in patient management of certain neurodegenerative disease such as Parkinson's disease and Multiple system atrophy (MSA) [14]. In this chapter we analyze the autonomic function tests of a patient with ChAc and discuss the mechanisms of sudden death in ChAc. These results may be helpful in developing management strategies in ChAc patients.

2 Case Report

2.1 Clinical Features

A 25-year-old woman presented to our hospital with involuntary movements of the extremities. Since age 19, she had insidiously developed involuntary movements of her lips and extremities. The symptoms had advanced to lip and tongue biting. At age 23, she had a single convulsive episode, and occasionally felt mild dizziness and reported blurred vision upon standing. On neurological examination, she was alert, well-oriented, had no gross memory defects and normal mentation. She slurred and stuttered slightly when talking. Cranial nerves appeared normal. Deep tendon reflexes were decreased in all four extremities. No pathological signs were observed. Mild muscular wasting was observed in the distal extremities. All the sensory modalities were normal. Very mild cerebellar signs were observed. There was constant grimacing and involuntary movements of the tongue and chorea-like movements were seen in the extremities.

Laboratory examinations showed high creatine phosphokinase (CK) levels of 640 U/L (normal 38–109) and mildly low β-lipoprotein levels of 232 mg dl^{-1} (normal 250–500). Other normal or negative laboratory investigations included; erythrocyte sedimentation rate, routine hematologic measures, urine sediment, serum electrolytes (K, Na, Cl, Ca, P), renal and hepatic functions, serum Cu and coeruloplasmin, thyroid and pituitary hormones and serum immunoglobulins (IgA, IgG, IgE, IgM). Investigation of red blood cell morphology in the peripheral blood revealed acanthocytosis of over 10%. The electroencephalogram was normal. X-rays of the skull, chest and spine were normal. Magnetic resonance imaging (MRI) studies of brain showed slight atrophy of cortex and caudate. Electromyography (EMG) showed reduced interference patterns and giant complex motor units of the interossei muscles bilaterally. Motor and sensory conduction velocities were low normal in median, tibial, sural and peroneal nerves. Her full-scale intelligence quotient was 98 using the Wecksler Adult Intelligence Scale (WAIS; verbal 98 and performance 98).

2.2 *Material and Methods*

The following autonomic function tests were performed according to Low's methods [11]. The room was controlled for temperature (23°C) and humidity (40%).

2.2.1 Heart Rate Response to Deep Breathing (HRdb) and the Valsalva Maneuver

Heart rate (HR) was continuously recorded via ECG leads connected to a cardiac monitor (Dyna Scope; Fukuda Denshi). Systolic (SAP), diastolic (DAP), and mean arterial pressures (MAP) were continuously monitored using the photoplethysmographic technique employing a Finapres monitor (Ohmeda, Englewood, CO).

For HRdb, the patient breathed deeply at a rate of 6 cycles min^{-1} and a total of eight cycles were recorded. The five largest responses were read from the computer (NEC98) using a cursor, averaged and the HR range (maximum-minimum) derived.

For Valsalva ratio (VR), the patient, rested and recumbent, was asked to maintain a column of mercury at 40 mmHg, for 15 s by blowing into a bugle (a small air leak was made in the bugle to prevent expiration against a closed glottis). The patient was instructed to breathe normally after the expiratory effort, and was allowed to rest for at least 3 min after stabilization of AP and HR before repeating the maneuver. The VR is the ratio of the maximal to the minimal heart rate in response to the Valsalva maneuver. Beat-to-beat BP responses to the Valsalva maneuver were also recorded and each of the four phases were quantified.

2.2.2 Orthostatic BP Recordings

Orthostatic blood pressure (BP) recordings were measured simultaneously using a sphygmomanometer cuff and with a continuous non-invasive BP device (Finapres monitor) to monitor beat-to-beat BP. Blood pressure was measured in the supine position, where a stable resting baseline was obtained, after which the patient was tilted to an angle of 80°head up for 5 min and then tilted back to the supine position.

2.2.3 Quantitative Sudomotor Axon Reflex Test (QSART)

A population of eccrine sweat glands was stimulated by the iontophoresis of acetylcholine using a constant current generator (Asahi Denshi Co., Osaka) and a stimulus of 2 mA for 5 min. Sweat output from a different compartment is evaporated off by nitrogen gas of low constant humidity and controlled flow rate and measured by a sudorometer (Hidrograph; K&S, Aichi). The response was recorded over the upper arm, thigh, distal leg and proximal foot on both the right and left sides.

2.2.4 Scoring of Autonomic Failure

The 10-point composite autonomic scoring scale (CASS) for autonomic function was used based on the autonomic reflex screen developed by Low [11].

3 Results: Autonomic Analysis of our Case (Tables 1, 2)

The heart period response to deep breathing was moderately decreased. The Valsalva ratio was mildly decreased. Analysis of beat-to-beat blood pressure in response to the Valsalva maneuver showed that the patient had an excessive fall in blood pressure in phase II_{late} but normal overshoot in phase IV. The patient had orthostatic hypotension (OH) and exhibited a heart rate increment upon tilt. The QSART was mildly abnormal in the lower extremities. The composite score for the patient was 6 points, indicating moderate autonomic failure.

Although not genetically confirmed, our patient showed typical signs and symptoms of ChAc: adult onset of symptoms, amyotrophy in the distal extremities, acanthocytosis of over 10%, progressive choreic movements of the extremities, tongue and lip biting, and an increased serum CK [6, 13, 18].

Table 1 CASS (composite autonomic scoring scale)

Recording sites	Latency (min; HN)	Volume (μlmin^{-1}; HN)
Lt Forearm	1.0 (0.9–2.9)	0.31 (0.21–4.39)
Rt Forearm	1.0 (0.9–2.9)	0.25 (0.21–4.39)
Lt Proximal leg	2.6 (0.5–2.2)	0.15 (0.20–2.36)
Rt Proximal leg	1.9 (0.5–2.2)	0.34 (0.20–2.36)
Lt Distal leg	1.7 (0.5–2.0)	0.19 (0.20–2.98)
Rt Distal leg	1.5 (0.5–2.0)	0.16 (0.20–2.98)
Lt Foot	2.7 (0.8–3.9)	0.11 (0.16–3.03)
Rt Foot	2.6 (0.8–3.9)	0.12 (0.16–3.03)
Adrenergic index: 2		
$BP_{BB}\ II_{late}$	−13 mmHg	(HN:>−5 mmHg)
Cardiovagal index: 2		
Valsalva ratio	1.42	(HN:>1.5)
HRdb	9.8	(HN:>18)

$BP_{BB}\ II_{late}$, Late phase II of beat-to-beat BP during Valsalva maneuver; HN hospital normal; *HRdb* Heart rate response to deep breathing

Table 2 Blood pressure alterations during tilt

Blood pressure (mmHg)	Rest	1 min	5 min
Systolic	155	120	131
Mean	121	95	112
Diastolic	98	87	104
Heart rate (beat min^{-1})	85	110	115

3.1 Heart Rate Response to Deep Breathing (HR_{DB})

The HR_{DB} is an indirect test of cardiovagal function and may be the most important of the HR responses since both efferent and afferent pathways are vagal. During deep breathing, the heart rate increases toward the end of inspiration and slows toward the end of expiration (HR_{DB}). Such HR variations are dependent on an intact vagus nerve. It has been suggested that a HR_{DB} measurement is more sensitive than the Valsalva ratio and is useful and sufficiently reproducible for early detection of subclinical parasympathetic cardiovascular autonomic dysfunction [5]. The heart rate response to deep breathing in our case was moderately decreased. This result suggests that our patient has parasympathetic cardiovascular autonomic dysfunction.

3.2 The Heart Response to the Valsalva Maneuver

The Valsalva ratio is the ratio of maximum HR:minimum HR during the Valsalva maneuver. The Valsalva ratio is reproducible in normal subjects and declines with increasing age. Low values may be seen in patients with lung and/or heart disease or peripheral autonomic neuropathy [7]. In our patient, the Valsalva ratio was mildly decreased. Although cardiomyopathy has been reported in rare cases of ChAc, it seems that this decreased Valsalva ratio is mainly due to autonomic neuropathy.

3.3 Beat-to-Beat BP Changes to the Valsalva Maneuver

There are four phases of blood pressure response to the Valsalva maneuver observed on photoplethysmographic recordings in normal subjects. Phase I, observed at the onset of straining, consists of a momentary increase in BP due to intrathoracic pressure compressing the aorta. Phase II is biphasic, consisting of an early phase and a late phase. In phase II_{early} there is an early progressive fall in systolic and diastolic pressure caused by continuous straining which impedes venous return to the heart and reduces cardiac output. The decreased cardiac output occurs despite tachycardia caused by withdrawal of parasympathetic influence. During phase II_{late}, as a result of increased efferent sympathetic discharges to muscle by mainly sympathetic α function, the total peripheral resistance increases and plasma epinephrine concentrations rise. With increased reflex vasoconstriction in place, within 4 s the fall of BP levels off and then rises toward baseline. In phase β, when the glottis is open and intrathoracic pressure returns to normal, there is a momentary (1–2 s) fall of mean arterial BP due to mechanical decompression of aorta. This phase is the inverse of phase I. There is a further burst of sympathetic activity during this phase. In phase IV, venous return is restored but the peripheral vessels remain constricted and cardiac output remains elevated due to sympathetic-mediated cardioacceleration, resulting in a rise of BP above the baseline value. This BP overshoot may persist for up to 1.5 min. This overshoot in BP stimulates the baroreceptors, causing a transient decrease in heart rate and a drop in BP back to baseline. The increased heart rate during the early phase of the Valsalva maneuver of at least 20–25 beats per minute above baseline is due mainly to sympathetic β function. This tachycardia is blunted or absent in patients with sympathetic nervous system dysfunction. The absence of BP overshoot in phase IV, or a fall in mean BP in phase II_{late} to below 50% of the resting mean pressure, is also consistent with sympathetic dysfunction [7].

The findings in our case indicated α-adrenergic failure, exhibited by the well-recognized pattern of a loss of late phase II and normal β-adrenergic function, since phase IV overshoot was present in the beat-to-beat BP response to the Valsalva maneuver. Sandroni et al. [15] observed the same phenomenon after α-blockade (phentolamine) administration in normal subjects.

3.4 Blood Pressure Response to Standing

As previously mentioned, the most important function of baroreceptors is maintenance of postural normotension when the body's position changes from supine to standing. Immediately upon standing, there is a fall of arterial pressure in the head and upper body, resulting in activation of baroreflexes and a strong sympathetic discharge, leading to arterial constriction in the splanchnic vasculature, limbs, kidneys, and lungs. The phenomenon of OH occurs in those individuals who are unable to maintain a normal BP upon standing. However, OH does not always indicate autonomic dysfunction. Some degree of OH is, in general, relatively common in patients beyond the age of 70 years [7]. Our patient is 25 years old, however, and her OH is much more likely due to sympathetic dysfunction.

3.5 *Quantitative Sudomotor Axon Reflex Test (QSART)*

QSART quantitatively measures axon-reflex mediated sweating in response to acetylcholine. Both the afferent and efferent pathways of this test are via sympathetic postganglionic fibers to eccrine sweat glands. The QSART provides a sensitive, reproducible, and accurate measurement of postganglionic sudomotor function. Abnormalities in QSART suggest postganglionic sympathetic dysfunction [12].

In our case, there was more QSART impairment in the distal sites compared to the proximal sites. We observed a length-dependent pattern of impairment suggestive of postganglionic sympathetic neuropathy.

3.6 *Composite Autonomic Severity Score*

CASS was determined by measuring the distribution and severity of postganglionic deficits, cardiovagal function and both peripheral α- and β-adrenergic function, and allotting 4 points for adrenergic and 3 points each for sudomotor and cardiovagal failure. This autonomic laboratory test is useful in grading the degree of autonomic failure in many neurological disorders such as Parkinson's disease, MSA, diabetic neuropathy and Guillain Barre syndrome (GBS) [8, 9, 14]. The composite score for our patient was 6 points (sympathetic score of 4 and parasympathetic score of 2), indicating moderate autonomic failure. Although the few previous reports describe primarily sympathetic autonomic dysfunction in ChAc, our case study suggests that there is also parasympathetic dysfunction.

4 Autonomic Dysfunction in ChAc

There are few reports studying autonomic function in patients with ChAc. Senda et al. [16] reported a case of ChAc with OH where the patient exhibited a heart rate increment upon tilt that is similar to our findings. In their case, the urinary noradrenaline and dopamine excretions increased, which led them to conclude that OH associated with ChAc might be due to a higher functional state of the catecholamine system and/or abnormal sensitivity of catecholamine receptors. Our case had postganglionic sympathetic dysfunction demonstrated by QSART, and α-adrenergic failure, exhibited by the well-recognized pattern of a loss of late phase II and normal β adrenergic function, since phase IV overshoot was present in the beat-to-beat BP response to the Valsalva maneuver. Both our case and Senda's case suggest that the underlying mechanism of autonomic dysfunction with ChAc is mainly that of peripheral adrenergic failure. We also consider the fact that cardiovascular dysfunction with ChAc may not only be due to autonomic neuropathy but also cardiomyopathy.

Patients with autonomic failure show an increase in mortality [12] in conditions such as Parkinson's disease, MSA, and in those who are bedbound [10, 14]. The presence of a lengthening of the QT interval in patients with diabetes correlates with abnormal findings on clinical autonomic tests and has been suggested to contribute to sudden death [2, 3]. In patients with ChAc, where autonomic symptoms are not so pronounced, autonomic studies are valuable in defining the severity and distribution of the autonomic deficits, in following the course of the disorder, and perhaps in the prevention of sudden death.

5 Cause of Sudden Death in NA

Previous reports have suggested that common causes of death in NA are aspiration pneumonia due to impairment in swallowing, cardiomyopathy, and suicide as a result of depression. Sudden death may be due to cardiomyopathy-related arrhythmia [4, 6]. It is also well known that NA patients have a hypersensitivity to dopamine and norepinephrine [6, 16]. Our case suggests postganglionic sympathetic neuropathy can be seen in ChAc. Based on previous reports and our results, we introduce the hypothesis that ChAc, and possibly the other core NA syndrome, McLeod syndrome, may show postganglionic sympathetic neuropathy in the early stages which results in denervation hypersensitivity of heart muscle to norepinephrine (Table 3). This denervation hypersensitivity may be due to an increase in

Table 3 Mechanism of sudden death in Neuroacanthocytosis syndromes

Early stage of NA
Post-ganglionic sympathetic neuropathy
+
(May be parasympathetic dysfunction also)
↓
Denervation hypersensitivity
↓
In cardiac muscle, noradrenaline receptor numbers and binding affinities increase.
↓
Fatal arrhythmia occurs
↓
Sudden death

norepinephrine receptor density and affinity on cardiac muscle and these abnormalities, may induce fatal arrhythmia. It is well known that patients with an overlap of sympathetic and parasympathetic autonomic dysfunctions develop a more malignant course than those with only sympathetic or parasympathetic dysfunction [14]. Our patient not only has sympathetic dysfunction but also parasympathetic dysfunction. This result suggests that parasympathetic dysfunction may play a part in the role of sudden death in ChAc patients.

6 Summary

1. We describe a 25-year-old woman with acanthocytosis, lip and tongue biting, chorea-like movements and increased levels of serum creatine phosphokinase.
2. This patient presented with OH. The heart rate response to deep breathing was moderately decreased and the QSART was mildly abnormal in the lower extremities. These results indicated that autonomic dysfunction in our patient was mainly due to a postganglionic lesion.
3. Although a few previous reports have described autonomic dysfunction associated with ChAc and have indicated that impairments are primarily sympathetic, our case study suggests that there is also parasympathetic dysfunction.
4. The composite autonomic score (CASS) for our patient was 6 points, which indicates moderate autonomic failure.
5. We hypothesize that ChAc, and possibly other NA syndromes, have postganglionic sympathetic neuropathy in the early stages, which results in denervation hypersensitivity to norepinephrine in heart muscle. This denervation hypersensitivity results in an increase in norepinephrine receptor density and affinity on cardiac muscle and these abnormalities may induce fatal arrhythmia.

References

1. American Academy of Neurology (1995). Assessment: clinical autonomic testing report of the Therapeutics and Technology Assessment Subcommittee. Neurology 48:873–880
2. Bellavere F et al. (1988) Prolonged QT period in diabetic autonomic neuropathy: a possible role in sudden cardiac death? Br Heart J 59:379–383
3. Chambers JB et al. (1990) QT prolongation on the electrocardiogram in diabetic autonomic neuropathy Diabet Med 7:105–110
4. Faillance RT, Kingston WJ, Nanda NC (1982) Cardiomyopathy associated with the syndrome of amyotrophic chorea and acanthocytosis Ann Intern Med 96:616–617
5. Genovely H, Pfeifer MA (1988) RR-variation: the autonomic test of choice in diabetes Diabetic Metab Rev 4:255–271
6. Gross K et al. (2006) Neuroacanthocytosis syndrome E-medicine Web MD: 1–16
7. Harati Y, Low PA (1990) Autonomic peripheral neuropathies: diagnosis and clinical presentation Curr. Neurol 10:105–176

8. Kihara M, Mitsui M, Nishikawa S, Nishimoto K, Takahashi M (1998) Comparison of electrophysiologic and autonomic tests in sensory diabetic neuropathy Clin Auton Res 8:213–220
9. Kihara M, Nakashima H, Takahashi M, Kawamura Y (1998) A dysautonomia case of Guillain–Barre syndrome with recovery: monitored by composite autonomic scoring scale. J Auton Nerv Syst 73:186–189
10. Kihara M, Takahashi M, Nishimoto K, Okuda K, Matsui T, Yamakawa T, Okumura A (1998) Autonomic dysfunction in elderly bedfast patients. Age Ageing 27:551–555
11. Low PA (1993) Composite autonomic scoring scale for laboratory quantification of generalized autonomic failure. Mayo Clin Proc 68:748–752
12. Low PA (ed) (1997). Laboratory evaluation of autonomic function. Clinical autonomic disorders, 2nd edn. Lippincott-Raven, Philadelphia, pp 179–208
13. Sakai T, Mawatari S, Iwashita H, Goto I, Kuroiwa Y (1981) Choreoacanthocytosis . Clues to clinical diagnosis. Arch Neurol 38:335–338
14. Sandroni P, Ahlskog JE, Fealey RD, Low PA (1991) Autonomic involvement in extrapyramidal and cerebellar disorders. Clin Auton Res 1:147–155
15. Sandroni P, Benarroch EE, Low PA (1991) Pharmacological dissection of components of Valsalva maneuver in adrenergic failure. J Appl Physiol 71:1563–1567
16. Senda Y, Koike Y, Sugimura K, Matsuoka Y, Takahashi A (1987) Chorea-acanthocytosis with catecholamine abnormality and orthostatic hypotension – a case report [Japanese]. Clin Neurol 27:898–903
17. Shepherd RFJ, Shepherd JT (1988) Control of blood pressure and circulation in man. In: Banister R (ed) Autonomic failure, 2nd edn. Oxford University Press, Oxford, pp 80–96
18. Sotaniemi KA (1983) Chorea-acanthocytosis. Neurological disease with acanthocytosis. Acta Neurol Scand 68:53–56
19. Spitz MC, Jankovic J, Killian JM (1985) Familial tic disorder, Parkinsonism, motor neuron disease, and acanthocytosis: a new syndrome. Neurology 35:366–370

Sleep Disorders in Neuroacanthocytosis

I. Ghorayeb(✉), L. Dolenc-Grošelj, J. Kobal, T. Pollmächer, A. Danek, and F. Tison

Abstract Sleep disorders are a well known feature of extrapyramidal disorders such as Parkinson's disease and Huntington disease. In neuroacanthocytosis (NA), data on the sleep-wake cycle deregulation are limited. Here, we report polysomnographic results for five patients with NA. Although limited by the small number, our results clearly point to a significant impact of the disease on the sleep architecture. Future studies with larger cohort of patients with NA are required to highlight specific sleep disorders features in this disease.

1 Introduction

McLeod syndrome and chorea-acanthocytosis (ChAc) are two of the heterogeneous neurological conditions grouped under the common term of "neuroacanthocytosis" (NA) [16]. ChAc is caused by mutations of a large gene on chromosome 9q21 and McLeod syndrome by mutations of the *XK* gene on chromosome Xp21 [16].

I. Ghorayeb
Service des Explorations Fonctionnelles du Système Nerveux-Hôpital Pellegrin, 33076 Bordeaux, France
imad.ghorayeb@u-bordeaux2.fr

The clinical features of McLeod syndrome and ChAc patients are similar. Chorea, which is defined as random, abrupt, irregular and involuntary movements, is the most characteristic movement abnormality observed in NA patients. It resembles the chorea observed in Huntington's disease (HD) patients, however, with prominent orofacial dyskinesias, often causing dysarthria and dysphagia. In addition, cognitive impairment, personality and behavioural changes are common [11]. Seizures are also frequently reported in these patients as well as psychiatric symptoms such as anxiety, paranoia, depression, obsessive behaviour and marked emotional instability [11]. Magnetic resonance imaging (MRI) shows symmetrical abnormalities in the basal ganglia, with atrophy and increased signal on T2-weighted sequences from the caudate and putamen nucleus [12], which are grossly similar to the MRI changes observed in HD [15]. Post mortem findings reveal atrophy of the caudate, putamen and to a lesser extent the globus pallidus and substantia nigra, with marked neuronal loss and astrocytic gliosis in caudate nucleus [8].

Sleep disorders in NA have been rarely reported [14], although they have been widely described in other neurodegenerative diseases of the basal ganglia such as Parkinson's disease (PD) [7], and to a lesser extent HD [6] and multiple system atrophy (MSA) [5]. Although the underlying mechanisms remain unclear, the neurodegenerative process appears to play a crucial role in the occurrence of such sleep-wake cycle disorganisation [4]. In patients with NA, it is not know if sleep disorders are a specific feature of the disease or whether they impact the quality of life of patients. The aim of our preliminary study was (a) to confirm previously reported polysomnographic (PSG) findings in NA [14] and (b) to identify specific PSG sleep pattern abnormalities in patients with NA.

2 Methods

A full-night PSG was performed in five patients with NA (three with ChAc; two with McLeod syndrome). Patients were recruited from different neurological centres in Europe (Ljubljana: two patients, Bordeaux: two patients, Munich: one patient). Patients were free of drugs which could interfere with nocturnal sleep for at least 2 weeks prior the PSG. Polysomnography included electroencephalogram (EEG), electro-oculogram (EOG) and electromyogram (EMG) of submentalis and right and left anterior tibialis. Nasal and oral airflow was recorded with thermistors, and thoracic and abdominal movements with a strain gauge. Arterial oxygen saturation was also continuously measured. PSG-synchronised video monitoring allowed movement analysis during the whole night for four out of the five patients. Sleep staging was manually analysed according to Rechtschaffen and Kale's criteria [13].

3 Results

Sleep parameters for the five patients (four males, one female, mean age 45.6 ± 8.8 years) are presented in Table 1 and 2. Most importantly, results showed great interindividual variability. Mean total sleep time (TST) was globally reduced (296 ± 167 min), but mean sleep latency was within normal range (14.3 ± 10.6 min). Mean duration of the time spent awake from sleep onset to final awakening (WASO) was relatively high (84 ± 54 min) and median sleep efficacy was also reduced 63 ± 25%. Frequent awakenings resulted in sleep fragmentation but with a relative preservation of mean percentages of different sleep stages (mean percentage of slow wave sleep = 28 ± 25% of TST, and mean percentage of rapid eye movement (REM) sleep = 19 ± 12% of TST).

Table 1 Polysomnographic features in five NA patients

	Patient No. 1	Patient No. 2	Patient No. 3	Patient No. 4	Patient No. 5
Diagnosis	McLeod	CHAC	McLeod	CHAC	CHAC
Sex	M	M	M	M	F
Age	48	31	55	46	48
Total period time (min)	499	542	488	342	341
Total sleep time (min)	399	513	297	114	154
WASO (min)	96	18	126	39	142
No. of awakenings	52	10	59	15	6
Sleep efficiency (%)	80	95	61	33	46
Sleep latency (min)	5	11	24	5	27
Sleep stages (min) and (% of TST)					
1	29	48	81	8	3
	7	9	20	7	2
2	264	229	107	24	24
	66	45	27	21	15
3+4	46	101	0	73	70
	12	20	0	64	45
5	59	134	52	8	57
	15	26	11	7	37
No. of sleep cycles	3	4	2	2	2
RBD (yes/no)	No	No	No	No	No
AI	20	2	–	–	–
PLMI	46	–	–	–	–
Excessive daytime somnolence in five NA patients					
EDS (yes/no)	Yes	No	No	No	No
Epworth scale	16	6	–	–	–
MSLT (min)	5	–	–	–	–

Table 2 Patients' polysomnography sleep parameters

Diagnosis	Age	TST (min)	SL (min)	Stage 1 (% TST)	Stage 2 (% TST)	Stages 3 and 4 (% TST)	REM sleep (% TST)	WASO (min)	SE (%)
McLeod	48	399	4.5	7	66	12	15	96	80
ChAc	31	513	11	9	45	20	26	18	95
McLeod	55	296	24	37	52	0	11	126	61
ChAc	46	114	5	7	21	64	7	39	33
ChAc	48	154	27	2	16	45	37	142	46

TST total sleep time, SL sleep latency, REM sleep rapid eye movement sleep, WASO wake after sleep onset, SE sleep efficacy

Video monitoring in four patients revealed reduction but not complete cessation of abnormal movements during sleep. These movements mostly occurred during sleep-wake transitions. REM behaviour disorder (RBD) was not observed in any of our patients.

Despite global reduced efficacy of night sleep in four out of the five patients, only one patient complained of excessive daytime somnolence (EDS). His Epworth sleepiness scale score was 16/24 and his multiple sleep latencies were less than 5 min which is indicative of severe daytime sleepiness. In this patient, PSG revealed a relatively mild obstructive sleep apnoea syndrome (OSAS) (mean apnoea/hypopnoea index = 20) and severe periodic limb movement syndrome (mean PLM index = 46).

4 Discussion

Our study demonstrates fragmented night sleep with poor sleep efficiency, short TST, and increased WASO in most of NA patients. These findings are not specific to NA but are in accordance with limited previous studies [10, 14]. Sleep disturbances are extremely common in extrapyramidal diseases, and encompass a wide variety of complaints including initial insomnia, sleep fragmentation, early awakenings and EDS. Complaints related to impaired sleep quality and efficacy have been extensively reported and documented in patients with PD but to date the underlying mechanisms not well deciphered [1]. In PD, RBD and restless legs syndrome (RLS) appear to be specific features of the disease which are more prevalent in patients with PD than in healthy controls [1]. Whether this is also the case for OSAS is still a matter of debate [3]. By contrast, in MSA, respiratory sleep-related disturbances are common findings and are probably related to the degeneration of the respiratory brainstem centres [4, 2].

Whether specific sleep disorders exist in NA has never been documented. Our results do not support the presence of these features in NA, however our small cohort does not allow us to draw definite conclusions. Interestingly, our findings parallel those reported in patients with HD regarding sleep fragmentation and decreased sleep efficacy

Neuroacanthocytosis is a rare neurological disease in which sleep was rarely studied. [6, 17]. In HD, these PSG abnormalities correlated in part with the degree of atrophy of the caudate nucleus [17]. In NA, the predilection for atrophy of the head of the caudate nucleus argues for a particular vulnerability of this part of the striatum in ChAc, and is in agreement with pronounced neuropsychological disturbances that are thought to rely on these regions [9]. It remains to be demonstrated whether, as in HD, the magnitude of sleep disturbances in NA parallel the degree of the caudate nucleus atrophy. To this end, larger PSG studies of patients with ChAc and McLeod syndrome are required to identify specific features of sleep disorders in NA and to better understand the underlying role of basal ganglia dysfunction in sleep.

References

1. Arnulf I (2006) Sleep and wakefulness disturbances in Parkinson's disease. J Neural Transm Suppl (70):357–360
2. Benarroch EE, Schmeichel AM et al. (2007) Depletion of putative chemosensitive respiratory neurons in the ventral medullary surface in multiple system atrophy. Brain 130(Pt 2):469–475
3. Diederich NJ, Vaillant M et al. (2005) A case-control study in 49 patients. Sleep apnea syndrome in Parkinson's disease. Mov Disord 20(11):1413–1418
4. Ghorayeb I, Bioulac et al. (2005) Sleep disorders in multiple system atrophy. J Neural Transm 112(12):1669–1675
5. Ghorayeb I, Yekhlef F et al. (2002) Sleep disorders and their determinants in multiple system atrophy. J Neurol Neurosurg Psychiatry 72(6):798–800
6. Hansotia P, Wall R et al. (1985) Sleep disturbances and severity of Huntington's disease. Neurology 35(11):1672–1674
7. Hardie RJ, Efthimiou J et al. (1986) Respiration and sleep in Parkinson's disease. J Neurol Neurosurg Psychiatry 49(11):1326
8. Hardie RJ, Pullon HW et al. (1991) Neuroacanthocytosis. A clinical, haematological and pathological study of 19 cases. Brain 114(Pt 1A):13–49
9. Henkel K, Danek A et al. (2006) Head of the caudate nucleus is most vulnerable in chorea-acanthocytosis: a voxel-based morphometry study. Mov Disord 21(10):1728–1731
10. Hori A, Kazukawa S et al. (1985) Electroencephalographic findings in neuroacanthocytosis. Electroencephalogr Clin Neurophysiol 61(5):342–348
11. Kartsounis LD, Hardie RJ (1996) The pattern of cognitive impairments in neuroacanthocytosis. A frontosubcortical dementia. Arch Neurol 53(1):77–80
12. Kutcher JS, Kahn MJ, et al. (1999) Neuroacanthocytosis masquerading as Huntington's disease: CT/MRI findings. J Neuroimaging 9(3):187–189
13. Rechtschaffen A, Kales A (1968) A manual of standardized terminology, techniques and scoring for sleep stages of human subjects. National Institute of Health Publications, Washington
14. Silvestri R, Raffaele M et al. (1995) Sleep features in Tourette's syndrome, neuroacanthocytosis and Huntington's chorea. Neurophysiol Clin 25(2):66–77
15. Simmons JT, Pastakia B et al. (1986) Magnetic resonance imaging in Huntington disease. AJNR Am J Neuroradiol 7(1):25–28
16. Walker RH, Jung HH et al. (2007) Neurologic phenotypes associated with acanthocytosis. Neurology 68(2):92–98
17. Wiegand M, Moller AA et al. (1991) Nocturnal sleep in Huntington's disease. J Neurol 238(4):203–208

Neurosurgery for Neuroacanthocytosis

F. Yokochi(✉) and P. Burbaud

Abstract Neuroacanthocytosis (NA) is a group of neurodegenerative diseases characterized by the presence of acanthocytes in peripheral blood and various types of involuntary movements which are often resistant to medical treatment. We report data concerning four NA patients who benefited from functional surgery; two had pallidotomies and two had deep brain stimulation (DBS) of the internal pallidum (GPi). The clinical outcome of pallidotomy was partial in the first patient, but clearly improved involuntary movements in the second patient without inducing side effects. In the patients who underwent DBS, GPi 40 Hz stimulation gave the best clinical benefit, improving chorea without an effect upon hypotonia. Higher frequency stimulation (130 Hz) was effective for dystonia but increased chorea, worsened dysarthria and induced drooling. Low frequency (10 Hz) GPi stimulation was ineffective. Therefore, in this small series of NA patients, both pallidotomy and pallidal stimulation were partially effective for relieving involuntary movements. However, in this progressive disease with a wide spectrum of symptoms, the decision to perform surgery must involve a full assessment of the clinical features of each patient.

F. Yokochi
Department of Neurology, Tokyo Metropolitan Neurological Hospital, 2-6-1 Musashidai, Fiuchru, Tokyo, Japan
fyokochi-tmnh@umin.ac.jp

R.H. Walker et al. (eds.), *Neuroacanthocytosis Syndromes II.*

1 Introduction

The term neuroacanthocytosis (NA) encompasses a heterogeneous set of neurodegenerative disorders associated with abnormal red blood cell morphology [6, 9, 12, 20, 26]. The commonest conditions are autosomal recessive chorea-acanthocytosis (ChAc), linked to a mutation in the *VPS13A* (formerly *CHAC*) gene localized on chromosome 9q21 [27], and McLeod syndrome (MLS), an X-linked multi-system disorder with absence of Kx antigen and weakened Kell glycoprotein antigens, due to mutation in the *XK* gene [26]. Both conditions share a common progressive clinical phenotype resembling that of Huntington's disease (HD), characterized by generalized choreatic movements, oral dyskinesias, tongue- and lip-biting, tics and dystonia [8, 26]. Other neurological signs such as epilepsy, neuropathy, muscular weakness, and cognitive or psychiatric changes are observed with variable frequency in ChAc and MLS. The movement disorders of NA are often severe and incapacitating, and may resist conventional drug therapy (i.e. neuroleptics and tetrabenazine).

Posteroventral pallidotomy has been reported to be effective for treating symptoms induced by long-term treatment of Parkinson's disease [15, 16, 19]. Its effectiveness in relieving levodopa-induced dyskinesia paved the way for the development of surgical treatment for other hyperkinetic disorders such as dystonia. Concurrently, deep brain stimulation (DBS) has emerged during the last 10 years as the treatment of choice for several types of movement disorders including Parkinson's disease [18, 21, 22], essential tremor [1], and dystonia [5, 31]. Compared to lesion strategies, this technique has the advantage of being a reversible procedure with low morbidity. The observation that DBS of the internal pallidum (GPi) was particularly effective in the treatment of hyperkinetic syndromes led to the idea that it could be potentially useful in the treatment of severe choreic syndromes [4, 23–25, 33, 34, 37]. However, there are few reports concerning the surgical treatment of involuntary movements in NA syndromes. Some benefit has been observed with pallidotomy [10], pallidal DBS [35] or thalamic stimulation [2], but the outcomes of these surgical treatments have not yet been clearly established.

In this review paper, we report a series of four NA patients treated by functional surgery for these severely debilitating hyperkinetic syndromes. Two strategies were used; pallidotomy in two patients and DBS of the GPi in the other two. The clinical benefits of these two approaches are compared.

2 Methods and Results

2.1 Pallidotomy

2.1.1 Patients

Patient 1 was a 41-year-old man with five siblings [10]. His elder brother was diagnosed as having NA and committed suicide at age 37. There was a consanguineous marriage in his family four generations previously. His initial symptom was clumsiness at age 33. He showed restlessness, discoordination when using chopsticks, and erratic behavior. Slurred speech, unsteady gait and involuntary movements of his extremities gradually became pronounced. At the age of 38, he was diagnosed as having NA on the basis of his choreoballistic movements, speech disturbance, ulcers on his lower lip, the presence of acanthocytes in his blood, and his family history. His involuntary movements were characterized by orofacial and lingual dyskinesias, and choreoballistic movements of the neck, trunk and extremities, and were more severe in the proximal than distal parts of his body. He could not remain standing for a long time even with support because of the severe flexion and extension movements of his trunk. The flexion of his knee joints and trunk became more exaggerated when he tried to stand or walk. He could not chew food because of his lingual dyskinesias, and swallowing was impaired. He had many ulcers on his lower lip from self-biting. He needed constant assistance by a caregiver in all activities of daily living (ADLs). Tests of cognitive function indicated slight dementia, but he showed no signs of disorientation. He suffered from severe dysarthria and communicated using only simple words. His other clinical symptoms included muscular hypotonia and areflexia of his extremities. Sensory examination showed a slightly impaired vibration sense in his lower limbs.

Magnetic resonance imaging (MRI) demonstrated symmetrical atrophy of the caudate nuclei and putamen, with high-intensity signal in the putamen on T2-weighted sequences (Fig. 1). The proportion of acanthocytes in his peripheral blood was less than 1%. HD was excluded by genetic testing. The consanguinity suggested the possibility of autosomal recessive disease, and the prominent orofacial dyskinesias were consistent with ChAc, however MLS was not excluded.

Patient 2 was a 37-year-old man. He had two siblings and his elder brother was diagnosed as having NA at 29 years and committed suicide at 33 years. There was no family history of consanguinity. Both parents had acanthocytes in their blood. His father had no clinical symptoms of NA, but his mother showed slight facial grimacing. His initial symptoms were foot dragging when walking and exaggerated gestures at 32 years of age. At age 33, he was diagnosed as having NA because of the presence of acanthocytes in his peripheral blood, chorea, and family history. He gradually developed other symptoms, including writing and speech disturbances, falling while walking, tongue-biting, and abnormal chewing and swallowing. He required complete assistance in his ADLs at age 35. He had slight orofacial, but not lingual, dyskinesias. He showed overextension and flexion of his trunk and neck, and ballistic swinging of his arms and legs, which was more violent proximally than distally. These movements were particularly exaggerated during sitting or standing, and it was difficult for him to remain in either of these positions. He sometimes hit his head against the wall and thus needed to wear a protective helmet. He would perform repetitive non-purposeful movements. For example, he would hold a pillow with both hands, and repeatedly put or pull it under his head. He would repeatedly hit table tops with his hand. This movement pattern was similar to that of ballism or negative myoclonus. His intelligence level was almost normal. His other clinical symptoms were muscle hypotonia and atrophy, hyporeflexia with no pathological reflexes, and a hoarse voice. There was no ataxia or sensory disturbance.

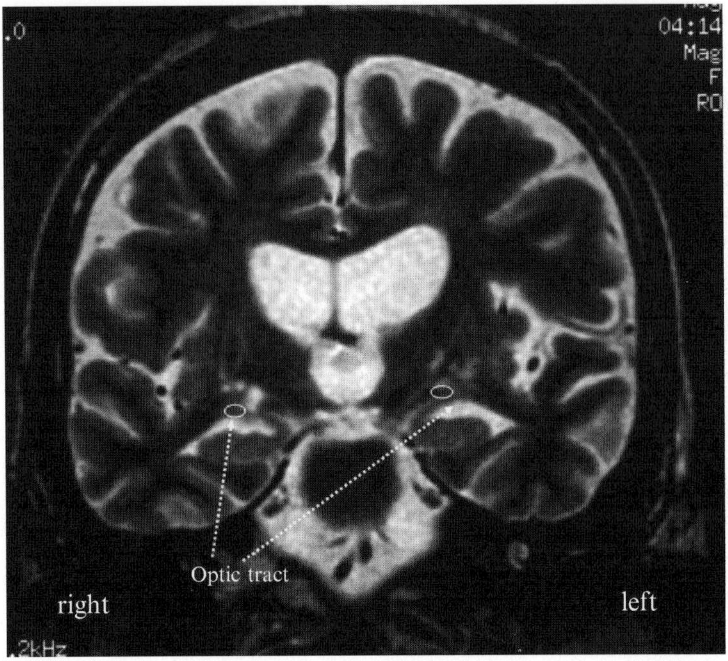

Fig. 1 MRI of patient 1. Symmetrical atrophy of the caudate nuclei and putamen, with high-intensity signal in the putamen on T2-weighted sequences. The thermal lesions were very small and were not located in the GPi. The right lesion involved the optic tract, and the location of the left lesion was unclear

MRI showed symmetrical atrophy of the caudate nucleus and putamen (Fig. 2) with high-intensity signals in the putamen on T2-weighted sequences. Figure 3 shows the surface EMG recordings of his choreoballistic movements and the performance of diadochokinesis with his left hand.

Genetic analysis showed a heterozygous mutation 4411C > T of *VPS13A*. The patient's mother was mildly affected and had acanthocytosis, as did his father. Heterozygous mutations have been reported to be associated with ChAc symptoms in a Japanese family with apparent autosomal dominant inheritance [28] similar to the family of patient 2. However, autosomal recessive inheritance cannot be completely ruled out as a mutation on the second allele cannot be excluded from the methods used.

2.1.2 Neurosurgical Procedures

Both patients underwent bilateral posteroventral pallidotomy under general anesthesia in two stages at an interval of 3 months. In both, the first operation was performed contralateral to the side with more severe involuntary limb movements,

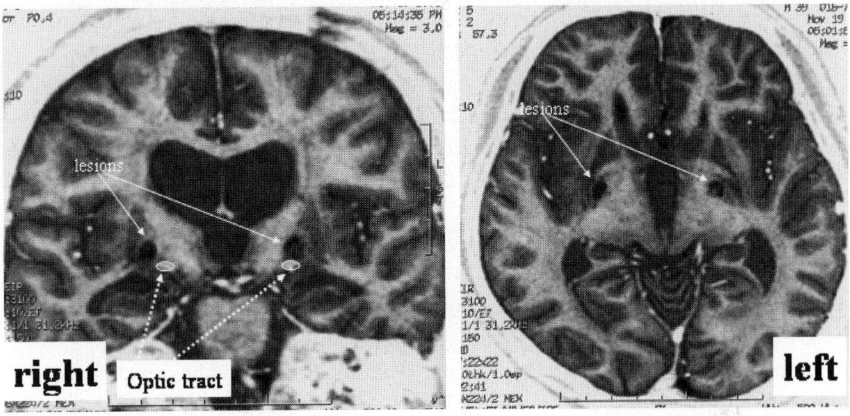

Fig. 2 MRI of patient 2. Symmetrical atrophy of the caudate nucleus and putamen. The left and right thermal lesions are located in the ventral GPi

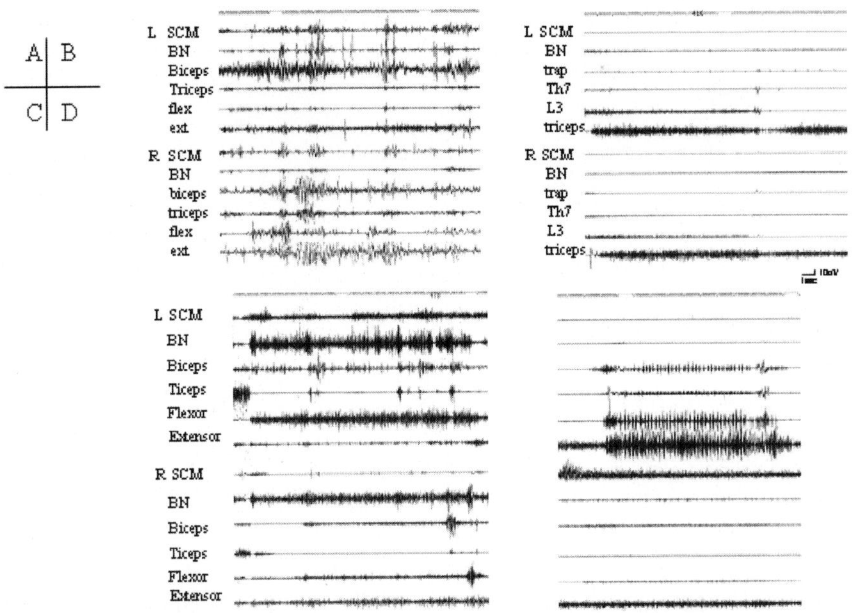

Fig. 3 Surface EMG recording of patient 2 before and 3 months after pallidotomy. A and B were obtained in a sitting position. Before pallidotomy (**a**) there were rapid and synchronous involuntary movements, which were not observed after pallidotomy (**b**). Before pallidotomy the patient was unable to perform diadochokinesis with the left hand (**c**), but after pallidotomy he was able to perform diadochokinesis without involuntary movements (**d**)

followed by the other side. Surgery was performed on patient 1 in 1995 and on patient 2 in 2001. The surgical procedure differed between the two patients.

Patient 1 was operated using the Sugita stereotactic frame without microrecording. The coordinates of the proposed target in the posteroventral part of the GPi as measured by ventriculography were 2 mm anterior to the midpoint of the intercommisural (AC-PC) line (Y), 4.8 mm inferior to the AC-PC line (Z) in the lateral view and 21 mm lateral to the midline of the third ventricle (X) in the AP view. The angle of the lesioning radiofrequency electrode was 60° to the AC-PC line. The location of the optic tract was confirmed using visual evoked potentials, and three thermal lesions were made starting at 2 mm dorsal to the optic tract, and at intervals of 1.5 mm proceeding dorsally from this. The thermal lesions were made at 70 °C for 50 s.

Patient 2 was operated using the Leksell stereotactic frame. The dorsal surface of the optic tract was identified as the tentative target, being immediately ventral to the GPi, and was determined by MRI. The coordinates of the tentative target were Y:1 mm anterior to the midpoint of the AC-PC line, Z:5 mm inferior to the AC-PC line and X:23 mm lateral to the midline. During the operation, neuronal activity was recorded by microrecording, and the specific location of the target was confirmed by neurophysiological and radiological methods. The distance to the optic tract was determined by recording the neural activity of the optic tract in response to a flashed light. Three thermal lesions made at 70 °C for 50 s using a radiofrequency electrode positioned away from the optic tract at an angle of 70° to the AC-PC line.

2.1.3 Results

Patient 1 showed a significant, albeit not completely satisfactory, improvement in his clinical condition after bilateral procedures, with reduction in lingual dyskinesia and involuntary movements of the trunk. After surgery, he was able to eat and stand with less assistance, although still with difficulty. Self-biting disappeared. The lesions as observed on MRI were very small and were not located in the GPi (Fig. 1). The right lesion involved the optic tract, and the location of the left lesion was unclear.

Six months after surgery, he committed suicide. He had not been noted to shows signs of depression. Before committing suicide, he did not complain of any defects or problems in his visual fields. The pathological findings confirmed the diagnosis of NA. The lesion on the right involved the optic tract.

For patient 2, following the right-side pallidotomy, the flexion and extension movements of the trunk improved and he could stand by supporting himself. The involuntary movements of his left hand and his tendency to hit his head against the wall both disappeared. Three months after bilateral pallidotomy, he was partially independent. He could eat without assistance in a sitting position. He could stand alone by holding on a bar, but it was difficult for him to remain standing and to walk without support because of contractures of the ankle joints. The non-purposeful limb movements and involuntary trunk movements gradually reduced in frequency

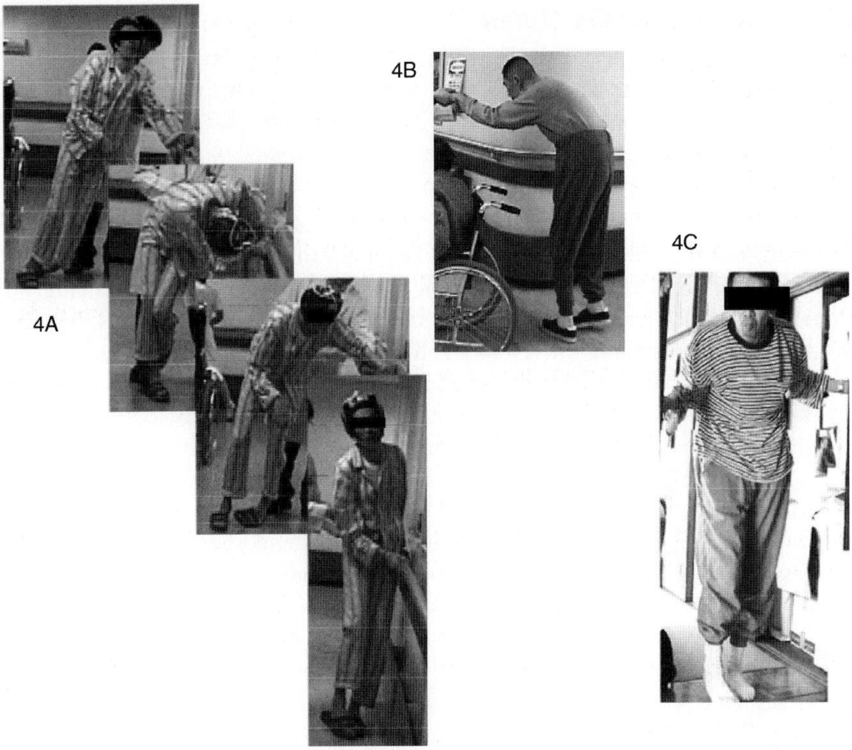

Fig. 4 Patient 2 standing before and after pallidotomy. Before pallidotomy patient 2 was unable to stand even with support because of his truncal movements (**a**). Three months after pallidotomy he could stand without involuntary movements with support (**b**). Four years after pallidotomy he could stand by himself (**c**)

until they disappeared completely. He continued to show an anteflexed posture while standing. His ADLs gradually improved until he was able to perform all indoor activities independently. There was no change in his cognitive functions. He had no difficulty in swallowing food or water, but he was bradykinetic. Four years after undergoing bilateral pallidotomy, he was able to sit, walk or stand and maintain these positions easily without any involuntary movements (Fig. 4). Five years after undergoing pallidotomy, he continued to perform his ADLs independently, and exhibited no involuntary movements.

On MRI, the left and right lesions in the GPi were located in the ventral part of the pallidum (Fig. 2). The surface electromyography (EMG) recorded before and after pallidotomy showed a marked effect of the surgery with a decrease in involuntary movements and improved ability to perform voluntary movements (Fig. 3).

2.2 Deep Brain Stimulation

2.2.1 Patients

Patient 3 was a 32-year-old man with an 8-year history of a choreatic-dystonic syndrome, dysarthria and generalized epileptic seizures who presented with recurrent distressing tasteless belching [29]. The frequency of belching had progressively increased over the previous year and had become extremely debilitating leading to severe depression. The patient also exhibited dramatic tongue-biting. Feeding was problematic and he had lost 20 kg during the previous year. Walking was disturbed by intermittent dystonia of the left foot and bilateral choreatic movements causing a jerky gait. Trunk flexion movements and a moderate back-arching dystonia were occasionally observed. Moderate cognitive impairment was obvious on clinical examination but neuropsychological testing was impossible because of dysarthria.

The patient had 10% of acanthocytes among the circulating red blood cells and a single intronic mutation in the *VPS13A* gene was detected, suggestive of the diagnosis of ChAc. Brain MRI, thoracic radiography and endoscopy of the upper gastrointestinal tract were normal. Videofluoroscopy showed normal deglutition and slight hypotonia of the inferior esophageal sphincter confirmed by manometry. EMG of the diaphragm demonstrated prolonged, recurrent, arrhythmic contractions of both diaphragmatic domes. Recording of respiratory muscles with surface electrodes revealed severely dyskinetic breathing frequently associated with violent belching (Fig. 5). Risperidone (up to 8 mg per day) was ineffective.

Patient 4 exhibited a progressive choreatic syndrome associated with muscular weakness at the age of 41 [13]. A brother and maternal uncle also had chorea. Investigations revealed 10% acanthocytes among the circulating red blood cells, compensated chronic hemolysis with splenomegaly, increased muscle enzymes and a sensory axonal neuropathy. Weak expression of red blood cell Kell antigens and a mutation in the *XK* gene were consistent with MLS. MRI revealed moderate atrophy of the caudate and putamen. Six years later the patient presented with progressive deterioration with severe generalized chorea, predominantly on the left, with hypotonia, postural instability causing repeated falls. He had moderate cognitive impairment (Mini mental state 28/30, Mattis scale 127/140). Tiapride (maximum dose 150 mg/d), risperidone (maximum dose 5 mg/d), and olanzepine (maximum dose 30 mg/d) were introduced without benefit, but with adverse sedative and cognitive effects.

2.2.2 Neurosurgical Procedures

For patients 3 and 4, the targets and entry point coordinates were calculated with regard to the AC-PC line reference system obtained through stereotactic 3D T1-weighted MRI as described elsewhere [7]. Under general anesthesia, per-operative electrophysiological recordings were performed using five parallel micro-electrodes

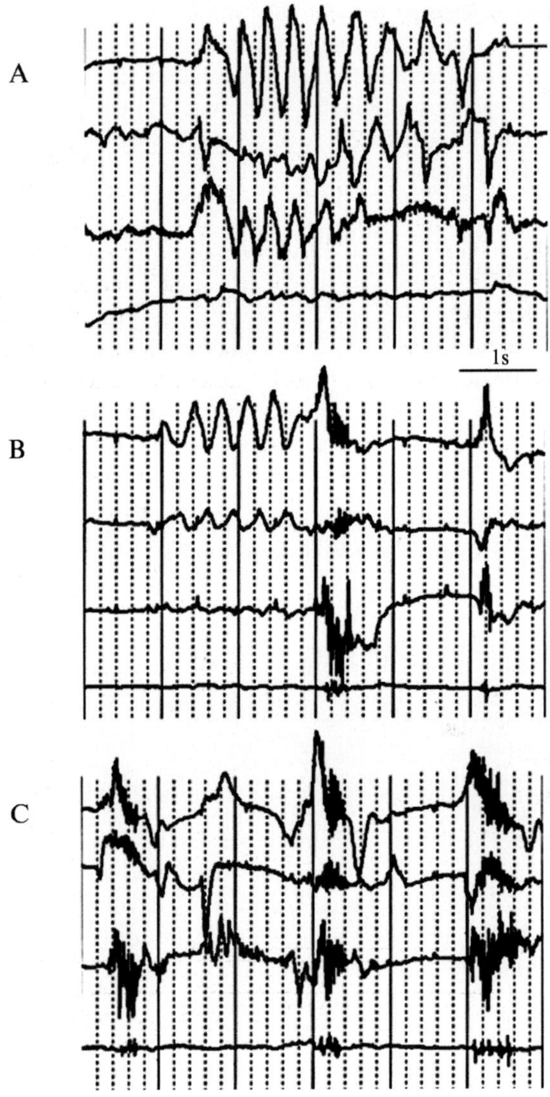

Fig. 5 EMG of axial symptoms in patient 3. Surface polygraphic recording of thoracic muscles recorded before surgery (from top to bottom; right intercostal, right paravertebral, left intercostal, left paravertebral). (**a**) Respiratory rhythms were disturbed by polypneic episodes of 2–3s. (**b**) Polypnea led to a series of violent belching indicated by the *white arrows*. (**c**) Belching appeared spontaneously in a series of violent spasms

(FHC, Bowdoinham, ME, USA) with an Alpha-Omega system (Micro-guide, Nazareth, Israël). At the end of each track, micro-stimulations were performed through the same micro-electrodes with a current intensity up to 10 mA in order to detect the per-operative occurrence of adverse effects. The final implantation sites of the definitive electrodes for chronic stimulation (DBS – 3387 electrode Medtronic®) were guided by the longest recording of GPi neuronal activity and the absence of visible side-effects after micro-stimulation. Electrodes were secondarily connected to a pulse generator (Kinetra, Medtronic®) internalized in the subclavicular region (Medtronic, Model 7425). A cerebral CT scan was performed three days after electrode implantation to identify electrode positions, and images were combined with the pre-operative 3D T1-weighted MRI. The post-operative location of electrode tracks was determined on a basal ganglia atlas consisting of the contours of basal ganglia structures traced and digitized from a brain specimen obtained at autopsy (Fig. 6). Three-dimensional surfaces constructed from these contours were co-registered with the patient's MRI by using specially designed algorithms [36].

2.2.3 Results

For Patient 3, initial stimulation settings involved the two lowest contacts with a voltage of 2.5 V, a frequency of 120 Hz and a pulse width of 90 μs. Within a few days after the onset of stimulation, a dramatic decrease in belching frequency and dyskinetic breathing was observed. Choreic and dystonic movements involving the lower limb and axial muscles completely disappeared, as did tongue-biting. However, after 2 months a partial relapse occurred, and belching and breathing dyskinesia reappeared, although tongue-biting did not. Changes in stimulation settings (voltage, frequency, contacts) were ineffective. MRI revealed that the right electrode was positioned at the border of the putamen and globus pallidus external segment (GPe). We therefore decided to re-operate on this patient in order to reposition the electrode. The first right electrode was left in place and a second electrode was introduced more medially on the same side. After the second implantation, contacts 0 and 1 (the two lower contacts) were located within the GPi on both sides. Contact 2 was at the border between the GPi and GPe and contact 3 within the GPe, bilaterally (Fig. 6). A double-blind evaluation was performed 3 months after this second intervention. Chorea, belching and dysarthria were improved by GPi 40 Hz stimulation compared to the "off" condition. The dystonia sub-score was slightly improved by 120 Hz GPi stimulation but at this frequency, chorea and dysarthria were worsened with drooling. There was no effect on gait whatever the settings used. GPi stimulation at 10 Hz was ineffective.

For patient 4, dysarthria was less pronounced and both the dystonia and choreic symptoms were improved at 40 Hz stimulation compared to the "off" condition. However, GPi 120 Hz stimulation clearly worsened chorea and induced drooling. There was no apparent improvement in hypotonia, and walking remained impossible without help whatever the stimulation parameters used. GPi stimulation at 10 Hz was ineffective. Contacts 0 and 1 were located within the GPi. Contact 2 was at the border between the GPi and GPe on the left, and in the GPe on the right (Fig. 6). Contacts 3 were within the GPe bilaterally. All of the contacts were located slightly more laterally in patient 1 than in patient 2 due to larger ventricles.

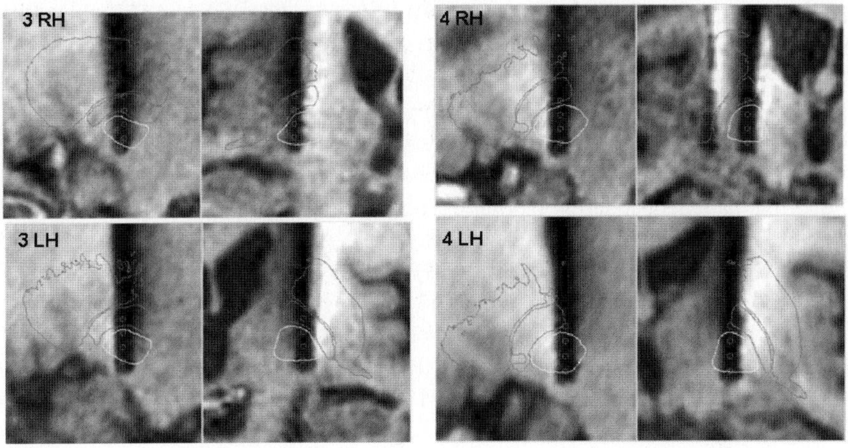

Fig. 6 Localization of the electrodes in patients 3 and 4. RH: right hemisphere; LH: left hemisphere. For each electrode a sagittal (*left*) and a frontal (*right*) view reformatted along the electrode trajectory is presented. Contacts for each electrode are indicated by small circles. The contours of the registered atlas comprise the putamen, and external and internal globus pallidus (from [13])

3 Discussion

These preliminary data obtained on a small series of patients reveal that functional surgery might be a useful tool in the treatment of choreatic and dystonic symptoms in NA syndromes.

The clinical outcome of pallidotomy in patient 1 was partial, however, his lingual dyskinesia and involuntary trunk movements improved and self-biting disappeared. After the surgery, he was able to eat and stand with less assistance, although still with difficulty. The clinical outcome of pallidotomy in patient 2 was better, as 3 months after his bilateral pallidotomy he became partly independent, and 5 years after undergoing pallidotomy he was able to sit, walk or stand easily without any involuntary movements. The difference between these two patients is likely to be due to the location and size of the thermal pallidotomy. In patient 2 the lesion was located in the sensory motor area of the GPi, i.e. the posteroventral area, whereas the lesion in patient 1 was located more anteriorly, and it is unclear as to whether there was actually any lesion of the GPi. In the treatment of levodopa-induced dyskinesia, careful anatomical analysis revealed a strong correlation between location of the lesion in the ventral GPi, its volume and the clinical improvement in contralateral limbs [16]. This is confirmed by data from high frequency stimulation (HFS) of the GPi. Stimulation with the most ventral contacts, located in the ventral part of the GPi, led to a complete arrest of the levodopa-induced dyskinesia [17].

In patients 3 and 4, bilateral GPi stimulation at 40 Hz gave the most reduction of chorea. This is in accordance with a recent finding on Huntington's disease using the same frequency [24]. In patient 3, belching appeared to be due to the combination of a hypotonic internal oesophageal sphincter and diaphragmatic dyskinesias [29]. GPi stimulation dramatically decreased the diaphragmatic dyskinesias, which is the most likely explanation for the improvement in belching [13]. The severe tongue-biting was responded to surgery independently of stimulation, probably through a pallidotomy effect. GPi 130 Hz stimulation worsened chorea and dysarthria, both patients becoming pallilalic with drooling, in line with a previous report showing that HFS did not improve chorea in an NA patient [35]. There was a modest effect on dystonia, but in our study dystonic symptoms were mainly observed in patient 3 and remained moderate.

The effect of GPe stimulation was more complex to interpret. Higher contacts were located at the border between the GPi and GPe thus some current might have reached the GPi, giving confusing data. Moreover, because of the angle of electrode track, these contacts were located in a medial part of the GPe leaving unstimulated a large part of the nucleus. Consequently, the effect of GPe stimulation in NA requires further investigation.

The mechanism by which pallidotomy or chronic stimulation of the GPi modifies chorea remains highly speculative. If HFS (130 Hz) is taken to have an inhibitory effect, as suggested by previous studies in parkinsonian patients [18, 21, 22], such GPi stimulation could worsen chorea, possibly by increasing the activity in the thalamo-cortical pathway. On the other hand, exogenous 40 Hz stimulation applied on the GPi could lead neurons to fire at a subnormal frequency allowing new dynamics in the cortico-subcortical network, thereby restoring a more physiological pattern of discharge, as previously proposed [11, 23]. However, contradictory data have been reported regarding the firing rates of GPi neurons in hyperkinetic syndromes. Whereas low discharge rates were initially found in a hemiballistic patient [32], high discharge frequencies comparable to those observed in parkinsonian patients have been recently reported in two choreic patients [30]. It is likely that further neuronal recordings, taking into account the nature of the anaesthesia used, are necessary to resolve this question.

Alternatively, pallidotomy, which totally suppresses the information passing through the pallido-thalamo-cortical pathway may improve chorea. This was illustrated by the good clinical outcome observed in patient 2 and is consistent with the previous reported case of improvement of back-arching movements in NA with HFS of the motor thalamus [2]. Finally, it seems that the therapeutic effect on chorea is obtained with several methods which have in common the capability of suppressing (with pallidotomy) or regulating (with DBS) an abnormal neuronal firing pattern. The choice between the two techniques could ultimately be dependent on the patient's preference or financial capability.

To conclude, although pallidotomy or pallidal DBS might be considered for the treatment of NA syndromes, it should be kept in mind that this progressive disease has a wide spectrum of symptoms. The decision to perform surgery involves full assessment of the risk of side-effects and the clinical features of each patient. The latter must also guide the choice of target and stimulation parameters. For example, DBS could have opposite effects on movement disorders such as chorea and dystonia that could limit the efficacy of a given target. An alternative could be to use several electrodes with different targets (e.g. GPi, GPe, motor thalamic nuclei) to treat specific symptoms (e.g. chorea, dystonia, hypotonia). In any case, study of a large series of patients with homogeneous clinical features is necessary to investigate this topic. This must be performed in the context of randomized, double-blind, controlled studies with optimal anatomo-clinical correlations.

References

1. Benabid AL, Pollak P, Gervason C et al. (1991) Long-term suppression of tremor by chronic stimulation of the ventral intermedius nucleus of the thalamus as a treatment of movement disorders. Lancet 337:403–406
2. Burbaud P, Rougier A, Ferrer X, Guehl D, Cuny E, Arne P, Gross Ch, Bioulac B (2002) Improvement of severe trunk spasms by bilateral high-frequency stimulation of the motor thalamus in a patient with chorea-acanthocytosis. Mov Disord 17(1):204–207
3. Burbaud P, Vital A, Rougier A, Bouillot S, Guehl D, Cuny E, Ferrer X, Lagueny A, Bioulac B (2002) Minimal tissue damage after stimulation of the motor thalamus in a case of chorea-acanthocytosis. Neurology 59(12):1982–1984
4. Choi SJ, Lee SW, Kim MC, Kwon JY, Park CK, Sung JH, Hong JT, Woo HK (2003) Posteroventral pallidotomy in medically intractable postapoplectic monochorea: case report. Surg Neurol 59(6):486–490
5. Coubes P, Roubertie A, Vayssiere N, Hemme S, Echenne B (2000) Treatment of DYT1-generalized dystonia by stimulation of the internal globus pallidus. Lancet, 355:2220–2221
6. Critchley EM, Clark DB, Wikler A (1968) Acanthocytosis and neurological disorder without betalipoproteinemia. Arch Neurol 18(2):124–140
7. Cuny E, Guehl D, Burbaud P, Gross C, Dousset V, Rougier A (2002) Lack of agreement between direct magnetic resonance imaging and statistical determination of a subthalamic target: the role of electrophysiological guidance. J Neurosurg 97(3):591–597
8. Danek A, Rubio JP, Rampoldi L, Ho M, Dobson-Stone C, Tison F et al. (2001) McLeod neuroacanthocytosis: genotype and phenotype. Ann Neurol 50(6):755–764
9. Ferrer X, Julien J, Vital C, Lagueny A, Tison F (1990) [Choreo-acanthocytosis]. Rev Neurol (Paris) 146(13):739–745
10. Fujimoto Y, Isozaki E, Yokochi F, Yamakawa K, Takahashi H, Hirai S (1997) A case of chorea-acanthocytosis successfully treated with posteroventral pallidotomy. Rinsho Shinkeigaku 37(10):891–894
11. Garcia L, D'Alessandro G, Bioulac B, Hammond C (2005) High-frequency stimulation in Parkinson's disease: more or less? Trends Neurosci 28(4):209–216
12. Gross KB, Skrivanek JA, Carlson KC, Kaufman DM (1985) Familial amyotrophic chorea with acanthocytosis. New clinical and laboratory investigations. Arch Neurol 42:753–756
13. Guehl D, Cuny E, Tison F, Bardinet E, Sibon Y, Ghorayeb I, Yelnick J, Rougier A, Bioulac B, Burbaud P (2007) Deep brain stimulation for movement disorders in neuroacanthocytosis. Neurology 68(2):160–161
14. Hebb MO, Garcia R, Gaudet P, Mendez IM (2006) Bilateral stimulation of the globus pallidus internus to treat choreathetosis in Huntington's disease: technical case report. Neurosurgery 58(2):E383
15. Jankovic J, Lai E, Ben-Arie L, Krauss JK, Grossman R (1999) Levodopa-induced dyskinesias treated by pallidotomy. J Neurol Sci 167(1):62–67
16. Kishore A, Panikar D, Balakrishnan S, Joseph S, Sarma S (2000) Evidence of functional somatotopy in GPi from results of pallidotomy. Brain 123 Pt 12:2491–2500

17. Krack P, Pollak P, Limousin P, Hoffmann D, Benazzouz A, Le Bas JF, Koudsie A, Benabid AL (1998) Opposite motor effects of pallidal stimulation in Parkinson's disease. Ann Neurol 43(2):180–192
18. Kumar R, Lozano AM, Kim YJ, Hutchison WD, Sime E, Halket E, Lang AE (1998) Double blind evaluation of subthalamic nucleus brain stimulation in advanced Parkinson disease. Neurology 51:850–855
19. Laitinen LV, Bergenheim AT, Hariz MI (1992) Leksell's posteroventral pallidotomy in the treatment of Parkinson's disease. J Neurosurg 76(1):53–61
20. Levine IM, Yettra M, Stefanini M (1960) A hereditary neurological disorder with acanthocytosis. Neurology 10:425
21. Limousin P, Pollak P, Benazzouz A, Hoffman D, Le Bas JF, Boussolle E, Perret JE, Benabid AL (1995) Effect on parkinsonian signs and symptoms of bilateral subthalamic nucleus stimulation. Lancet 345:91–95
22. Limousin P, Krack P, Pollak P, Benazzouz A, Ardouin C, Hoffmann D et al. (1998) Electrical stimulation of the subthalamic nucleus in advanced Parkinson's disease. N Engl J Med 339(16):1105–1111
23. Montgomery EB Jr. (2004) Deep brain stimulation for hyperkinetic disorders. Neurosurg Focus 17(1):E1
24. Moro E, Lang AE, Strafella AP, Poon YY, Arango PM, Dagher A, Hutchison WD, Lozano AM (2004) Bilateral globus pallidus stimulation for Huntington's disease. Ann Neurol 56(2):290–294
25. Nakano N, Uchiyama T, Okuda T, Kitano M, Taneda M (2005) Successful long-term deep brain stimulation for hemichorea-hemiballism in a patient with diabetes. Case report. J Neurosurg 102(6):1137–1141
26. Rampoldi L, Danek A, Monaco AP (2002) Clinical features and molecular bases of neuroacanthocytosis. J Mol Med 80(8):475–491
27. Rubio JP, Danek A, Stone C, Chalmers R, Wood N, Verellen C, Ferrer X, Malandrini A, Fabrizi GM, Manfredi M, Vance J, Pericak-Vance M, Brown R, Rudolf G, Picard F, Alonso E, Brin M, Nemeth AH, Farall M, Monaco AP (1997) Chorea-acanthocytosis: a genetic linkage to chromosome 9q21. Am J Hum Genet 61(4):899–908
28. Saiki S, Sakai K, Kitagawa Y, Saiki M, Kataoka S, Hirose G (2003) Mutation in the CHAC gene in a family of autosomal dominant chorea-acanthocytosis. Neurology 61(11): 1614–1616
29. Sibon I, Ghorayeb I, Arne P, Tison F (2004) Distressing belching and neuroacanthocytosis. Mov Disord 19(7):856–859
30. Tang JK, Moro E, Lozano AM, Lang AE, Hutchison WD, Mahant N et al. (2005) Firing rates of pallidal neurons are similar in Huntington's and Parkinson's disease patients. Exp Brain Res 166(2):230–236
31. Vidailhet M, Verceuil L, Houeto JL, Krystowiak P, Benebid A, Cornu P, Lagrange C, Tezenas du Montcel S, Dormont D, Grand S, Blond S, Detante O, Pillon B, Ardouin C, Agid Y, Destée A, Pollak P (2005) Bilateral deep-brain dystonia of the globus pallidus in primary generalized dystonia. N Engl J Med 352:459–467
32. Vitek JL, Chockkan V, Zhang JY, Kaneoke Y, Evatt M, DeLong MR et al. (1999) Neuronal activity in the basal ganglia in patients with generalized dystonia and hemiballismus. Ann Neurol 46(1):22–35
33. Wang Y, Turnbull I, Calne S, Stoessl AJ, Calne DB (1997) Pallidotomy for tardive dyskinesia. Lancet 349(9054):777–778
34. Weetman J, Anderson IM, Gregory RP, Gill SS (1997) Bilateral posteroventral pallidotomy for severe antipsychotic induced tardive dyskinesia and dystonia. J Neurol Neurosurg Psychiatry 63(4):554–556
35. Wihl G, Volkmann J, Allert N, Lehrke R, Sturm V, Freund HJ (2001) Deep brain stimulation of the internal pallidum did not improve chorea in a patient with neuro-acanthocytosis. Mov Disord 16(3):572–575

36. Yelnik J, Damier P, Demeret S, Gervais D, Bardinet E, Bejjani BP et al. (2003) Localization of stimulating electrodes in patients with Parkinson disease by using a three-dimensional atlas-magnetic resonance imaging coregistration method. J Neurosurg 99(1):89–99
37. Yianni JJ, Nandi D, Bradley K, Soper N, Gregory R, Joint C, Stein J, Scott R, Azziz T (2004) Senile chorea treated by deep brain stimulation: a clinical, neurophysiological and functional imaging study. Mov Disord 19(5):597–602

Multidisciplinary Neurorehabilitation in Chorea-Acanthocytosis: A Case Study

J. McIntosh

Abstract With change of the "no cure, no help" attitude towards patients with long-term neurological disability, neurorehabilitation is increasingly being seen as an option. The journey can be helped through negotiation between the patients, their families and professionals to set goals, to enable readjustment and to explore the potential for the person living with the disorder.

1 Introduction

A multidisciplinary therapeutic approach in rehabilitation is not new. It is common to see a group of professionals working together towards shared goals with a patient who has had a stroke and suffered the neurological consequences. The view is one of recovery, compensation, perhaps capitalising on natural recovery and preventing, or at least minimising, the long-term consequences. It is not as common to

J. McIntosh
Wolfson Neurorehabilitation Centre and Atkinson Morley Wing (Neurosciences), St. George's Healthcare NHS Trust, Wimbledon, London, UK
Jacq.McIntosh@stgeorges.nhs.uk

R.H. Walker et al. (eds.), *Neuroacanthocytosis Syndromes II.*

view a progressive disorder or long-term condition in the same way. However, things seem to be changing with the development in the UK of guidelines for the treatment of people with chronic, progressive neurodegenerative diseases such as dementia, multiple sclerosis and Parkinson's disease. The "no cure, no help, no hope" attitude is diminishing, yet despite this it is not always easy for people with these disorders to get expert rehabilitation therapy.

In neuroacanthocytosis (NA), as with many other neurological diseases, there have been many recent advances in understanding pathophysiology and developing treatments. In this chapter 1 demonstrate the potential for assistance for people and families living with such chronic progressive conditions. It is vital for patients and families to have access to professionals who can negotiate goals with them, putting the patient at the centre of their care to ensure that "living with" is not "suffering with".

Although the paradigm involved is not "rocket science", it does involve more than just common sense. The paradigm requires recognition that as a disease progresses and the needs of the person with it change, constant reviews, support and adjustments may be necessary. Those working in the area of neurorehabilitation have the appropriate skills to assist the family and patient with this process. Much can be done even in the absence of a definitive cure. Each person with a progressive neurological condition has a unique personality, intellect, education, culture, skills, relationships and opinions. To provide appropriate care requires the recognition of the need for support and services tailored to these individual qualities. In addition, the person with the condition and family may well know or have ideas on how their needs can be met, needing only the tools and facilitation to do so.

There is much in the dementia literature discussing personhood. Recent work discusses how social and interpersonal factors come into play and challenges the view that all the difficulties are due to catastrophic neurodegenerative changes. It is argued that this view is a narrow conception that diverts attention away from the inadequacy of our social arrangements, and even that this narrow view has resulted in an imbalance in research.

2 The National Service Framework for Long-Term Conditions

In March 2005 the UK Government, through the Department of Health, published The National Service Framework for Long-Term Conditions [6]. Much of the work we do at the Wolfson Neurorehabilitation Centre is recommended in these guidelines. The guidelines recommend a new model through self-care, disease management and individual case management, with emphasis placed on giving people choice and tailoring services to the individual's needs. Included in the requirements is; early and specialist rehabilitation to meet continuing and changing needs, and access to a comprehensive range of rehabilitation, advice and support; and recognition that whilst cures are still being sought the physical, sensory, communication, cognitive, behavioural, emotional and psychosocial effects of a long-term condition should be and can be addressed and services provided. (It needs to be noted though the UK Government did not grant any extra monies along with these guidelines.) To fulfil this requires are a person-centred approach, prompt diagnosis, referral and treatment, rehabilitation, adjustments, social reintegration and long-term support for the person with the condition, and their families and carers.

With regard to rehabilitation the Framework states that this can prevent deterioration and secondary complications. Indeed, rehabilitation does more – it supports adjustment, advocates for support, experiments with strategies, enables choice and compensations, monitors change, explains and educates when done well.

3 The Approach of the Wolfson Neurorehabilitation Centre

By describing the therapeutic intervention at the Wolfson Neurorehabilitation Centre in relation to a person with chorea-acanthocytosis (ChAc) I hope to highlight the many areas of consideration. The World Health Organisation Classification of Disability [16] influences any multidisciplinary team working in neurodisability. Disability should be viewed and reviewed in terms of impairment, activity and participation. The primary aim of the Wolfson Neurorehabilitation Centre is to enable people with neurological disability to negotiate their pathway through education and treatment, in tandem with their families, independence and choice being primary concerns.

The Wolfson is a large purpose-built regional rehabilitation centre with staff comprising a wide range of professionals, including doctors, nurses, therapists and neuropsychologists. We provide services for neurological and cognitive rehabilitation. Medical staff includes specialists in neurology and rehabilitation medicine and visiting specialists in neuropsychiatry and ophthalmology. The types of neurological conditions for which patients may be referred include single-incident neurological disorders, e.g. stroke, brain or spinal trauma, or anoxic damage.

We work to reduce the levels of impairment arising from neurological disorders and to promote activities and participation using the skills of a multidisciplinary team to achieve patient-centred goal planning. We work to facilitate social reintegration and provide support to families and carers. The social model view of disability, as described by Byng [2], encourages us to think of creating authentic involvement and establishing user control. We liaise with general practitioners, hospital and community therapy teams, and social services to support and advise on rehabilitation management. We provide an educational and training resource for local district rehabilitation services. We aim to provide continuing professional development for all our staff, and to encourage audit and research. The primary aim of the Wolfson is to enable individuals to maximise their potential and thereby live as independently as possible.

4 The Multidisciplinary Team

Participation as part of a multidisciplinary team (MDT) is accepted by health care professionals in rehabilitation. Multidisciplinary and interdisciplinary, though used synonymously, do not mean the same thing. MDT working has been referred to "wedges of pie" [5] whereby a group of professionals work in parallel towards a common goal.

"Interdisciplinary rather refers to a team working which involves a partnership between a *team* of health professionals and a *client* in a participatory, collaborative and coordinated approach to *share decision making* around health issues" [14]. This has been described as analogous to a hand; joined and working in unison though having separate digits. This reflects more what we at the Wolfson aim to do. The team needs to have roles and responsibilities with boundaries but not barriers when working together.

The occupational therapist (OT) addresses the following with the client; personal care, e.g. washing and dressing; domestic tasks, e.g. cooking, washing up; leisure activities, e.g. computers, gardening; work activities, e.g. posture, memory strategies; and vocation, e.g. work skills, and cognition.

The physiotherapist concentrates on; balance, movement, muscle weakness, muscle and joint stiffness and pain, assessing the ability to transfer, mobility, and hydrotherapy.

The social worker has a role in discharge planning, assessment of community care needs, welfare benefits advice, housing assistance, liaison with other agents, and acting as advocate.

The medical team's concerns are; ongoing medical care, medication review, education, discussion and review of medical issues, and referrals for specialist opinion.

The speech and language therapist's (SLT) main responsibilities are; assessing and supporting communication abilities and needs in a range of situations; assessing swallowing; providing therapy; advocacy; promoting awareness of the nature of communication and swallowing disorders with the patient and carers.

The nurse's main responsibilities are; to work with client and family to increase independence in personal care with a 24-h approach; advice regarding skin care, diabetes, continence and general health promotion; liaising with district nurses and general practitioners, teaching self-help strategies; dietary advice; understanding medication; methods of lifting and handling.

The neuropsychologist's main responsibilities are; assessing and treating cognitive, emotional and behavioural problems, adjustments, and family issues.

Various characteristics have been described as being important for goal planning and team work to be effective in providing care; "…– a critical factor is to create an environment of interdisciplinary care. Such interdisciplinary working involved acknowledging the overlap between professional disciplines, with a greater sharing of duties and responsibilities …" [14]; "The most effective teams consist of a cohesive group of professionals who have different, but complimentary skills, and who trust and respect the expertise of other team members" [1]; "Successful teams may possess combinations of skills that no single individual demonstrates alone" [5].

The healthcare professional also benefits from an integrated approach and feeling part of the team, and from an increased awareness that a complex problem requires a complex, multifaceted solution; feeling "actively involved in [the patient's] care …… [with an i]mpact on motivation, perceptions of control and freedom in decision making" [3].

The challenges are that being part of an MDT may be overwhelming and confusing, with loss of clarity of roles, and concern that a specific aspect may be missed. This can be particularly difficult for medical personnel who still seem to be expected to have all the answers. There is a need to change preconceived ideas with respect to roles and recovery. This approach can break down barriers, provide a learning opportunity about the condition, develop a high level of negotiation skills, model techniques and carry over into more effective management.

5 Goal Planning

Goal planning is a flexible framework, within which each client's strengths, needs and wants can be assessed and managed. This enables each admission to be smoothly coordinated by the MDT involving the client, their family and friends.

The *process* of goal planning aims to increase a client's and family's involvement in their rehabilitation programme and to reduce their dependency on the MDT and the "institution". The process should be a transparent service, clear to all those involved, ensuring an effective and efficient admission for the client, and resulting in their seamless transition back into the community.

Goal areas which are expressed by clients and then discussed and negotiated with the MDT are unique and can be multifaceted. Goal planning is a simple process that takes a high level of skill and time to negotiate and work with usefully. It should not be seen as a predetermined series of steps or tasks, but a flexible framework within which an individual can be guided forwards towards their future. Goals are negotiated, set, worked upon and reviewed. Goal planning is used as an outcome measure. McMillan and Sparkes [13] discuss the usefulness of goal planning as providing basic data on outcome and audit information. A goal should be stated in terms of the most important area(s) of the client's life and should incorporate a state which can be reached in a reasonable time [11].

When setting a goal we need to leave our preconceived ideas to one side and not anticipate the patient's focus. A patient who is in a wheelchair and adjusted to that way of life may not have a goal of walking but of independence and function in the wheelchair.

Goal setting needs to be part of a process of adjustment, acceptance and hopefully improvement. Thus, an MDT approach to goal planning enables the goal to be the driving force, offers a consistent approach, makes sense and links to real-life.

6 Case Report

6.1 Clinical Presentation

Patient A presented at 28 years with a primary diagnosis of ChAc and was referred to the rehabilitation centre by her neurologist. The aim was to determine the level of disability and to explore ways to maintain independence. As the disease progressed her level of activity and participation declined. At early stages in the disease she exhibited mild executive function deficits related to basal ganglia damage. This reduced planning organisation, caused disinhibition, affected her perception of reality, and gave her difficulty with shifting tasks. All of these features affected communication [9].

Patient A presented with choreiform and stereotypic involuntary movements of her upper limbs, poor coordination and clumsiness of gait. She had classic lingual dystonia, expelling food from her mouth, with some dysfluency and tongue clicks. She had mild cognitive changes, specifically with visual memory. The MDT aimed to identify the problems for Patient A and what could be done though rehabilitation to reduce or compensate for these.

6.2 Assessment

The patient's problems were identified on her first admission. She was introduced to the ideas of goal setting; long-term disability; compensation rather than cure; and disability management and strategies. She was helped in understanding her impairments and their consequences when taking part in activities, and the limitations on participation in her everyday life. These were challenging and unexpected choices and considerations, especially for a young woman with already reduced cognitive function.

In working to accepting her condition with the MDT, the following issues were identified; a struggle to live independently as formerly; swallowing difficulties with occasional choking (no dinner parties); involuntary clicking noises (too loud to go to the theatre); slurred speech; unsteady gait with decreased balance and falling; inability to walk long distances (unable to go out alone); difficulty with fine finger movements (impairing keyboard skills); changes in cognition and memory difficulties (no longer able to work); and foot drop. The MDT identified one further problems which could be a big barrier, specifically her lack of insight into the difficulties. At early stages it was necessary to maintain a very delicate balance between being overwhelmed by her limitations in light of a progressive neurodegenerative condition, and setting up false unrealistic hope. She was keen to return to independent living, and although the MDT acknowledged this, it was always highlighted that this would be difficult.

The strengths identified on the first admission were; a positive attitude, engaging and attendance of all therapy, highly supportive family, no dependents, good general health, high level of education and understanding, and multiple interests.

The goals set were the following:

- To have explored all possible strategies to achieve the best compensated gait
- To have explored possible eating strategies to achieve the best compensated eating
- For Patient A to write an informational leaflet about NA for friends and family so that she would have a better understanding of her problems and why they were happening
- To acquire knowledge of management strategies to help cope with her illness
- For Patient A to explore methods of relaxation
- To have identified cognitive strategies to achieve independence for:
  - self medication
  - managing appointments
  - remembering names
  - organising meal provision

6.3 Level of Achievement

The MDT was initially involved in evaluating Patient A's independence with a view to her returning to her former lifestyle. She was seen regularly by the SLT, physiotherapist, OT, psychologist and the medical team. She had initially wanted to work at improving her level of impairment, with the idea, as many people with long-term conditions have, that exercises will return them to function.

Therapy focussed on her understanding of her illness, her abilities and the impact on her current and future lifestyle. Specific strategies were tried which she was able to learn but not necessarily generalise to outside situations. For example, strategies for reducing lingual projection of food were learnt, but took tremendous effort to remember and institute in everyday eating. She was able to organise information into a folder, but it remained difficult for her to use organisational strategies to remember to do at all times. In terms of physiotherapy, whilst she was able to maintain trunk stability for short distances, i.e. 40 m, she was unable to maintain this with additional tasks. Some strategies have been leant and put into use. Many techniques Patient A has taught herself and maintained, e.g. reducing involuntary noises of her tongue by holding her tongue in certain positions, maintaining her ability to walk outside by recognising and identifying when she has problems with gait and instigating a new walking strategy. Strategies have been discussed with Patient A and her mother, and they will need future help to monitor these and make new adjustments.

6.4 Admission Summary

Throughout her stay at the Wolfson Centre, Patient A was extremely enthusiastic and cooperative, continually trying to evaluate therapy and its effects for and upon her. She rarely showed signs of the frustration which she was obviously feeling, although she found it hard at times that she could not find answers to problems. Her insight and awareness were impaired as to the nature and severity of some of her difficulties with eating, walking and general functioning. Patient A's level of productivity and level of expectation decreased. She could not expect to do as much in a day as prior to her illness, thus it was necessary to build up a lifestyle, and research work and leisure opportunities to establish a daily/weekly routine.

Patient A was able move towards a more appropriate and realistic view of her abilities and expectations, and decided on the last week of admission that a return to independent living would be difficult. This was an important shift for her.

It was recommended that the patient and her family seek assistance in exploring future choices and support. She was referred to a community rehabilitation team to maintain progress. The initial stay of 8 weeks, Monday to Friday, enabled important goals to be achieved.

6.5 Follow-up

Patient A was strongly supported by her family, although a shortage of resources in the community was identified. She continued to live with her parents and to work a little at home. One year later she was reviewed by the Wolfson and returned for another inpatient stay of 3 days/week over 10 weeks. This was recommended to enable linking intensive therapy with home.

At this time her condition had progressed. The problems identified were the following:

- Increasing dependence on parents for social and financial aspects of daily living
- Decreasing ability to work
- Decreasing independence in mobility
- Difficulty eating and swallowing
- Decreasing cognitive ability and primary difficulties of memory
- Reduced balance and equilibrium reactions
- Difficulty grading movement during functional activity due to reduced proximal stability

Her needs for the second stay were to specify in terms of activity and participation realistic goals to maintain an appropriate level of independence. It was important that the consequences of neurological illness should not preclude her participation in many aspects everyday life. This can be achieved by the MDT helping to explore the multifaceted nature of the problem and working with the client to enable the goal.

One year later Patient A was admitted for a short stay of 4 weeks, with progressive difficulties. The MDT worked with her again to identify her difficulties, and set new goals for this admission. On this admission her insight and expectations of rehabilitation were appropriate and she became the leader, recognising what she could get from us and what she might be able to explore with us. This put her family at the centre as they were becoming experts, but needed more understanding of the disorder, its progression and areas where help was needed.

The following problems were identified by the MDT with Patient A on her third stay:

- Decreased safety in the kitchen
- Decreased safety on walking
- Problems carrying items when walking
- Not going out
- Memory difficulties
- Decreased opportunities to communicate
- Decreased communication skills
- Difficulty eating normal consistency food
- Unable to eat out socially

The goals set on her third admission were the following:

- To continue to cook safely at home
- To explore strategies to help working memory
- To understand cognitive difficulties
- To maintain handwriting
- To maintain some ability to work at home
- To have a trial of callipers
- To plan home environment and ongoing support
- To explore leisure activities
- To eat out
- To establish a regime for best eating and eat out on one occasion
- To explore opportunities to communicate with others

The problems which were worked through at this time were the following:

- Barriers to old life
- Reality of illness
- Patient A and therapists' struggle to agree on the reality of the illness
- Support needed
- Desire to work at impairment level
- Results of first stay
- Some insight into her situation … unable to return to independent living
- Safety awareness
- Effort needed to improve eating
- Adjustment
- Maintenance

6.6 Specific Orolingual Issues

Not wanting to "throw the baby out with the bath water" the team recognised the need to work specifically on one of the impairments, i.e. the distressing lingual dystonia and oral clicks that were becoming very marked. She was losing food from her mouth constantly during meal times and the involuntary noises were becoming very loud. She was referred to a neurophysiologist at the neighbouring hospital, and it was decided to try botulinum toxin injections into the genioglossus muscle.

The first signs of the disorder had been dysfluent speech and involuntary vocalisations – grunts, clicking noises and bruxism. She then became aware of problems chewing and swallowing food, with rare episodes of choking, and avoided eating out or with strangers. She had recently bitten her lip for the first time. She has experimented with different textures of food and developed various trick manoeuvres. Much work was done with the SLT aimed at bringing eating under volitional control with some success, leading to closed mouth and inhibited tongue movement using video feedback. This behavioural therapy was successful to a point but was hard work for her.

The role of the tongue in swallowing is paramount for oral preparation of food; to manipulate, place, collect food, to draw together and hold the bolus in a groove, to taste and feel the food, texture and temperature, and to approximate with the palate to prevent early escape into the pharynx. The tongue is the main contributor in propulsion of the bolus by peristaltic action and anterior elevation to allow food into pharynx. With a large bolus the tongue divides it into smaller boluses by humping to the hard palate. During swallowing the tongue continues manipulation and control of the contents of the oral cavity. All of these elements were being affected in Patient A.

The results of the botulinum toxin were impressive, showing reduction in lingual dystonia and improvements in eating, which started to wear off after 8 weeks as expected (Figs. 1–5). She had two injections 8 weeks apart. However, she decided not to have further injections due to the pain of the injection and she returned to the strategies which she had previously learnt. Behavioural therapy showed improvements in most objective measures, but in general was not as effective as it required great effort and was only beneficial under supervision of the therapist.

7 Conclusion

The approach of the MDT in this patient and family showed benefits in terms of functional and psychological adaptation. The process of structured goal-setting enabled clarification of the aims of therapy, and was useful in highlighting discrepancies between the patient's somewhat unrealistic goals, and what was more likely to be achieved. The process also gave the patient and her family an enhanced and active role in the therapy process, and an increased autonomy in the face of this

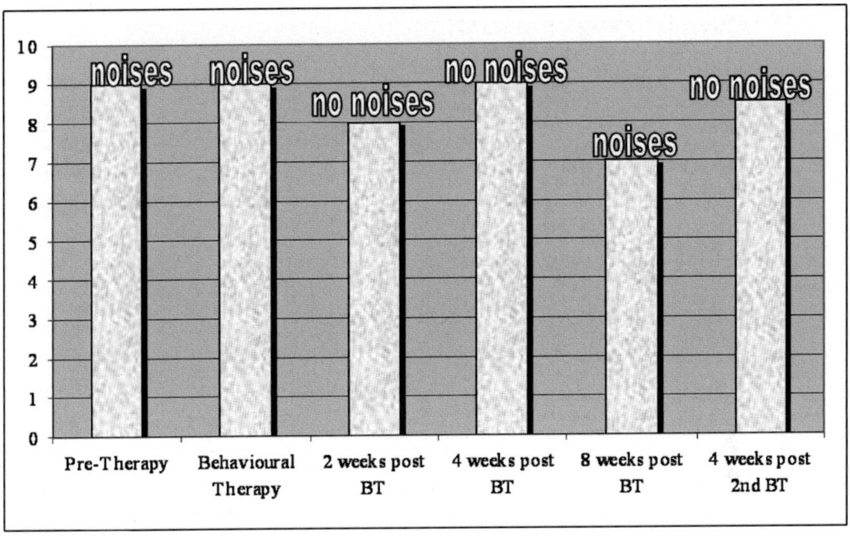

Fig. 1 Intelligibility of speech as rated on an arbitrary scale out of ten by speech and language therapist (SLT). Involuntary noises (oral clicks) decreased following botulinum toxin (BT) injections

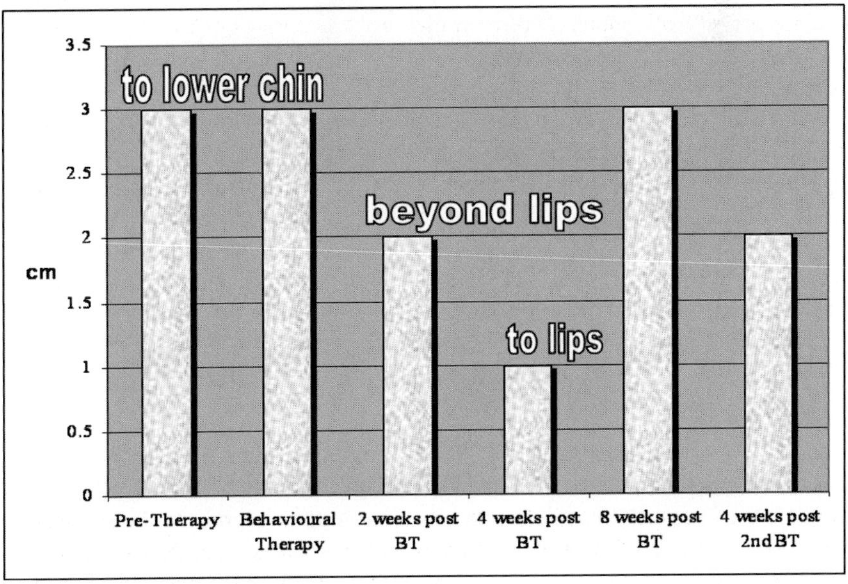

Fig. 2 Distance of tongue protrusion. The distance of involuntary tongue protrusion decreased following botulinum toxin injections, as an indication of effective injection

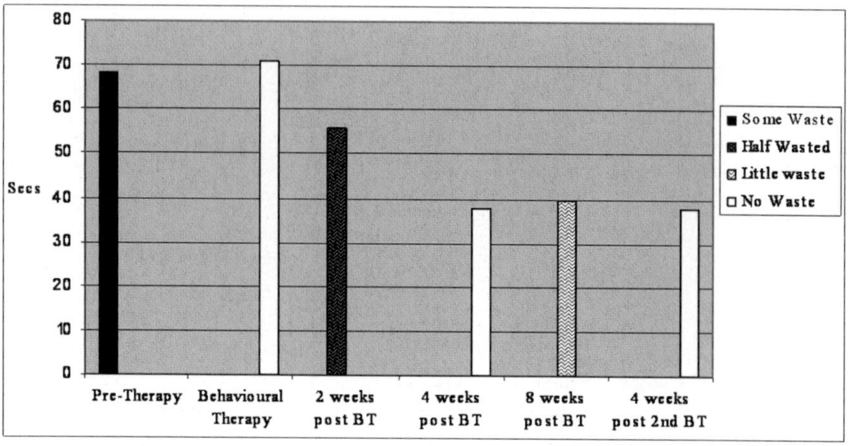

Fig. 3 Average time to eat one mouthful of food. Time taken to eat decreased and the amount of food wasted was reduced following botulinum toxin injections, although marked benefit was also seen with behavioural therapy

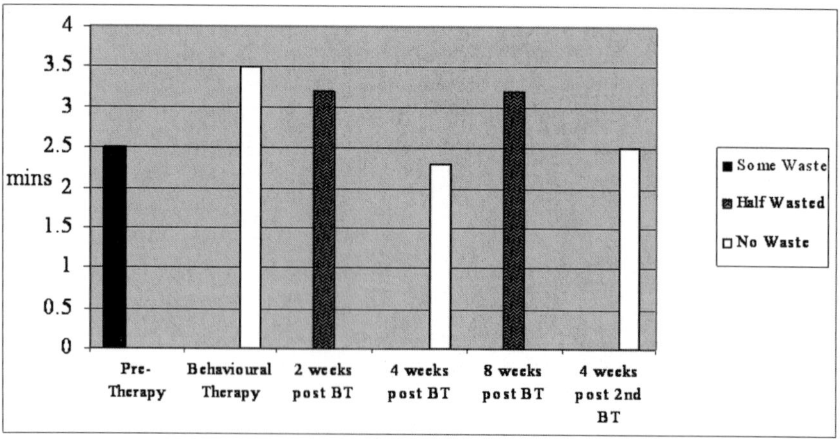

Fig. 5 Average time to eat five mouthfuls of food. The time to eat five mouthfuls was reduced following injection. A comparable effect was seen with behavioural therapy but required great effort

potentially overwhelming situation. Although not quantified, it is likely that these factors increased their ability to cope psychologically.

Review of the patient and her family is an ongoing process. In light of the expected changes in impairments with this progressive disorder, the patient and her family will be encouraged to return with suggestions and expectations of the team. The MDT will in partnership with Patient A and her family continue to negotiate if and how best we might set useful goals. As Patient A's dependence increases there

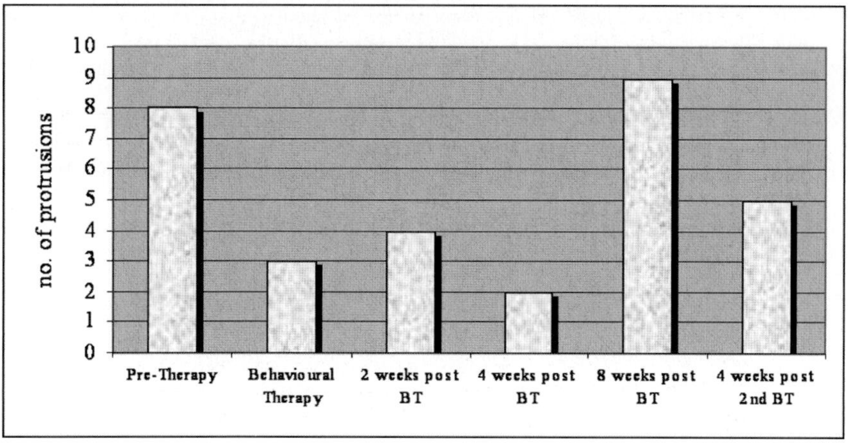

Fig. 4 Tongue protrusions during five mouthfuls. Number of tongue protrusions during eating five mouthfuls. A reduction is seen following injections. A comparable effect was seen with behavioural therapy but required great effort and supervision

may be a need to evaluate her decreasing environment. Communication aids will also need to be considered. If a short inpatient stay is indicated, the outcome will be measured by goal achievement. The guiding principle will continue to be to put Patient A at the centre of the team, and through team work to optimise her function and independence.

References

1. Baldwin, Tsukuda (1994) The interdisciplinary team approach to long term disability. Conference Proceedings, 2nd OT Congress
2. Byng S, Felson J, Ducham (2005) Social model philosophies and principles: their applications to therapies for aphasia. Aphasiology 19(10/11):906–922
3. Conneeley A (2004) Interdisciplinary collaborative goal planning in a post-acute neurological setting; a qualitative study. Br J Occup Ther 67(6):248–255
4. Coombes R (2005) Health Serv J 115:18
5. Crawford, Price (2003) Team working in palliative care as a model of interdisciplinary working. Ther Mag
6. Department of Health (March 2005) The national Service Framework for long-term conditions
7. Freeman JA, Ford H, Mattison P, Thompson AJ, Clark F, Ridley J et al. (2002) Developing MS healthcare standards: evidence-based recommendations for service providers. The MS Society and the MS professional Network, London, pp E1+2
8. Iles V (1997) Open University Press
9. Keif, Kazuiak (2002) Aphasiology 16(3)
10. Leigh PN, Abrahams S, Al Chalabi A, Ampong MA, Goldstein LH, Johnson J et al (2003) The management of motor neurone disease. J Neurol, Neurosurg Psychiatry 74:Suppl-iv47. R2 High Direct

11. McIntosh J, Sparkes C (1997) Goal setting: can the process be empowering? RCSLT Bulletin, Issue 548
12. McMillan T, Oddy M Service provision for social disability and handicap after acquired brain injury. Chapter 12
13. McMillan T, Sparkes C (1999) Goal planning and neurorehabilitation: the Wolfson neurorehabilitation centre approach. Neuropsychol Rehabil 9(3/4):241–251
14. Orchard et al (2002) Creating a culture of interdisciplinary collaborative practice. Med Educ Online 10(11):1–13
15. Royal College of Physicians (2000) National clinical guidelines for stroke
16. WHO (2000) Health Services Management

Part VI
The Way Forward

"Virtual Neuroacanthocytosis Institute": A Look Forward

A. Danek(✉) and B. Bader

Abstract Five years after the initial symposium in Seeon, Germany, international cooperation in the field of neuroacanthocytosis is well established. Improved recognition of the individual diseases now forms the basis for standardized observation in the context of a web-based collection of cases that may be uncommon locally, but are nevertheless amenable to systematic investigation if accumulated across and between continents.

The neuroacanthocytosis submodule (www.euro-hd.net/html/na/submodule) within the European Huntington's Disease Network should eventually provide the basis for observational treatment studies and may transform into a veritable "Virtual Neuroacanthocytosis Institute."

1 Meeting the Agenda of 2002

Development of an agenda to improve the understanding of the neuroacanthocytosis (NA) syndromes concluded the proceedings of our first symposium, yet many of the questions posed still remain a challenge [7]. Further work will particularly be necessary with respect to the acanthocytic shape change. Uncovering its causes in the easily accessible red cells promises to reveal the mechanisms responsible for neuronal damage in NA. Two areas in which there are new data

---

A. Danek
Neurologische Klinik, Ludwig-Maximilians-Universität, D-81366 München, Germany
danek@lmu.de

R.H. Walker et al. (eds.), *Neuroacanthocytosis Syndromes II.*

are regarding the tissue distribution of the NA proteins and the ChAc mouse model [5, 19, 20, 24].

Most progress, however, has concerned the mutual exchange among researchers and among patients under the lead of the "Advocacy for Neuroacanthocytosis Patients" (see introductory chapter by Walker et al.), as demonstrated by the follow-up symposia, including a meeting planned for 2008 in London and Oxford, and the existence of the present volume. We have succeeded in making information about the field more easily accessible, both through written reviews [9, 11, 12, 17, 27, 28] and invited oral presentations, most notably at the 2007 World Congress on Huntington's Disease. The educational success is tangible from the ever increasing numbers of inquiries regarding the chorein Western blot [13] that we offer as a diagnostic aid, with financial support from the "Advocacy for Neuroacanthocytosis Patients" [6].

2 Global Occurrence of NA Syndromes

Application of the chorein Western blot has supported a diagnosis of chorea-acanthocytosis (ChAc) in thirty patients during the last year and led the way to the alternative diagnosis of McLeod syndrome (MLS) in additional cases. Among the MLS cases is an extended family from Chile that offers a new opportunity to systematically assess the phenotype of MLS [23], in addition to the large families known from New Zealand and Switzerland. MLS so far has also been reported from the USA, the UK, France, Germany, Brazil, and Japan.

New cases of ChAc, as diagnosed with the chorein blot, have come from Germany, Italy, Spain, Switzerland, UK, USA, Sri Lanka, Brazil, and Australia. They add up to the considerable number of single cases with a suggestive clinical picture from all over the world, which were often reported only locally [1, 2, 3, 4, 14, 15, 18, 22, 25, 26, 29]. In addition we note the Japanese cases (see chapter by Hirose) and the patients in whom *VPS13A* mutations were detected [8]. The current map of global distribution of ChAc (Fig. 1) shows absence of cases mainly from Africa and the former Soviet Union – most likely due to inaccessibility of the reports in local literature or because of underdiagnosis.

With the help of the questionnaire previously used in MLS [10] we collected data on 106 patients (62 males, 44 females) in whom mutations of *VPS13A* were found [8]. Results show that the onset of the disorder peaks at around the age of 35 years. There is a wide age range in ChAc, yet overall the disease appears to manifest about ten years earlier than MLS (Fig. 2).

Observations of this nature help to put single case reports in perspective. The presumed diagnosis of ChAc in an 86 year old Japanese woman with acanthocytosis seen only once upon conventional red cell staining [16] does not appear likely from the age distribution described above. MLS was reportedly excluded. In keeping with the clinical impression, ChAc was not confirmed by DNA analysis (Saiki, personal communication).

Fig. 1 Approximate number of cases of ChAc with respect to country of origin, based on clinical reports and on cases with *VPS13A* mutations or chorein non-expression in the red cell membrane Western blot

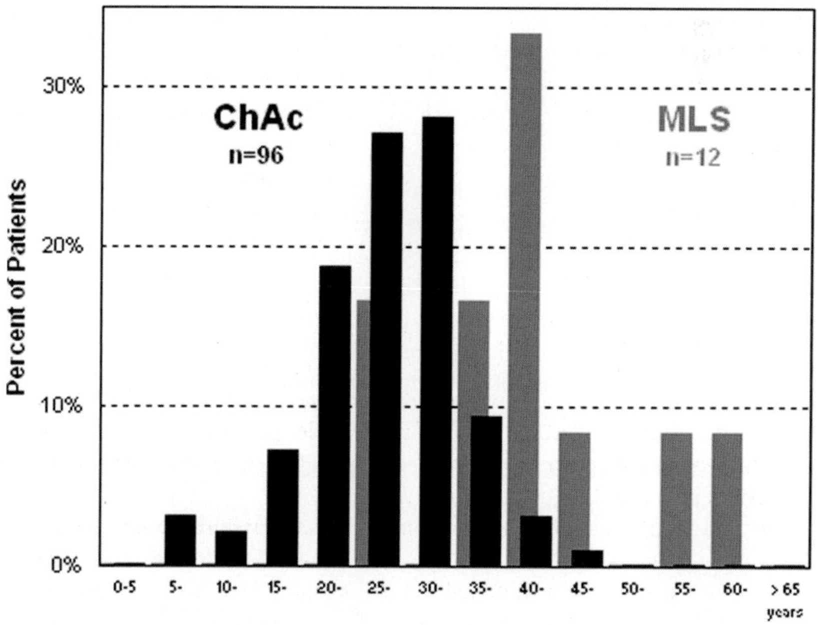

Fig. 2 Age at onset in 96 patients with ChAc. ChAc starts about ten years earlier than McLeod syndrome (12 cases from [10])

3 Building a World-Wide Database

Given the number of at least 200 accessible cases, a structured collection of clinical observations and systematic analysis of data will define the natural course of the NA syndromes and may provide a basis for treatment studies. The data registry should be internet-based and easily accessible so that physicians worldwide can share information on their patients. Data entry into the case record forms must be standardized to maintain quality and comparability across centers. To achieve this, an NA submodule has been established within the European Huntington's Disease Network (EHDN), where procedures to maintain data confidentiality are already established (Fig. 3).

In addition to the instruments used for Huntington's disease, special rating scales are needed in NA, e.g. to document dysphagia or involvement of the peripheral nervous system.

Since the movement disorders of NA defy easy classification, it is of great interest to build a video collection so that the motor abnormalities may then be categorized by a panel of experts. A standard protocol for filming has recently been agreed upon; the sample video can be downloaded from the submodule webpage. Further developments may include a standardized neuropsychology battery [11] as well as neuroimaging protocols for MRI volumetric studies (see chapter in this book by Henkel et al.) or PET transmitter imaging (see chapter by Leenders and Jung) to make data comparable among the patients scattered throughout various countries.

The most important aspect, however, from a clinical point of view, is the gathering of detailed information about past treatments. Experiences with botulinum toxin injections were recently collected among the patients (via neuroacanthocytosis@ yahoogroups.com) but this approach must be extended using a systematic protocol. Similarly, the effects of deep brain stimulation ought to be compared between the few patients who so far have undergone the procedure. Drug treatments such as sulpiride, tetrabenazine, tiapride, levetiracetam [21] or other substances should be documented, as should be approaches such as physiotherapy, dysphagia treatment, and psychotherapy. Accumulation of an extensive set of therapeutic experiences may then be followed by the design of observational treatment studies that in such uncommon medical conditions are only feasible if close international collaboration is achieved. Collectively, these activities effectively constitute a "Virtual Neuroacanthocytosis Institute."

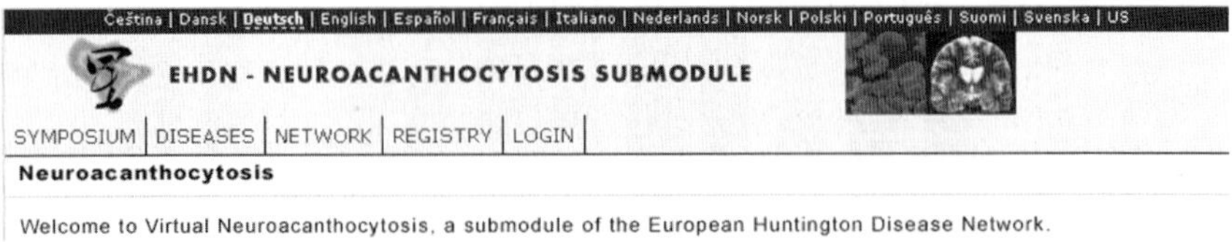

Fig. 3 Screenshot of the NA submodule portal (www.euro-hd.net/html/na/submodule) within the webpages of the European Huntington's Disease Network

References

1. Balhara YPS, Varghese ST, Kayal M (2006) Neuroacanthocytosis: presenting with depression. J Neuropsychiatry Clin Neurosci 18:426
2. Bohlega S, Al Jishi A, Dobson-Stone C et al. (2003) Chorea-acanthocytosis: clinical and genetic findings in three families from the Arabian peninsula. Mov Disord 18:403–407
3. Bonakis A, Agathonikou A, Sarrigiannis P et al. (2001) Neuroacanthocytosis presenting with epilepsy in two siblings. Seventh Mediterranean Epilepsy Conference
4. Boughammoura-Bouatay A, Karmani M, Chebel S et al. (2005) Neuroacanthocytose et épilepsie. Rev Méd Interne 26:756–758
5. Claperon A, Hattab C, Armand V et al. (2007) The Kell and XK proteins of the Kell blood group are not co-expressed in the central nervous system. Brain Res 1147:12–24
6. Danek A, Bader B (2007) Chorein Western blot instructions. http://www.nefo.med.uni-muenchen.de/~adanek/Chorein_Blot.pdf
7. Danek A (2004) Research agenda in neuroacanthocytosis. In: Danek A (ed) Neuroacanthocytosis syndromes. Springer, Dordrecht, The Netherlands, pp 269–275
8. Danek A, Dobson-Stone C, Velayos-Baeza A et al. (2006) The phenotype of chorea-acanthocytosis: a review of 106 patients with *VPS13A* mutations. Mov Disord 20:1678
9. Danek A, Jung HH, Melone MAB et al. (2005) Neuroacanthocytosis: new developments in a neglected group of dementing disorders. J Neurol Sci 229–230:171–186
10. Danek A, Rubio JP, Rampoldi L et al. (2001) McLeod neuroacanthocytosis: genotype and phenotype. Ann Neurol 50:755–764
11. Danek A, Walker RH (2005) Neuroacanthocytosis. Curr Opin Neurol 18:386–392
12. Dobson-Stone C, Rampoldi L, Velayos-Baeza A et al. (2006) Chorea-acanthocytosis [ChAc, Choreoacanthocytosis]. http://www.geneclinics.org/profiles.chac
13. Dobson-Stone C, Velayos-Baeza A, Filippone LA et al. Chorein detection for the diagnosis of chorea-acanthocytosis. (2004) Ann Neurol 56:299–302
14. Gökçay F, Çolakoğlu Z (2003) Koreakantositoz 3 olgu sunumu [Turkish]. Türk Nöroloji Der 9:103–109
15. Harirchian MH, Maghbooli M, Shirani A (2006) A case of choreoacanthocytosis with marked weight loss: impact of orolingual dyskinesia. Neurol India 54:296–297
16. Ikawa M, Yoneda M, Kuriyama M (2005) A case of chorea-acanthocytosis with onset at age 86 [Japanese]. Clin Neurol (Tokyo) 45:603–606
17. Jung HH, Danek A, Dobson-Stone C et al. (2007) McLeod neuroacanthocytosis syndrome. http://www.geneclinics.org/profiles/mcleod
18. Kayahan B, Özdemir F, Bora E et al. (2005) Neuroacanthocytosis: a case report [Turkish]. Turkiye Klinikleri J Med Sci 25:576–580
19. Kurano Y, Nakamura M, Ichiba M et al. (2007) In vivo distribution and localization of chorein. Biochem Biophys Res Commun 353:431–435
20. Lee S, Sha Q, Wu X et al. (2007) Expression profiles of mouse Kell, XK, and XPLAC mRNA. J Histochem Cytochem 55:365–374
21. Lin FC, Wei LJ, Shih PY (2006) Effect of levetiracetam on truncal tic in neuroacanthocytosis. Acta Neurol Taiwan 15:38–42
22. Liu SP, Li DN, Wang SZ et al. (2001) Familial neuroacanthocytosis [Chinese]. Chin J Neurol 34:16–18
23. Miranda M, Castiglioni C, Frey BM et al. (2007) Phenotypic variability of a distinct deletion in McLeod syndrome. Mov Disord 22:1358–1361
24. Mizuno E, Nakamura M, Agemura A et al. (2007) Brain-specific transcript variants of 5′ and 3′ ends of mouse *VPS13A* and *VPS13C*. Biochem Biophys Res Commun 353:902–907
25. Tiftikcioglu BI, Dericioglu N, Saygi S (2006) Focal seizures originating from the left temporal lobe in a case with chorea-acanthocytosis. Clin EEG Neurosci 37:46–49
26. Türe S, Bilgin R, Gedizlioglu M (2006) Neuroacanthocytosis: report of 2 cases [Turkish]. Parkinson Hast Hareket Boz Der 9:38–43

27. Walker RH, Danek A, Dobson-Stone C et al. (2006) Developments in neuroacanthocytosis: expanding the spectrum of choreatic syndromes. Mov Disord 21:1794–1805
28. Walker RH, Jung HH, Dobson-Stone C et al. (2007) Neurologic phenotypes associated with acanthocytosis. Neurology 68:92–98
29. Zheng BR, Han H, Qi ZZ et al. (1989) A case report of chorea-acanthocytosis [Chinese]. Harbin Med Univ Bull 23:126–127

Index

Printing: Krips bv, Meppel, The Netherlands
Binding: Stürtz, Würzburg, Germany